MW01048264

NORTHERN LIGHTS AGAINST POPs

Northern Lights against POPs

Combatting Toxic Threats in the Arctic

EDITED BY

DAVID LEONARD DOWNIE

TERRY FENGE

Published for the Inuit Circumpolar Conference Canada

by

McGill-Queen's University Press

Montreal & Kingston · London · Ithaca

© McGill-Queen's University Press 2003
ISBN 0-7735-2448-7 (cloth)
ISBN 0-7735-2482-7 (paper)

Legal deposit second quarter 2003
Bibliothèque nationale du Québec

Printed in Canada on acid-free paper.

McGill-Queen's University Press acknowledges the support of the Canada Council for the Arts for our publishing program. We also acknowledge the financial support of the Government of Canada through the Book Publishing Industry Development Program (BPIDP) for our publishing activities.

National Library of Canada Cataloguing in Publication

Northern lights against POPs: combatting threats in the Arctic / edited by David Leonard Downie, Terry Fenge.
Includes bibliographical references.
ISBN 0-7735-2448-7 (bnd)
ISBN 0-7735-2482-7 (pbk)

1. Persistent pollutants — Health aspects — Arctic Regions. 2. Organic compounds — Health aspects — Arctic Regions. 3. Persistent pollutants — Environmental aspects — Arctic Regions. 4. Organic compounds — Environmental aspects — Arctic regions. 5. Stockholm Convention on Persistent Organic Pollutants (2001) 6. Arctic regions — Environmental conditions. I. Downie, David Leonard II. Fenge, Terry, 1950- III. Inuit Circumpolar Conference.
TD190.5.N67 2003 363.738'4'09719 C2002-905849-X

This book was typeset by Dynagram Inc. in 10/13 Sabon.

DEDICATION

The circumpolar region has for too long been seen by many in southern capitals as a world on the fringe, populated by Indigenous peoples destined for assimilation into larger, more vibrant, and technologically modern societies. This need not, and should not, be the case. Arctic Indigenous peoples are committed to defending their ways of life and the values and traditions that underlie them. Life on the land and eating what the land provides are essential components of Indigenous peoples' culture. Their survival is important, for in a very real way the state of the Arctic and its Indigenous peoples indicates to us the environmental health of the globe. It is for this reason that we dedicate this book to the Indigenous peoples of the circumpolar Arctic.

Contents

Foreword

The Stockholm Convention on Persistent Organic Pollutants, adopted on 22 May 2001, is a great achievement in humanity's efforts to protect itself and the environment from the threats posed by chemical pollution. For the first time, countries worldwide have agreed to legally binding obligations to eliminate or severely restrict production and use of pesticides and industrial chemicals and take actions towards eliminating the release of chemical by-products.

Persistent organic pollutants represent a subset of the roughly 70,000 to 100,000 chemicals that are currently on the market. In general, industrial chemicals and pesticides provide important benefits to societies by helping them to produce more food for growing populations, find cures for diseases, and develop products that make our lives easier, safer, and more productive.

However, some chemicals present unacceptable risks to human health and the environment, and in such cases governments must take action to protect our peoples and our environment. This is the case with the twelve persistent organic pollutants included in the Stockholm Convention. These chemicals persist in the environment for a long time, accumulate in the tissues of living organisms, travel long distances, and can poison humans and wildlife. As a result, persistent organic pollutants are everywhere and can cause damage wherever they are.

Scientific consensus that global action was necessary to address persistent organic pollutants took years to reach. It has been four decades since Rachel Carson, in her book *Silent Spring*, demonstrated that humans and wildlife were in great peril from certain chemicals, including seven that are now included in the Stockholm Convention. Many governments have since banned these chemicals nationally. When, however, high levels of a number of the persistent organic pollutants were found in the blood and breast milk of Inuit people of northern Canada and in people and wildlife in other remote locations, it became clear that persistent organic pollutants can travel thousands of kilometres from their sources. We are learning, too, as with persistent organic pollutants and other environmental matters like climate change, that the Arctic is an indicator of global environmental health.

Clearly, no country acting alone can adequately protect its citizens or its environment from the threat of persistent organic pollutants. This was understood by the Intergovernmental Forum on Chemical Safety in 1996, when it was concluded that international action, including a global legally binding instrument, is required to reduce the risks to human health and the environment arising from the release of the twelve specified persistent organic pollutants (aldrin, chlordane, DDT, dieldrin, dioxins, endrin, furans, heptachlor, hexachlorobenzene, mirex, PCBs, and toxaphene).

On this scientific basis, the United Nations Environment Programme launched the negotiations of a global treaty to address persistent organic pollutants. With the understanding that the process would affect not only the eventual substance of the treaty but also the future commitment of all stakeholders to its implementation, the negotiations were conducted in an open and transparent manner that welcomed and encouraged the participation of environmental organizations, public-interest groups, industry groups, and academic institutions. In total, 350 such organizations took part in the negotiations, many of which became part of the International POPs Elimination Network. Indigenous peoples' organizations played a special role in voicing a moral dimension to the treaty, a role we would encourage them to play in other global environmental issues as well. As a result, the Stockholm Convention is highly regarded by the governments that adopted it as well as by the full range of stakeholders who participated in its development.

The true test of the effectiveness of the Stockholm Convention will be its ability to protect human health and the environment from persistent organic pollutants. It is well designed to do so. The convention requires the elimination of production and use of nine chemicals. The use of one chemical, DDT, will be limited to disease vector control according to World Health Organization standards until economically feasible and effective alternatives are discovered. Of the unintentionally produced persistent organic pollutants, most notably the dioxins and furans, governments are to develop action plans and other measures to reduce their releases. For persistent organic pollutant stockpiles and wastes, governments are to develop strategies and guidelines for their identification; and their storage, transport, and disposal are to be undertaken in an environmentally sound manner consistent with the Basel Convention on Transboundary Movement of Hazardous Wastes and Their Disposal and other relevant international agreements. The convention also includes criteria and a procedure for adding new chemicals that enable it to address newly identified chemical risks for which global action is needed.

The challenge now is to implement the Stockholm Convention. One hundred and fifty-one governments signed the convention, indicating their commitment to take action to rid the world of persistent organic pollutants. We now urge all governments to ratify the treaty quickly so that it will enter into force as soon as possible, which will contribute to creating a world free of these chemicals. Implementing the treaty will involve setting in place the necessary legal and administrative infrastructure, the first step of which is the development of national implementation plans. Resources will also be necessary to fund projects that result in real reductions in releases of persistent organic pollutants. This will require a strong commitment on the part of governments, intergovernmental organizations, non-government organizations, and international financial institutions.

The Stockholm Convention and the process that developed it are evidence of the growing importance of multilateral environmental agreements in the twenty-first century. These agreements are becoming keystones of sustainable development by helping to ensure that globalization does not result in increased environmental degradation. The Stockholm Convention, together with the Basel Convention and the Rotterdam Convention on Prior Informed Consent Procedure for Certain Hazardous Chemicals and Pesticides in International Trade, provides a regime for international governance of chemicals throughout their life cycles. Our experience with these multilateral environmental agreements shows that we can achieve environmentally sound production, use, and disposal of chemicals in a manner that promotes sustainable development.

With the creation of the Stockholm Convention, humanity has the opportunity to make the world a safer and healthier place to the benefit of current and future generations. The time has come to capitalize on this opportunity by effectively implementing the convention. We have every confidence that this will happen.

David Anderson
Chair, UNEP Governing Council
Minister of the Environment,
Canada

Klaus Töpfer
Executive Director
United Nations
Environment Programme

Acknowledgments

This book is the result of vision, commitment, support, and plain hard work by many people. We ran the risk of giving offence during the preparation of *Northern Lights against POPs*, pressing hard, sometimes relentlessly, on the authors. Now that the book is published, we run a similar risk through inadvertently omitting those who deserve our recognition and thanks for helping to translate an idea into a book.

All the authors featured in this book gave unstintingly of their time and energy. We were warned of the difficulties of dealing with multiple authors, many very senior in universities, government agencies, or other institutions. The warnings proved false. Each author not only prepared the paper we requested, but did so with good grace and in good humour, lightening our task considerably. We thank them all.

This project received critical financial support from the Walter and Duncan Gordon Foundation, the EJLB Foundation, the Salamander Foundation, and the George Cedric Metcalf Charitable Foundation. Few Canadian foundations support work in the territorial North and fewer still support Aboriginal peoples' organizations, yet these foundations understood the national and international importance of the story we proposed to tell and took a chance that we could and would complete the project. This book is testament to their vision.

Michael Francis and Edie Van Alstine, our copy editors, enthusiastically and most capably shepherded the individual manuscripts to completion. We cannot over-emphasize their contribution and we extend our gratitude and admiration for their expertise, commitment, and support.

Our publisher also deserves recognition. When first approached, Philip J. Cercone, executive director and senior editor, McGill-Queen's University Press, was immediately enthusiastic and promised ongoing support. We were particularly pleased that a well-known Canadian university press would publish this book, for Canadians played a significant role in the POPs story from research through to negotiation of international conventions.

We would like to thank David Anderson and Klaus Töpfer for writing the foreword and, more importantly, for their work in addressing the dangers of toxic pollutants in the Arctic and around the world. We also thank Jim Willis, John Whitelaw, David Ogden, Bo Wahlstrom, Agneta Sunden, and others at UNEP Chemicals – both for their diligent work on the Stockholm Convention and for the logistical assistance they provided to various aspects of our project. David Stone, chair of Canada's Northern Contaminants Program, was both supportive and very helpful, as was Stephanie Meakin, our close colleague with the Inuit Circumpolar Conference Canada.

We would like to acknowledge the support of our home institutions – the Inuit Circumpolar Conference and Columbia University. Each allowed us the latitude needed to devote time to this project. For research assistance and other support we would also like to thank the Columbia Earth Institute, Victoria Elman, Rafael Flor, Mike Kraft, Jayne Laiprasert, and Noam Lupu.

Finally, we would like to express our profound gratitude to our families and friends – to whom we owe much and thank far too little. For David, this book would not have been possible without his wife, Laura, and children, Lindsey and William. Terry enjoyed the support of his wife, Wendy, and children, Arwen, Elanor, and Gareth.

David Leonard Downie, Columbia University, New York City, U.S.A
Terry Fenge, Inuit Circumpolar Conference (Canada), Ottawa, Canada

October 2002

Introduction

> The world can tell us everything we want to know. The only problem for the world is that it doesn't have a voice. But the world's indicators are there. They are always talking to us.
>
> Quitsak Tarkiasuk[1]

In the mid-1980s, scientists began to find surprisingly elevated levels of certain toxic chemicals in the blood and lipid tissues of Inuit and then other Indigenous peoples in northern Canada. Many doubted that toxic residues from pesticides, insecticides, fungicides, industrial chemicals, and waste combustion could be found so far from the closest possible emission source – hundreds and sometimes thousands of kilometres away. Over the next thirteen years, however, detailed research in Canada and other circumpolar countries determined that certain toxic chemicals, known as persistent organic pollutants (POPs), were indeed transported over long distances to the Arctic and that they bioaccumulated and biomagnified in the Arctic ecosystem. Moreover, these POPs were being ingested in dangerous quantities by northern Indigenous people when consuming "country food," particularly marine mammals.

In 1997, Canada's Northern Contaminants Program (NCP) and the Arctic Monitoring and Assessment Programme (AMAP – part of the eight-nation Arctic Environmental Protection Strategy) published exhaustive, government-endorsed assessments of transboundary contaminants in the North. These and other studies solidified scientific and political understanding of the presence and threat posed by the long-range transport of toxic chemicals, setting the stage for important international policy discussions.

The most significant results were two binding international treaties that seek to eliminate particular POPs: the 1998 Århus Protocol on Persistent Organic Pollutants, which covers Europe, North America, Asiatic Russia, and central Asian countries of the former Soviet Union and is part of the 1979 UNECE Convention on Long-range Transboundary Air Pollution (CLRTAP); and the global 2001 Stockholm Convention on Persistent Organic Pollutants. Both treaties ban or limit the production, use, release, and trade of particularly toxic POPs, establish scientifically based criteria and specific procedures for establishing controls on additional POPs, and seek to prevent the development and commercial introduction of new POPs.

Of course, negotiating the accords, while a critical first step, is insufficient without further action. Much more must be done to address the POPs issue; both POPs agreements require additional national ratifications to enter into force. More research is needed on the health effects of contaminants, particularly contaminant mixtures. Other chemicals should be examined as candidates for inclusion among those chemicals slated for elimination. And providing adequate financial and technical assistance so that developing countries and countries with "economies in transition" can fulfil their treaty obligations remains a critical challenge.

Nevertheless, as policy-makers, activists, scholars, and students, we must pause to reflect upon, analyze, and give credit to the accomplishments to date. Important international agreements on toxic chemicals have been concluded where only four years ago they did not exist. If fully implemented, the regional Århus protocol and the global Stockholm Convention will contribute significantly to making our planet a healthier place. Both agreements owe a great deal to many "northern lights" – scientists working in the Arctic, particularly in Canada and Sweden; several Arctic-nation governments and their leadership; and northern Indigenous peoples, with their committed participation, patience, and moral authority. The efforts of these people, in their battles against toxic chemicals at the top of the world, will benefit us all.

THE ARCTIC, ITS PEOPLES, AND ITS ENVIRONMENT: A DEBUT IN GLOBAL POLICY

The Arctic rarely impinges on the consciousness of decision-makers in more southern capitals. Until recently, this region lay frozen in the ideological divide between east and west. The collapse of the Soviet Union provided the opportunity – the political space, if you will – for nation-states, Indigenous peoples, and other actors to examine issues of importance to the region from a fresh perspective. With surprising speed, new circumpolar institutions dealing with public health, environmental protection, scientific inquiry, Indigenous peoples, and political governance, such as the International Arctic Science Committee (IASC) and the Arctic Environmental Protection Strategy (AEPS), helped to bring the Arctic dimension of global issues to the attention of international bodies. The Århus POPs protocol and the Stockholm Convention were clearly influenced by this new political activity.

That Arctic perspectives, articulated in part by Indigenous peoples, have had an impact on major international environmental debates is surprising. Indeed, *Agenda 21*, the global blueprint for sustainable development adopted at the 1992 Earth Summit in Rio de Janeiro, does not mention the Arctic. Yet, in relation to POPs and public health, the Arctic has acted as an "indicator region" for the globe, alerting decision-makers of the need for urgent action. The region's Indigenous peoples, particularly Inuit, have been, in a very real sense, canaries in a global POPs coal mine.

Unfortunately, the same may also be true for climate change. The Intergovernmental Panel on Climate Change (IPCC) has noted repeatedly that the effects of global climate change will be particularly severe and most pronounced in the latitudes of the Arctic. Northern Indigenous peoples, drawing upon their traditional knowledge of the environment's subtle rhythms and nuances, have been reporting impacts of climate change, including melting glaciers, decreasing permafrost, and changes in the distribution, abundance, and behaviour of harvested animals, for some years.

To Indigenous peoples in the Arctic, POPs and climate change are not solely environmental or public health issues: they are threats to long-enduring cultures based upon hunting, fishing, trapping, herding, and gathering that therefore raise fundamental issues of human rights. If eating muktuk (whale blubber) laced with POPs is injurious to health – particularly that of a fetus – should Inuit abandon this age-old practice, and in so doing relinquish their hunting culture? Could they then still identify themselves as Inuit? What are the risks of eating country food compared with the risks of modifying or even abandoning this diet? What are the risks of abandoning one's cultural heritage?

Inuit leaders often characterize Inuit – when on the land hunting, fishing, and trapping – as "guardians of the environment." This view has merit. Nobody is better equipped to warn of environmental changes with potentially global impact than Indigenous peoples drawing upon first-hand information and traditional knowledge handed down from generation to generation by hunters and Elders.

The POPs story is partly one of translating science into policy. It is also a story of Arctic Indigenous peoples defending their cultures and economies in international negotiations among states. This may surprise casual observers; indeed, many governments, institutions, and individuals view Indigenous peoples as powerless victims of change. But this is not how Inuit see themselves. Nor is it the ethos reflected in the operations of their representative organizations.

The Inuit Circumpolar Conference, Saami Council, Russian Association of Indigenous Peoples of the North, Aleut International Association, Arctic Athabaskan Council, and Gwich'in Council International, all "permanent participants" in the eight-nation AEPS and, since 1996, the Arctic Council, were, in one way or another, involved in the POPs story. Their advocacy on POPs, first and foremost to protect their cultures, has had beneficial impacts for us all.

Taking their place in a rapidly globalizing world, Arctic peoples have begun to use advocacy to address new international agreements and economic activities. Energy and mineral resources in this region have begun to attract attention in a world quickly consuming supplies of both. There is serious interest in using the Arctic Ocean and the Northeast and Northwest passages for general cargo shipping. Evidence continues to accumulate concerning environmental degradation caused by activities beyond the region's control, fuelling the need to increase scientific research and monitoring and to turn political attention to this huge and still poorly understood region.

NORTHERN LIGHTS AGAINST POPS: AN OUTLINE

Persistent organic pollutants are a set of extremely toxic, long-lasting, chlorinated, organic chemicals that can travel long distances from their emission source and accumulate in animals, ecosystems, and people. They include pesticides such as aldrin and DDT, industrial chemicals such as PCBs, and unintentionally produced by-products of combustion such as dioxins and furans. Scientists have studied the impact of these chemicals since the 1960s; however, in the last decade, concern over the broad dangers that POPs pose to the environment and human health has increased significantly. Their tendency to bioaccumulate in the food chains in northern latitudes threatens Indigenous peoples who rely upon country foods.

Scientists, government officials, and northern Indigenous peoples have significantly influenced both our understanding of the threats these chemicals pose and the dynamics of policy discussions aimed at controlling them. Still, most policy-makers, the public, and students know relatively little about POPs, the struggle to begin dealing with them through national and international policy, and the compelling roles played in that struggle by northern scientists, policy-makers, and Indigenous peoples.

This book explains the threat posed by POPs to the environment and to the health of Indigenous peoples in the Arctic. It examines the leadership of these peoples, and of some of their governments, in pushing for a meaningful policy response and presents the lessons that can be learned from this important process. In telling the POPs story, at times from a northern perspective and while its lessons are still fresh, we hope that *Northern Lights against POPs* will not only inform but also provide insight and guidance to lend support for further research and political action.

The authors in this volume are leading scientists, activists, policy-makers, and scholars. Many played instrumental roles in the development of POPs science and policy. None of these chapters has been published before, and we are pleased with the authors' authoritative discussions of important scientific, political, and legal issues.

As editors, we made no attempt to constrain the opinions or the style of the authors. Indeed, because so many of the authors have played, and continue to play, leadership roles on this issue, we encouraged them to write from both personal and professional perspectives to reflect, to project, to express opinions, and to raise hopes and fears. Thus, we are delighted with the mix of professionally referenced, scholarly discussions with personal essays. We recognize that some chapters touch on subjects of continuing controversy. Although we do not endorse or agree with every conclusion or personal observation in this book, we passionately respect the expertise of each author and the importance of having his or her views recorded.

Chapters in the book's first section examine the development and current understanding of "POPs science." We sought leading experts who could speak authoritatively to the presence and impacts of POPs in the Arctic, their long-range transport to the region, and the research and monitoring programs needed to increase our understanding of this fragile ecosystem and the impact of these chemicals on it. We also urged several authors to look back on their own roles in the development of knowledge regarding the presence and impact of POPs in the North, while asking others to examine important issues about those questions that still linger.

Research teams led by Eric Dewailly and Harriet Kuhnlein author the first two chapters and set the scientific scene – especially in Canada's Arctic. Both Eric Dewailly and Harriet Kuhnlein have made historical and fundamental contributions to our understanding of POPs in the North. Written with key colleagues, their chapters canvas the scientific literature, report on ongoing research on transboundary contaminants in the Arctic, specify which contaminants are of most concern, and report on the dietary and public health consequences of POPs to Arctic Indigenous peoples. Drs Dewailly and Kuhnlein also discuss some of the history of the field, reflecting on particular puzzles and controversies they encountered along the way.

Russel Shearer, a Canadian government official with broad experience in this field, and Siu-Ling Han, co-author of *Highlights of the Canadian Arctic Contaminants Assessment Report: A Community Reference Manual*, describe the work of Canada's Northern Contaminants Program (NCP) in the next chapter, following which Lars-Otto Reiersen, Simon Wilson, and Vitaly Kimstach of the Arctic Monitoring and Assessment Programme (AMAP), examine this unique circumpolar co-operative program. These chapters outline the research response by Arctic states to the POPs problem and the comprehensive assessment and monitoring programs they sponsored. Importantly, Indigenous peoples participated actively in both research programs, learning quickly about contaminants science. If the first set of chapters underscores the critical breakthroughs and pioneering importance of active scientists, this second set highlights the scientific and political importance of large-scale, government-supported systematic research and monitoring. To our minds, authors of both chapters also argue persuasively, if implicitly, that increased environmental monitoring in the Arctic should be a global priority.

Long-range transport of POPs to the Arctic is now an accepted phenomenon, but important questions remain concerning the ability to pinpoint particular emission sources and transport routes. In a potentially controversial chapter, Barry Commoner and his colleagues link emissions of dioxins from named sources in the United States and southern Canada – and even Mexico – to Nunavut in the Canadian Arctic. The chapter does not attempt to predict levels of dioxins in northern communities, and it is important to note that PCBs, DDT, and certain other POPs are of greater immediate concern than dioxins to public health authorities in the North. Indeed, Canadian research shows that dioxin levels in the Arctic are low. Nevertheless, by attempting to calculate the percentage contribution of total dioxins to a receptor location from individual and named emission sites, this chapter graphically illustrates the long-range transport phenomenon and points the way to future research. It is also an example of the type of research that may equip receptor communities in the North with sufficient information to question offending emitters in the South.

In the second section, chapters by Selin, Downie, Bankes, Fenge, Huntington and Sparck, Buccini, and Watt-Cloutier address international POPs policy from a variety of perspectives and disciplines. Henrik Selin, a post-doctoral scholar at the Massachusetts Institute of Technology, discusses the development of the Århus POPs protocol to the CLRTAP, a subject that he has studied extensively. In addition to providing an excellent overview of the development and the content of the protocol, both of which influenced the Stockholm Convention, Dr Selin's work details the important leadership roles Canada and Sweden played in the development of POPs science and policy.

David Downie, a scholar of international environmental agreements at Columbia University, attended the global POPs negotiations. His chapter describes and analyzes the Stockholm Convention in detail and provides a brief history of the development of global chemical policy. He traces the evolution of the Stockholm Convention from chapter 19 of *Agenda 21* through the establishment and activities of the International Forum on Chemical Safety and other international organizations, the negotiation of the 1995 Washington Declaration on Protection of the Marine Environment from Land-based Activities, the 1998 Rotterdam Convention on Prior Informed Consent, and the 1998 Århus POPs protocol to the CLRTAP.

Nigel Bankes served as a member of Canada's delegation to the global POPs negotiations during a sabbatical year from duties as a leading scholar of environmental law at the University of Calgary. His chapter closely examines the Stockholm Convention in comparison to other multilateral environmental agreements (MEAs) and ongoing evolutions in international environmental law.

Terry Fenge, Strategic Counsel to the chair of ICC, details the roles played by Arctic Indigenous peoples in the negotiation of both the CLRTAP and the Stockholm Convention. With a career in northern and Aboriginal issues and as a key participant in these processes, Dr Fenge brings special insight to this important subject. His central conclusion that Arctic Indigenous peoples exerted significant influence and punched above their weight raises interesting avenues for scholars and policy-makers alike and expresses optimism for more inclusive and productive international debates in the future.

Henry Huntington, an Alaska-based researcher and consultant, and Michelle Sparck, an Inuk who serves as special liaison between several Alaskan Indigenous organizations and the state and federal political processes, view the POPs debate from Alaska. They suggest that U.S. policy in this area suffered from a lack of POPs research in the Alaskan Arctic. As a result, and because Arctic perspectives on global issues traditionally do not come easily to the Department of State, the United States found itself initially hampered during the regional and global negotiations.

John Buccini, chair of the global POPs negotiations, presents his first detailed, written reflections on that work. Having also worked on POPS for many years as a senior official in the Canadian Department of the Environment, Dr Buccini has comprehensive views on the subject. Tracing the growing concern over POPs in international policy development, he pinpoints the key events, processes, tradeoffs, and personalities that led to the Stockholm Convention and offers his personal observations and cogent analysis. His conclusions concerning the causal factors that most shaped the successful negotiations bring together key elements of all the preceding chapters. These include the critical influence of advancing scientific knowledge; the gradual development of a broad political consensus produced by that advancing knowledge; the sustained involvement of stakeholders throughout the process; an open and transparent negotiation process; the critical impact of previous work done by the Intergovernmental Forum on Chemical Safety (IFCS), UNEP, and other organizations; and the role of environmental organizations and northern Indigenous peoples in publicizing both the POPs threat and the importance of taking international action to address it.

In the final chapter, Sheila Watt-Cloutier, an Inuk from Nunavik (northern Quebec) and the elected chair of the Inuit Circumpolar Conference, offers a personal recollection of her participation in the global POPs negotiations. Watt-Cloutier reflects on the role of individuals of good will who respected and gave weight to Indigenous peoples' efforts and interventions. We note the certainty and steadfastness with which she and her colleagues approached the issue, understanding its importance to the environment and public health of Inuit and, indeed, of all our peoples. It is apt that Watt-Cloutier has the final word, for Arctic Indigenous peoples, particularly the Inuit, are the northern lights that continue to shine against POPs.

> Without the hunter-gatherers, humanity is diminished and cursed; with them, we can achieve a more complete version of ourselves.
>
> Hugh Brody[2]

NOTES

1 Miriam McDonald, Lucassie Arragutainaq, and Zack Novalinga, *Voices from the Bay: Traditional Ecological Knowledge of Inuit and Cree in the Hudson Bay Bioregion* (Ottawa: Canadian Arctic Resources Committee and Environmental Committee of the Municipality of Sanikiluaq 1997), p. vii.

2 Hugh Brody, *The Other Side of Eden: Hunters, Farmers, and the Shaping of the World* (New York: North Point Press 2001), p. 300.

Acronyms and Abbreviations in the International POPs Arena

AAC	Arctic Athabaskan Council
AMAP	Arctic Monitoring and Assessment Programme
BAT	Best available techniques
BEP	Best environmental practice
CARC	Canadian Arctic Resources Committee
CBD	Convention on Biological Diversity (1992)
CCD	Convention to Combat Desertification
CEIT	Countries with economies in transition
CFCs	Chlorofluorocarbons
CINE	Centre for Indigenous Peoples' Nutrition and Environment
CLRTAP	Convention on Long-range Transboundary Air Pollution
COP	Conference of the Parties
CORINAIR	Core Inventory of Air Emissions
CYFN	Council of Yukon First Nations
DDT	Dichlorodiphenyl trichloroethane
DIAND	See INAC
EC	European Community
EEA	European Environment Agency
ELV	Emission limit values
EMEP	Co-operative Programme for Monitoring and Evaluation of the Long-range Transmission of Air Pollutants in Europe
ESM	Environmentally sound manner
EU	European Union
FAO	(United Nations) Food and Agriculture Organization
FCCC	Framework Convention on Climate Change (1992)
G-77	Group of 77
GCI	Gwich'in Council International
GEF	Global Environment Facility

HCB Hexachlorobenzene
HCH Hexachlorocyclohexane
HELCOM Helsinki Commission
ICC Inuit Circumpolar Conference
IEA International Energy Agency
IEN Indigenous Environmental Network
IFCS Intergovernmental Forum on Chemical Safety
IGO Intergovernmental organization
ILO International Labour Organization
INAC Indian and Northern Affairs Canada (Department of)
INC Intergovernmental Negotiating Committee
IOMC Inter-Organization Programme for the Sound Management of Chemicals
IPCC Intergovernmental Panel on Climate Change
IPCS International Programme on Chemical Safety
IPEN International POPs Elimination Network
IUCN International Union for Conservation of Nature and Natural Resources (also: the World Conservation Union)
MEA Multilateral environmental agreement
MSC-E Meteorological Synthesizing Centre-East
MSC-W Meteorological Synthesizing Centre-West
NGO Non-government organization
NIP National implementation plan
OECD Organisation for Economic Co-operation and Development
PAHS Polycyclic aromatic hydrocarbons
PARCOM Paris Commission
PCBs Polychlorinated biphenyls
PCP Pentachlorophenol
PIC Prior informed consent
POPs Persistent organic pollutants
PRC POPs Review Committee
PRTR Pollutant Release and Transfer Register
RAIPON Russian Association of Indigenous Peoples of the North
REIO Regional economic integration organization (e.g., the European Union)
SCCP Short-chain chlorinated paraffins
SEPA Swedish Environmental Protection Agency
TDI Tolerable daily intake
TRI Toxic Release Inventory
UNCED United Nations Conference on Environment and Development
UNDP United Nations Development Programme
UNECE United Nations Economic Commission for Europe

UNEP	United Nations Environment Programme
UNIDO	United Nations Industrial Development Organization
UNITAR	United Nations Institute for Training and Research
UN	United Nations
U.S. EPA	United States Environmental Protection Agency
WHO	World Health Organization
WWF	World Wildlife Fund

SOME OF THE KEY PARTICIPANTS IN NEGOTIATION OF THE 2001 STOCKHOLM CONVENTION ON PERSISTENT ORGANIC POLLUTANTS

Jim Willis (right) of UNEP Chemicals, led the negotiations secretariat, providing, in Chair Buccini's words, the "strong and professional secretariat support for both the chair and the delegates [that] is a key ingredient in the negotiation process."
Photo courtesy of International Institute for Sustainable Development

Greenpeace, one of the many NGOs present at the negotiation sessions, urged delegates to include more chemicals in the convention and to hasten the negotiations.
Photo courtesy of International Institute for Sustainable Development

Dr Klaus Töpfer, former minister of the Environment for Germany and current UNEP executive director, received the Inuit carving "Mother and Child" as a gift from ICC. Seizing the significance of the gesture, he quickly turned it over to Chair John Buccini, who ensured that it sat at the head table thereafter.
Photo courtesy of International Institute for Sustainable Development

Sheila Watt-Cloutier (centre), then president of ICC Canada and vice-president of ICC International, passionately presented the views of the Canadian Arctic Indigenous Peoples Against POPs (CAIPAP) at INC-3 in Geneva. She called for elimination of POPs and for financial and technical assistance and new funding for developing countries and countries with economies in transition through a reformed Global Environment Facility or through a multilateral fund (left: Cindy Dickson, Council of Yukon First Nations; right: Stephanie Meakin, technical advisor to CAIPAP).
Photo courtesy of International Institute for Sustainable Development

Speaking for the World Wildlife Fund, Cliff Curtis, Director, Global Toxics Programme, supported the goal of ultimate elimination and stressed that import or export of POPs should be consistent with the Basel Convention and for environmentally sound disposal only.
Photo courtesy of International Institute for Sustainable Development

Chair John Buccini holds aloft the Inuit carving given to him at INC-2. Sheila Watt-Cloutier, then president of ICC Canada and vice-president of ICC International, commented that "John Buccini told me ... that when fatigue threatened, he would only have to look to the carving for further energy and strength..."
Photo courtesy of International Institute for Sustainable Development

SECTION ONE

Persistent Organic Pollutants: Global Poisons Threaten the North

1
POPs, the Environment, and Public Health

ERIC DEWAILLY AND CHRISTOPHER FURGAL

INTRODUCTION: THE DISCOVERY OF ORGANIC CONTAMINANTS IN ARCTIC PEOPLES

In June 1985, the Public Health Research Unit of the Laval University Medical Research Centre was implementing a large provincial survey in Quebec to monitor for possible breast-milk contamination by PCBs and other chlorinated organic contaminants.[1] The research team had discussed including populations in Nunavik, James Bay, and along the lower North Shore of the St. Lawrence River in the study but for practical and budgetary reasons had decided not to include these remote communities. Moreover, these communities were located in what was thought at that time to be a pristine environment far from industrial and agricultural activities and pollution. The research collection of 536 breast-milk samples in twenty-two hospitals started in 1988; pilot sampling had begun in 1986.[2]

In June 1986, a midwife from Puvignirtuk, Nunavik, offered to collect breast-milk samples in this eastern Hudson Bay community. Having a "blank" control from a clean environment was considered to be a good idea and of value to the study for comparison with the other samples. This sampling started in September 1987, and the first analysis of the samples showed extraordinary results. When the first gas chromatograph profiles were printed and viewed, the chemists were stunned: not only was the number of different chemicals present in the samples extremely high but the concentrations of these chemicals were elevated far beyond what anyone had expected. The researchers' initial reaction was that something had contaminated the samples; however, after further analyses of the milk samples and controls, it was obvious to us that the breast milk of the women in Puvignirtuk was highly contaminated by organochlorines.[3]

Meanwhile, in September 1985, Drs Kinloch and Kuhnlein had begun a study in the community of Broughton Island (now part of Nunavut) and, in

1988, reported a high intake of PCBs and high PCB blood concentrations among the residents as a result of exposure through the consumption of marine mammals.[4]

The existence of high levels of organic contaminants in the Canadian Arctic food chain – and consequently in Inuit people – has been studied since 1987–1988 and is now confirmed. Adult and newborn exposure to these chemicals has been described in Nunavik, Nunavut, Northwest Territories, and, more recently, in other circumpolar countries. This chapter discusses our understanding of this exposure, its implications, and the challenges for environmental public health in the Canadian and circumpolar North.

ORGANOCHLORINES IN THE ARCTIC

The primary group of environmental contaminants that has been identified in the Arctic food chain and that poses a risk to human health is the organochlorines, including substances such as polychlorinated biphenyls (PCBs) and chlorinated pesticides. Many of these contaminants are used in places distant from the North and are transported via atmospheric and oceanic currents to the Arctic, where they are deposited in the environment through precipitation and other forms of natural deposition. They then find their way into the terrestrial, marine, and freshwater food chains. These substances are characterized by, among other things, their high lipophilicity and persistence in the physical environment, traits that allow them to bioconcentrate in fatty tissues of animals and biomagnify as they move up the food chain. With diets that include animals occupying the top levels of Arctic food chains, northern Indigenous peoples are more exposed to these contaminants than are populations living in more southern regions of the globe.

Sources of exposure: Recent dietary assessments in the North

Dietary surveys conducted in the Canadian North have now given us a great deal of information about how people are exposed to these environmental contaminants. The most recent survey, conducted by Dr Kuhnlein and colleagues at the Centre for Indigenous Peoples' Nutrition and the Environment (CINE) in five Inuit regions (Inuvialuit, Kitikmeot, Kivalliq, Qikiqtaaluk (Baffin), and Labrador), revealed that about fifty-five per cent of Inuit consume a daily diet containing heavy metals such as mercury, lead, and cadmium. This diet also contains PCBs and chlorinated pesticides, often in quantities exceeding established tolerances.[5]

The main source of POPs in the Inuit diet is marine mammal fats, including narwal blubber, beluga oil, and walrus blubber. The CINE study also found some regional variations in the sources of contaminants. Specifically, higher intakes of contaminants were observed in Baffin communities, where there are traditionally higher intakes of marine mammal species (e.g., seals, beluga whale, walrus). The major source of POPs for Baffin region communities appears to be narwal muktuk, whereas in the Inuvialuit region and among Kitikmeot and Kivalliq Inuit the main source is beluga muktuk. On the north coast of Labrador, the largest source of POPs intake is through consumption of lake trout. Similar forms of dietary surveys conducted in Nunavik have shown that close to eighty per cent of the intakes of chlorinated pesticides and PCBs in men and women originate from the consumption of beluga skin and fat (muktuk), ringed seal fat, and, to a lesser extent, fatty fish such as Arctic char, Atlantic salmon, and lake trout. These dietary surveys, although time consuming and often costly, have been critical in identifying the specific sources of human exposure to POPs in the Arctic.

WHAT IS THE IMPACT: STUDYING HEALTH EFFECTS OF POPs IN THE ARCTIC

With a better understanding of the levels of POPs in Arctic residents and an identification of the sources of exposure, the next logical step in the research has been to determine if these substances are having any effect on the health of Arctic peoples. "Potential" effects are known from other studies conducted around the world and from extrapolations from laboratory investigations, but a great deal of time and effort has been – and is still – required to determine if exposures at the levels present in the Arctic are having negative implications for human health.

Health effects: Immune system and infections

When the breast milk of Nunavik Inuit women was found to have POPs concentrations five to ten times greater than concentrations found in the breast milk of women in southern Quebec, the environmental health research group from Laval University and Nunavik regional health authorities were called upon to manage this unusual and unexpected situation. Seeking advice and support for potential actions, we first contacted experts at the World Health Organization (Europe) and rapidly learned that the available scientific knowledge was inadequate to formulate a good risk assessment and management solution for this problem in this northern region. At the time, some experts favoured banning or restricting breastfeeding until an infant had reached one month of age, whereas others were less convinced of the real risk to nursing infants. The lack of consensus among the medical and scientific communities made dealing with this issue very difficult; nevertheless, to protect public health, something had to be done.

Previous studies conducted in Michigan[6] and North Carolina[7] suggested that *in utero* exposure to PCBs was probably more critical than post-natal exposure. However, with respect to health impacts for the mother, a major difference exists between post-natal exposure via breastfeeding and prenatal exposure (during pregnancy). There is little a mother can do to decrease her body burden of contaminants and reduce the transfer of contaminants to her fetus, but the obvious option to decrease post-natal exposure for the baby is to stop breastfeeding. Following this rationale, we immediately began a review and evaluation of possible health risks related to contaminated breast milk. Secondly, we established priorities among potential health outcomes gleaned from this review to help focus our efforts for action. For this, we looked at the health of Inuit infants to see whether there were any conditions that were currently public health problems (a high incidence or causing mortality) and to discern which of those conditions might be partially a result of exposure to environmental contaminants.

In 1988 we knew that several organic contaminants displayed immunotoxic properties in both laboratory animals and humans. In cases of accidental exposure to PCBs and polychlorinated dibenzofurans (PCDFs) in Taiwan ("Yu-Cheng disease"),[8] infants born to Yu-Cheng-afflicted mothers were found to have more episodes of bronchitis or pneumonia during their first six months of life than infants from the same neighbourhoods who were not exposed to these substances. The researchers speculated that the increased frequency of pulmonary diseases could be resulting from a generalized immune disorder induced by transplacental or breast-milk exposure to dioxin-like compounds and, more likely, PCDFs.[9] In later studies, eight- to fourteen-year-old children born to Yu-Cheng mothers were found to be more susceptible to middle-ear diseases than their matched controls.[10] In view of this evidence, and knowing that Inuit infants in Nunavik were suffering from high infection rates early in life, we decided to track Inuit infants during their first year of life to see if highly exposed babies were more at risk of infections than those in the same population who were exposed to lower rates of these contaminants.

The Quebec Ministry of Health funded an epidemiological study to investigate the potential immunological effects of organochlorine exposure among infants in Nunavik. Implemented in 1988, the study investigated whether organochlorine exposure was associated with the incidence of infectious diseases and immune dysfunction among Inuit infants. The research concluded in 1991, and initial results confirmed the previously found high POPs concentrations in 107 human milk samples collected throughout the region of Nunavik (in Hudson Bay, Hudson Strait, and Ungava Bay communities). The number of infectious disease episodes in ninety-eight breast-fed and seventy-three bottle-fed infants was recorded during their first year of life and concentrations of organochlorines were measured in early breast-milk samples from their mothers.

Biomarkers of immune system function (lymphocyte subsets, plasma immunoglobulins) were assessed in venous blood samples collected from the infants at three, seven, and twelve months of age. None of the immunological parameters included in the study were associated with perinatal POPs exposure.

The study found that otitis media was the most frequent disease among these infants: 80% of breast-fed and 81.3% of bottle-fed infants experienced at least one episode during their first year of life. During the second follow-up period, the risk of otitis media increased with prenatal exposure to the organic contaminants *p,p'*-DDE, HCB, and dieldrin. Furthermore, the relative risk of recurrent otitis media (three episodes) increased with prenatal exposure to these compounds. The study also found that no clinically relevant differences existed between breast-fed and bottle-fed infants with respect to the biomarkers of immune function and that prenatal organochlorine exposure was not associated with these biomarkers. From this study, we concluded that prenatal organochlorine exposure could very well be a risk factor for acute otitis media in Inuit infants.[11]

From 1991 to 1993 we conducted a variety of communication activities to inform the Inuit population and the Nunavik health staff about the study and these preliminary results. During the same period, we started to prepare an assessment of the effects of prenatal exposure on the central nervous system of the fetus. This assessment led to the cord/maternal blood monitoring program conducted in Nunavik from 1993 to 1996, and during this three-year period the monitoring was extended to the Northwest Territories and finally to other circumpolar countries between 1994 and 1997 under the Arctic Monitoring and Assessment Programme (AMAP). The initial studies conducted in the Canadian Arctic showed that the primary possible health effects from environmental exposure to these contaminants were on the immune system and with the incidence of infections; however, the international literature published at that time focused largely on impacts to child neurodevelopment.

Health effects: Neurobehavioural development

A number of different forms of studies have increased our knowledge of the potential neurodevelopment impacts of organic contaminants. Some of these studies followed industrial incidents that generated very high levels of exposure in a population, whereas others have studied effects of longer term, low-level exposure through cohort investigations. Both forms of research have added significantly to our understanding of potential effects on pre- and postnatal exposure to POPs and have established the basis for our investigations in the Arctic.

The developmental toxicity of heat-degraded PCBs was first recognized in Japan in the late 1960s and in Taiwan in the late 1970s. Following similar industrial accidents in both countries, infants born to women who had consumed rice oil contaminated with mixtures of PCBs and PCDFs were observed to have skin rashes and showed poorer intellectual functioning during infancy and childhood.[12]

Effects of prenatal exposure to background levels of PCBs and other POPs from environmental sources have been studied since the 1980s in prospective longitudinal studies in Michigan, North Carolina, the Netherlands, Oswego (New York), and Germany. Results show effects including impacts on behavioural functioning, slowed growth rates, and negative implications for intellectual functioning up to age eleven. In these studies, PCB exposure was associated with less optimal newborn behavioural function. Adverse neurological effects of exposure to PCBs have been found in infants up to eighteen months of age in the Netherlands study.[13] In Michigan and the Netherlands, higher cord serum PCB concentrations were associated with lower birth weight and slower infant growth rates.[14] In Michigan, prenatal PCB exposure was associated with poorer visual recognition memory in infancy,[15] an effect that was more recently confirmed in the Oswego study.[16] In North Carolina, deficits in psychomotor development up to twenty-four months of age were seen in the most highly exposed children.[17] In Michigan, prenatal PCB exposure was linked to poorer intellectual functioning at four and eleven years of age,[18] a finding recently confirmed in the Netherlands study at forty-two months.[19] In a German cohort of 171 children, negative associations between PCB concentrations in maternal milk and mental/motor development were reported at all ages, with the differences becoming statistically significant after thirty months of age. After thirty months, greater PCB concentrations in mothers' milk were associated with lower mental and motor scores of these children.[20]

Although much larger quantities of PCBs are transferred to nursing infants by breastfeeding than prenatally across the placental barrier, virtually all of the adverse neurobehavioural effects reported to date have been linked specifically to prenatal exposure. This indicates that the fetus is particularly sensitive to these substances, and therefore the period of development in the womb is perhaps the stage at which humans are most vulnerable to the effects of organic contaminants.[21]

Knowing this, one might logically think that we have a good understanding of the effects we might expect to see in the Arctic and that there is greater need for action than for more research. While the need for action has been evident for some time, the decision to conduct a new epidemiological study on neurodevelopmental effects of POPs in the Arctic was made based on the results found outside the Arctic, as well as on considerations of the specificity of the Arctic "situation." The nature of the issue, including our current understanding of organic contaminants in the Arctic, is unique and warrants some elaboration.

First, the mixtures of contaminants found in the Arctic food chain (i.e., no single traditional food item contains only PCBs, but rather a mixture of organic contaminants, heavy metals, and other contaminants) can have an effect that differs from the sum of the effects of each component of the mixture. Further, Arctic seafood contains a mixture of contaminants that may differ from the mixtures found in other parts of the world. The importance of these differences and their effects on human health are not known. It should also be noted that there are similarities; for example, the PCB congener profile found in human tissue in the Arctic is similar to that found in southern Canada. Simply stated, the effects of chemical mixtures on human health need to be better understood.

Second, the possibility that nutrients common to many of the seafood items (n-3 fatty acids, selenium, etc.) could modify or counteract the toxicity of these contaminants is highly probable. This probability appears to be specific to populations such as the Inuit and other northern Aboriginal groups in Canada that consume large amounts of fish. These contaminant–nutrient interactions need to be better understood to improve the management of matters such as nutritional advice and for the protection of human health.

Third, it is important to recognize that human genetic variability may affect susceptibility to the effects of such pollutants. Gene–environment interactions might explain why some populations or individuals are more susceptible than others to exposure to organic contaminants. Since few genetic studies have been conducted in the Arctic, and considering that Aboriginal people have a distinct genetic background, the impact of genetic differences needs to be evaluated.

Fourth, many health endpoints such as those related to neurobehaviour are multifaceted; socioeconomic and lifestyle factors may very well contribute – to various degrees – to the etiology of these diseases, and we know that these conditions are often quite different in the Arctic than elsewhere and even vary considerably among circumpolar Arctic regions. Recent and ongoing studies to examine these interlaced issues have been attempting to unravel the complex mystery of chemical mixtures, interactions, and genetic influences to better understand the effects of contaminant exposure in the circumpolar North. One such prospective longitudinal study, involving Nunavik and Greenlandic mothers and infants, began in 1997. As this study is still in its phase of statistical analysis, no results have yet been published concerning the neurobehavioural effects of prenatal exposure to POPs. Results are expected to be released to Nunavik communities at the end of 2002.

Health effects: Reproduction

Typical organochlorine mixtures found in highly exposed human populations contain a large variety of compounds, including substances with estrogenic, anti-estrogenic, or anti-androgenic capacities. We may therefore expect complex real-life mixtures, composed of numerous compounds that can interact with different human receptors, to result in impaired male fertility in adults. One study investigating this topic will take place in Greenland starting in autumn 2002. Other potential impacts related to human reproduction are implications for pregnancy. A number of studies have investigated this in the Arctic and elsewhere.

The 1989–91 Nunavik cohort study on the health effects of Inuit newborns exposed to organochlorines,[22] found a statistically significant association between male newborn height and exposure to PCBs and PCDD/FS even after adjusting for other variables. More recently, the initial results of the Nunavik cohort study were reported by Dr Muckle. After controlling for potential confounding variables with the health outcomes measured in the project, researchers in this study associated higher cord–plasma concentrations of PCBs with shorter durations of pregnancy and lower birth weights. These effects remained significant despite the protective effect of high n-3 fatty acid exposure on duration of gestation and on birth weight. Additionally, the magnitude of the negative association of PCBs with birth weight was stronger than the associations with prenatal exposure to alcohol and tobacco. These findings are the first of their kind in an Arctic population and are consistent with results of previous epidemiological studies conducted in populations exposed to PCBs through consumption of PCB-contaminated food. As well, this study went one step further as it examined the hypothesis of protective effects of n-3 fatty acids against the negative effects of PCB exposure on birth outcomes for the first time. The results indicate that the negative effects of prenatal PCB exposure on birth weight and duration of pregnancy can still be demonstrated despite the beneficial effects of n-3 fatty acids.[23] This study and similar projects contribute significantly to our understanding of the impacts of some of these chemical–nutrient interactions on human health and neurodevelopment. Further, because the study was conducted among an Arctic population it is even more useful in formulating public health action in the circumpolar North.

Other health effects: Chronic exposure to POPs and possible links to cancer and osteoporosis

Results from earlier studies on humans generally supported the existence of a relationship or suggested a possible link between the risk of developing breast cancer and exposure to organochlorines, more specifically exposure to *p,p'*-DDE, the main metabolite of DDT.[24] In contrast, however, more recent studies involving larger sample sizes yielded negative results to this effect.[25] In particular, the studies by Hunter and Høyer using a nested case control study design failed to observe a relationship between *p,p'*-DDE or PCB plasma concentrations and the risk of breast cancer. However, Høyer did find that high plasma concentrations of dieldrin were associated with breast cancer risk. Previous studies focused solely on the risk of developing a new breast cancer; however, some researchers have suggested that hormonally active organochlorines might also modulate cancer growth and prognosis.[26]

What is the relevance of all this data to people in the North? Should we not study these relationships among highly exposed Arctic populations to see if there is a link between observed rates of cancer and exposure to these contaminants? At this time, the value of studying the relationship between cancer and POPs among Inuit is questionable. First, the small size of the populations living in the Arctic limits the use of epidemiological approaches. The number of cancer cases (not in proportion to the population, but in total) is small and case control studies are extremely difficult to conduct; only cross-sectional and cohort designs are possible. Second, the incidence of and mortality rates for most cancers, especially hormone-associated cancers, are still lower in the Arctic than in southern latitudes. It may, however, be possible to avoid some of these limitations by conducting circumpolar studies in populations of several of the Arctic countries. Research is now in the planning phase to study these and other relationships between health outcomes and chronic exposure to various POPs.

Another important health outcome potentially associated with exposure to some POPs is osteoporosis, a multifactorial chronic disease that may progress silently in individuals for decades until characteristic fractures occur later in life. To evaluate the prevalence of risk factors for osteoporosis fracture and, more specifically, environmental factors and their association with bone mass in menopausal Inuit women, a study was conducted in Nuuk, Greenland, in September 2000. Nineteen per cent of the Inuit women involved in the study were identified as being at high risk for osteoporosis fracture, compared with 7.2% of the women in the province of Quebec. Age, body mass index, former use of oral contraceptives, current use of hormone replacement therapy, and exposure to PCBs (anti-estrogenic congeners) accounted for thirty-six per cent of the variance in these results. The explanation behind this relationship between PCBs and osteoporosis is based on the fact that some PCB congeners share some structural similarities with dioxins and can act in the same biochemical way. Dioxin-like compounds elicit a broad spectrum of anti-estrogenic activities and may in fact reduce bone density.

Some attention has therefore been turned towards the relationships between environmental exposures to some organic pollutants and the risks of osteoporosis and breast cancer among Arctic residents.

ASSESSING THE RISK TO THE ARCTIC POPULATION

Based primarily on laboratory, but also on some epidemiological evidence, national health agencies have established various guidelines to recommend at what levels action should be taken to immediately minimize exposure to substances such as PCBs out of concern for potential negative human health effects. These guidelines incorporate uncertainties in their calculations, are in some cases extrapolations from laboratory studies to human populations, and are based on only single contaminant exposure scenarios. Based on these and other limitations, their applicability and use in decision-making situations regarding exposure among human populations in the Arctic is debated. The guidelines do, however, provide a reference level upon which to compare populations (analytical and research methods being the same) and do raise various questions regarding when to act and what measures to take to minimize risk for the protection of public health.

In the Northwest Territories and Nunavut, forty-three per cent of Inuit women involved in the cord-blood monitoring program had blood PCB concentrations above the "level of concern" established by Health Canada for women of child-bearing age (5 µg/L); however, eighty-seven per cent of these were less than the level of concern for men and post-menopausal women (20 µg/L), and none exceeded 100 µg/L. Health Canada sets 100 µg/L as the "level of action" – the level at which health authorities are recommended to take immediate action to minimize exposure regardless of gender or age.[27] Further, it must be kept in mind that the northern Canadian population is quite diverse in many respects, including the levels of exposure to PCBs that exist among Inuit regions. Examining the percentage of individuals exceeding the 5 µg/L PCB blood guideline, we see a regional variation in populations, with more individuals in Baffin (seventy-three per cent) exceeding the guideline than in Kivalliq (fifty-nine per cent) or Nunavik (fifty-nine per cent). The corresponding values for Métis/Dene and Caucasian populations in the NWT/Nunavut regions were found to be 3.2% and 0.7%, respectively.[28]

Just as there is variation among the northern Canadian population, there are considerable differences within and between levels of exposure in different circumpolar countries. Among women of child-bearing age in Greenland, ninety-five per cent of those in Disko Bay, fifty-two per cent of those in Ilulissat, eighty-one per cent of those in Nuuk, and eighty-one per cent of those in Ittoqqortoormiit (Scoresbysund) were found to exceed the level of concern (lower limit five µg/L for PCBs such as Aroclor 1260).[29] In Ittoqqortoormiit, twelve per cent of pregnant women exceeded the Canadian level of action for exposure to PCBs, while fifty-two per cent of non-pregnant women exceeded this blood guideline. These markedly higher proportions of the population exceeding the (Canadian) Level of Concern reflect the considerably higher PCB levels in Inuit in Greenland.

As stated earlier, many public health guidelines are based on laboratory tests; few are based solely on epidemiological evidence. Because of the various confounding factors and the potential interactions between different chemicals

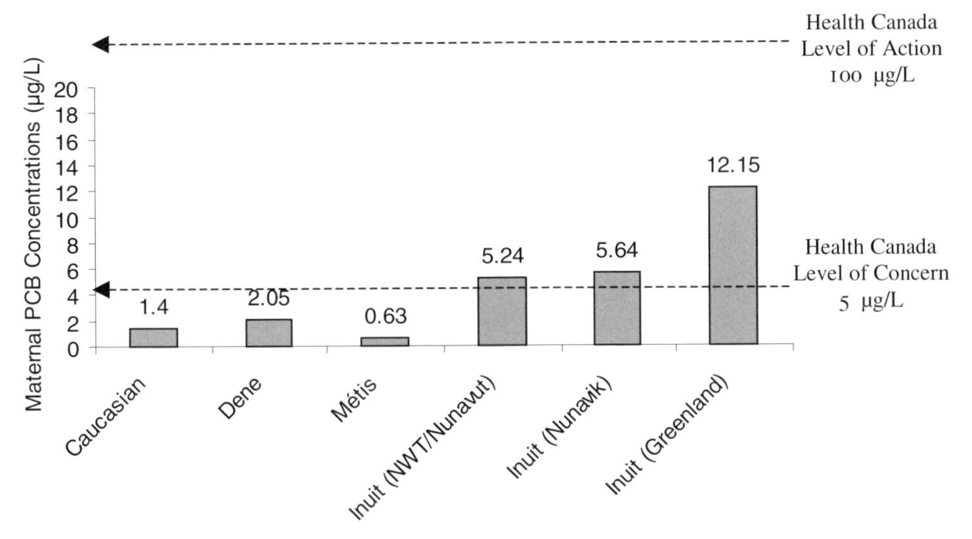

Figure 1.1 Mean (geometric) Maternal Blood PCB (as Aroclor 1260) Concentrations among Various Canadian and Greenlandic groups and Health Canada Guidelines Level of Concern and Level of Action. Ranges around means are not shown (based on data from J. Walker et al., "Organochlorine Levels in Maternal and Umbilical Cord Blood Plasma in Arctic Canada," *Science of the Total Environment* (in press); J.V. Van Oostdam et al., "Human Health Implications of Environmental Contaminants in Arctic Canada: A Review," *Science of the Total Environment* 230, nos 1-3 (1999): 1-82; E. Dewailly et al., *Integration of 12 years of Data in a Risk and Benefit Assessment of Traditional Food in Nunavik* (Ottawa: Indian and Northern Affairs Canada 1998).

and between chemicals and nutrients present in the diets of northern people, epidemiological evidence is critical to refine our understanding of levels of exposure in these populations that affect human health. However, with the exception of studies of the neurodevelopment effect conducted in the Faroe Islands and the immune system studies in Nunavik, Canada, very few major environmental epidemiological studies have been conducted in the circumpolar North. There may be a variety of reasons for this: conducting Arctic studies is extremely difficult because of the remoteness of communities, the cultural context, climate, small population size, and the many potential social and behavioural confounders of this work. Yet the specificity of the Arctic situation raises the question: to what extent can results and conclusions from epidemiological studies conducted outside the Arctic apply to this region?

COMPLICATING THE PROCESS: COMBINED EFFECTS

Clouding our understanding of the effects of contaminants on human health, and thus our management of the issue, is the simultaneous exposure of Arctic residents to a variety of contaminants present in the food chain. Because POPs are composed of numerous compounds, most of which accumulate in the food chain and in humans, it is difficult to determine which compound is responsible for any single observed effect. Any assessment of risk is therefore of limited relevance for regulators and public health professionals. Further complicating the issue is evidence that Arctic residents are exposed to a variety of different forms of chemicals as well. Just as there are many different substances in the group we refer to as "POPs," Arctic residents are also exposed to different forms of chemicals from the same food source; in the case of muktuk consumption, the resulting exposure is to both methylmercury (MeHG) and a mixture of POPs. These mixtures show some interactive effects, which further complicates our understanding of the effects of contaminants on human health in the North.

In a study conducted in the Faroe Islands, the association between cord-blood PCB and cord-blood mercury levels suggests a possible confounding relationship. While no effects of PCBs were apparent in children with low exposure to mercury, PCB-associated deficits were seen among the highest exposure groups, indicating a possible interaction between these two neurotoxic substances.

Additionally, the existence of these mixtures and their chemical behaviour further confuses the discussion around the applicability or use of exposure guidelines for northern populations. The exposure profile found in the Arctic may, because of the diffuse properties of contaminants, differ significantly from those reported at mid-latitudes where local sources may be primary contributors to the chemical mixture present in the environment. As well, lifestyle influences such as hunting and fishing seasons could influence patterns of exposure, and steady state and peak exposures may have different toxic consequences. Also, Arctic residents' consumption of wild plants and animals, many having both a mixture of environmental contaminants and a suite of nutrients that influence or counteract contaminant toxicity, makes this anything but a simple problem. Importantly, Aboriginal people have a specific genetic background that could influence their susceptibility to toxic agents, a fact that is often excluded from the discussion but should not be forgotten.

In general, many health endpoints are multifactoral and environmental stressors may contribute to varying degrees to the etiology of these diseases. Contaminants are likely to play a modest role compared with the roles played by lifestyle and genetic factors in the etiology of most diseases. However, exposure to contaminants is preventable and one can argue that their presence in the remote Arctic regions is unethical and unjust. Therefore the question remains: At what level and how should we act based on our current understanding of this issue and the potential implications for environmental public health?

TAKING ACTION: SHOULD WE ACT LOCALLY OR INTERNATIONALLY?

It is well accepted that solutions must be found to fight against the presence of these environmental contaminants in the Arctic food chain. One of the most logical and sustainable solutions is to ban production and use of these chemicals. Even though international conventions (for POPs and mercury) are at various stages of development, ratification, and implementation to achieve this goal, we also know that, because of the persistent nature of these chemicals, it will take several years to see a substantial improvement in the environment even after these emissions cease. Ethically, we must ask the question of the need, feasibility, and efficacy of immediate dietary intervention to decrease population exposure. Without a doubt, one of the most vulnerable groups for POPs exposure is pregnant women and their fetuses. POPs (especially PCBs) and methylmercury are toxic for the fetus and may affect the development of the central nervous and immune systems at this stage of human development. To date, and with varying degrees of success, some regional actions have been taken to reduce exposures to these substances.

In the Faroe Islands, dietary advice was released to reduce consumption of pilot whale meat during pregnancy and thus decrease prenatal exposure to methylmercury. This information program proved to be extremely efficient; methylmercury exposure among pregnant women dropped by a factor of ten and ultimately reached background concentrations observed in general urban populations.[30] However, the Faroe Islands situation is not comparable to that in the high Arctic. In the Faroe Islands, exposure was considered to be almost entirely from a single source (pilot whale meat) and associated with what was considered at that time a single harmful contaminant: it was believed that only mercury was associated with negative health effects. However, mercury has a relatively short half-life, and thus abstention from consumption can show a very fast reduction in blood mercury levels. It was not until later that PCBs were reported to have a possible interaction with this mercury exposure. These factors led the public health authorities to propose stopping pilot whale consumption among pregnant women as an effective and efficient means to reduce exposure.

This option would not necessarily be effective in the Canadian Arctic. In this region, there are multiple contaminants (metals and POPs) coming from the consumption of various food items and no restrictive dietary advice could be proposed without risking nutritional and cultural disruption that could have negative health consequences. Any public health action based on just one of these factors, without an appreciation of the context and potential impacts of restrictive dietary advice in Canadian Arctic communities, is likely to be met with significant resistance and limited success. Through our work with public health authorities in Nunavik, we have learned one potential strategy through which to address some of the complexities and potentially negative effects of restrictive dietary advice. The option of promoting the consumption of nutritionally rich, contaminant-free species during pregnancy is currently being assessed in Nunavik through a pilot risk-management program.

One of the mandates of the Nunavik Nutrition and Health Committee (NNHC), the recognized and authorized body for the region on health and environment issues, is to provide advice to the regional health authorities on policies relating to nutrition and health. Information gathered by researchers working in Nunavik and provided to the committee has shown that Arctic char is a potential healthy "country food" to promote to women during pregnancy. Char would provide significant benefits while reducing exposure to some important environment contaminants present in other country food species (e.g., mercury and POPs in marine mammal meat and blubber). Positive effects for the mother and developing fetus from the consumption of char include the reduction of contaminant exposure and potential increase of exposure to some important nutrients (e.g., omega-3 fatty acids, which have been shown to be valuable for fetal development). However, because of the differing half-lives of these substances, such a program is more effective in reducing exposure to mercury than to POPs. As intakes of mercury in this region are primarily from beluga meat (thirty-five per cent), lake trout (fifteen per cent), and seal liver (eighteen per cent), we believe that an increase in char consumption will decrease these major sources of mercury intake and thus maternal and fetal exposure to this substance. Based on known average levels of consumption, we have estimated that with increased char consumption the daily mercury dose will be decreased by eighty per cent and that, with a mercury half-life of two months, maternal blood or hair concentrations will be decreased by seventy per cent after six months.

This project, which starts in 2002 and will be completed in 2003, is built on the need to provide appropriate (culturally and socially acceptable and desired), healthy, and viable (economically) alternatives when providing health advice to reduce contaminant exposures in northern communities. Arctic char is an important source of protein, which is needed for normal growth, repair of body tissues, and the production of antibodies to fight disease. It is also an excellent source of vitamin D and phosphorus, both of which are essential for the development of strong bones and teeth. Additionally, Arctic char contains significant amounts of selenium, which acts as an antioxidant and may reduce the risk of mercury poisoning. Finally, research has shown that among northern fish species Arctic char is the best source of omega-3 fatty acids. In fact, the amount promoted in the pilot study (three meals of Arctic char per week @ 230 g/meal) provides as much as ten grams of DHA (docosahexanoic acid) and

EPA (eicosapentanoic acid) per week (1,430 mg a day). These fatty acids protect against some heart diseases, high blood pressure, and some types of cancers and have been associated with healthy child development. Moreover, Arctic char is one of the preferred traditional foods in Nunavik, and research in this region has shown that ability to procure traditional foods is a limiting factor to their consumption. Research also demonstrates that health advice related to consumption habits is more closely followed when viable alternatives are provided; the NNHC is now testing free access to char for pregnant women as a pilot project. The committee's experience suggests that a positive approach – promoting the consumption of one species rather than banning another – to reducing contaminant exposure is far less socially disruptive.

CONCLUSIONS

The public health management of exposure to environmental contaminants in the Arctic food chain is a complex matter. Although contaminants found in northern traditional foods may pose certain public health risks, these foods also constitute a valuable source of several key nutrients, some of which may act to reduce the risk to contaminant exposure. In the risk assessment of exposure to levels found in the Arctic it is reasonable to conclude that the traditional diet in the Arctic includes substances that have a negative influence on health. Although fetuses and children are the most susceptible groups potentially affected by POPs, further studies are needed to address hormone-related chronic diseases such as breast cancer and osteoporosis.

Whenever possible, the assessment of the risks posed by contaminants in the Arctic should be based on epidemiological evidence; however, for substances currently identified in northern food chains, this evidence is sparse or nonexistent. Nevertheless, we should seriously consider results from the two available cohort studies on neurological disorders associated with prenatal PCB and methylmercury exposure (Faroe Islands) and observations of immune dysfunction in children prenatally exposed to POPs (Nunavik). As the dietary exposure to contaminants includes a mixture of many different substances, it is not reasonable (and may not even be possible) to consider the risk of a single substance in isolation. The exposure levels observed in the Faroe Islands can also be found in other places in the circumpolar Arctic (e.g., Greenland); therefore it seems likely that the negative effects could also be found at these other locations. Public health authorities have the responsibility to decide on a suitable undertaking to reduce the exposure to contaminants among humans and to consider especially the possible negative effects that changes in lifestyle habits, such as diet, may cause. Any reduction of intakes of contaminants among northern populations as a result of decreased consumption of traditional foods would likely be at the expense of the nutritional benefits they provide. Thus, access to viable, appropriate, and healthy diet alternatives must be considered when taking action. However, the search for nutritional alternatives may prove to be problematic in some Arctic regions and among some segments of the population.

At this time, the scientific community cannot assess the health impacts of contaminant exposure among northern populations with absolute certainty. Direct action to eliminate this potential health hazard without causing problems related to a loss of the benefits provided by traditional foods and their associated activities is difficult, at best. Further work towards better understanding and refining our abilities to ascertain the exact impact of these substances on human health is ongoing through various research programs. Local and regional public health initiatives, such as the pilot project of the NNHC, are developing and assessing risk-minimization scenarios using the best knowledge available. Meanwhile, international conventions for the control of many of these substances are at various advanced stages and are the ultimate solution for this global problem.

The consumption of traditional foods in the North presents certain benefits and certain risks. Inuit know the role that these foods play in their daily lives and that traditional food is central to the social, mental, and spiritual health of individuals and communities throughout the circumpolar North. For these reasons, we still believe that Inuit should continue to eat traditional food as much as they desire, and thus benefit from the enormous advantages provided by these gifts from the land and sea.

NOTES

1 The environmental health research group is part of the Public Health Research Unit of the Laval University Medical Research Centre (CHUQ), located in Quebec City. Ten researchers in community medicine, epidemiology, toxicology, nutrition, psychology, and anthropology and thirty research assistants and students contribute to our research program. Most of the material presented in this chapter draws on work by this group that has been supported by the Northern Contaminants Program (INAC). Portions of the work discussed here are also published in past and the most recent reports of the Arctic Monitoring and Assessment Programme and in the *Canadian Arctic Contaminants Assessment Report* (I and II).

2 E. Dewailly et al., "Health Risk Assessment and Elaboration of Public Health Advices Concerning Food Contaminants in Nunavik" (Ottawa: INAC 1996).

3 E. Dewailly et al., "High Levels of PCBs in Breast Milk of Inuit Women from Arctic Quebec," *Bulletin of Environmental Contamination and Toxicology* 43, no. 1 (1989): 641–6.

4 H.V. Kuhnlein and D. Kinloch, "PCBs and Nutrients in Baffin Island Inuit Foods and Diets," *Arctic Medical Research* 47, supplement 1 (1988): 155–8.

5 H.V. Kuhnlein et al., "Assessment of Dietary Benefits/Risks in Inuit Communities" (Montreal: CINE, McGill University 2000).
6 S.W. Jacobson et al., "The Effects of Intrauterine PCB Exposure in Visual Recognition Memory," *Child Development* 56 (1985): 853–60.
7 W.J. Rogan et al., "Neonatal Effects of Transplacental Exposure to PCBs and DDE," *The Journal of Pediatrics* 109, no. 2 (1986): 335–41.
8 K.J. Chang et al., "Immunologic Evaluation of Patients with Polychlorinated Biphenyl Poisoning: Determination of Lymphocyte Subpopulations," *Toxicology and Applied Pharmacology* 61 (1981): 58–63.
9 W. Rogan et al., "Congenital Poisoning by Polychlorinated Biphenyls and their Contaminants in Taiwan," *Science* 241 (1988): 334–36.
10 W.Y. Chao et al., "Middle-ear Disease in Children Exposed Prenatally to Polychlorinated Biphenyls and Polychlorinated Dibenzofurans," *Archives of Environmental Health* 52 (1997): 257–62.
11 E. Dewailly et al., "Susceptibility to Infections and Immune Status in Inuit Infants exposed to Organochlorines," *Environmental Health Perspectives* 108 (2000): 205–10.
12 Y.C.J. Chen et al., "Cognitive Development of Yu-Cheng ("Oil Disease") Children Prenatally Exposed to Heat-degraded PCBs," *Journal of the American Medical Association* 268 (1992): 3213–18; M.L. Yu et al., "*In Utero* PCB-PCDF Exposure: Relation of Developmental Delay to Dsymorphology and Dose," *Neurotoxicology and Teratology* 13 (1991): 195–202.
13 M. Huisman et al., "Neurological Condition in 18-month-old Children Perinatally Exposed to Exposure to Polychlorinated Biphenyls and Dioxins," *Early Human Development* 43 (1995): 165–76.
14 G.G. Fein et al., "Prenatal Exposure to Polychlorinated Biphenyls: Effects on Birth Size and Gestational Age," *The Journal of Pediatrics* 105 (1984): 315–20; J.L. Jacobson, S.W. Jacobson, and H.E.B. Humphrey, "Effects of Exposure to PCBs and Related Compounds on Growth and Activity in Children," *Neurotoxicology and Teratology* 12 (1990): 319–26; S. Patandin et al., "Effects of Environmental Exposure to Polychlorinated Biphenyls and Dioxins on Birth Size and Growth in Dutch Children," *Pediatric Research* 44 (1998): 538–45.
15 S.W. Jacobson et al., "The Effects of Intrauterine PCB Exposure"; J.L. Jacobson, S.W. Jacobson, and H.E.B. Humphrey, "Effects of *in utero* Exposure to Polychlorinated Biphenyls and Related Contaminants on Cognitive Functioning in Young Children," *The Journal of Pediatrics* 116 (1990): 38–45.
16 T. Darvill et al., "Prenatal Exposure to PCBs and Infant Performance on the Fagan Test of Infant Intelligence," *Neurotoxicology* 21, no. 6 (2000): 1029–38.
17 B.C. Gladen et al., "Development after Exposure to Polychlorinated Biphenyls and Dichlorodiphenyl Dichloroethene Transplacentally and Through Human Milk," *The Journal of Pediatrics* 113 (1988): 991–5; W. Rogan and B.C. Gladen, "PCBs, DDE, and Child Development at 18 and 24 Months," *Annals of Epidemiology* 1 (1991): 409–13.

18 J.L. Jacobson, "Effects of *in utero* exposure"; J.L. Jacobson and S.W. Jacobson, "Intellectual Impairment in Children Exposed to Polychlorinated Biphenyls *in utero*," *New England Journal of Medicine* 335 (1996): 783–9.

19 S. Patandin et al., "Effects of Environmental Exposure to Polychlorinated Biphenyls and Dioxins on Cognitive Abilities in Dutch Children at 42 Months of Age," *The Journal of Pediatrics* 134 (1999): 33–41.

20 J. Walkowiak et al., "Environmental Exposure to Polychlorinated Biphenyls and Quality of the Home Environment: Effects on Psychodevelopment in Early Childhood," *Lancet* 358 (2001): 1602–7.

21 N. Ribas-Fito et al., "Polychlorinated Biphenyls (PCBs) and Neurological Development in Children: A Systematic Review," *Journal of Epidemiology and Community Health* 55, no. 8 (2001): 537–46.

22 E. Dewailly et al., "Health Status at Birth of Inuit Newborn Prenatally Exposed to Organochlorines," *Chemosphere* 27 (1993): 359–66.

23 G. Muckle et al., "Affective and Behavioural Effects Associated with PCBs, Mercury and Lead Exposure Among the Northern Quebec Inuit Children," Paper presented at the International Neurotoxicology Conference, Little Rock, Arkansas, 18–21 November 2002. The current mean mercury maternal hair concentration in Nunavik is 4.5 ppm. At the end of this one-year study we expect to see an average of 1.3 ppm and thus very few mothers with levels exceeding the threshold value for neurodevelopment effects. For POPs such as PCBs, because the half-life is considerably longer (average is four years), we expect only minimal effects; for example, for the PCBs that currently average 1.1 ppm (lipid basis) in maternal blood, a reduction of 80% of the dose will decrease the body burden by only 8.2% after six months of increased consumption of Arctic char.

24 N. Krieger et al., "Breast Cancer and Serum Organochlorines: A Prospective Study among White, Black, and Asian Women," *Journal of the National Cancer Institute* 86 (1994): 589–99.

25 D.J. Hunter et al., "Plasma Organochlorine Levels and the Risk of Breast Cancer," *New England Journal of Medicine* 337 (1997): 1253–8; L. López-Carrillo et al., "Dichlorodiphenyltrichloroethane Serum Levels and Breast Cancer Risk: A Case-control Study from Mexico," *Cancer Research* 57 (1997): 3728–32; K.B. Moysich et al., "Environmental Organochlorine Exposure and Postmenopausal Breast Cancer Risk," *Cancer Epidemiology, Biomarkers, & Prevention* 7 (1998): 181–88; P. van't Veer et al., "DDT (dicophane) and Postmenopausal Breast Cancer in Europe: Case-control Study," *British Medical Journal* 315 (1997): 81–5; A.P. Høyer et al., "Organochlorine Exposure and Risk of Breast Cancer," *Lancet* 352 (1998): 1816–20.

26 A. Demers et al., "Risk and Agressiveness of Breast Cancer in Relation to Plasma Organochlorine Concentrations," *Cancer Epidemiology, Biomarkers, & Prevention* 9 (2000): 161–6.

27 J. Walker et al., "Organochlorine Levels in Maternal and Umbilical Cord Blood Plasma in Arctic Canada," *Science of the Total Environment*, in press.

28 J.V. Van Oostdam et al., "Human Health Implications of Environmental Contaminants in Arctic Canada: A Review," *Science of the Total Environment* 230, nos 1–3 (1999): 1–82.
29 B. Deutch and J.C. Hansen, "High Human Plasma Levels of Organochlorine Compounds in Greenland. Regional Differences and Lifestyle Effects," *Danish Medical Bulletin* 47, no. 2 (2000): 132–7.
30 P. Weihe and P. Grandjean, "Effects of Methylmercury in Humans," Paper presented at the AMAP conference and workshop, "Impacts of POPs and Mercury on Arctic Environments and Humans," Tromsø, Norway, 20–24 January 2002.

2
Canadian Arctic Indigenous Peoples, Traditional Food Systems, and POPs

HARRIET V. KUHNLEIN, LAURIE H.M. CHAN, GRACE EGELAND, OLIVIER RECEVEUR

INTRODUCTION

Information on PCB contaminants in sea mammal food items in the mid-1980s triggered both concern and action by Arctic peoples about toxic hazards in their traditional food resources. This chapter presents major findings from several POPs-related projects conducted by the Centre for Indigenous Peoples' Nutrition and Environment (CINE) at McGill University.

In 1985, Health Canada's Medical Services Branch decided to investigate dietary intake of traditional food resources among Inuit on Baffin Island, who were thought to be the greatest consumers of sea mammal species. One community was selected based on local advice and community harvest data generated by the Baffin Regional Inuit Association (BRIA). A pilot study consisting of dietary data collection was conducted, and the Medical Services Branch collected convenience blood samples in the community. Following the pilot study, the community joined researchers at McGill University in undertaking a thorough, year-round dietary intake study to understand the extent of traditional food consumption in the total diet. The protocol included various interviews and food sample collections conducted bi-monthly.[1]

During the course of this study, the Dene Nation in Yellowknife also requested investigations into contaminants in their traditional food system, particularly in fish in the Mackenzie River consumed in communities downstream from the oil and gas complex at Norman Wells. Two Dene/Métis communities requested investigations similar to the food and dietary intake studies on Baffin Island, with participation of the same McGill University team.[2]

These two studies set the stage for understanding the complex patterns of exposure to several organochlorines and heavy metals in a variety of species in traditional Arctic food systems and the extent of exposure depending on level of consumption by segments of the populations. The studies would also contribute knowledge of the considerations of nutrient and cultural benefits of traditional food use within Native peoples' dietary patterns balanced against the risks from contaminants in the same food. During the 1990s, CINE conducted three large studies with similar objectives with funding from the Northern Contaminants Program. They included more than forty Arctic communities in Dene/Métis and Yukon First Nations and Inuit territories.[3] Several smaller community studies, laboratory studies, and education initiatives were, and continue to be, carried out.

CINE

CINE was established in 1992 in response to a need expressed by Aboriginal peoples in Canada for participatory research and education to address their concerns about the integrity of their traditional food systems. Aboriginal leaders worked together to lobby for funds and to establish a working structure for CINE's activities. Discussions about establishing the centre, begun in 1988, resulted in funding through the Arctic Environmental Strategy, managed by Indian and Northern Affairs Canada (INAC), as an initiative of Canada's *Green Plan*. Staff recruitment and preparation of physical space at McGill University began in March 1992, and CINE officially opened in autumn 1993. The CINE governing board includes representatives from the Assembly of First Nations, Council of Yukon First Nations, Dene Nation, Inuit Circumpolar Conference, Inuit Tapiriit Kanatami, Métis Nation of the Northwest Territories, and the Mohawk Council of Kahnawake. As the Aboriginal community geographically closest to the university, the Mohawk Council of Kahnawake serves as the host of CINE.[4]

CINE is a permanent multidisciplinary research and education resource with an international outlook. The forces of environmental and cultural change that impact on traditional food systems, nutrition, and health of Indigenous peoples have global similarities and significance. The centre operates at arm's length from government and works closely with Indigenous peoples' communities on topics related to their traditional food systems.

Aboriginal peoples, McGill administration, and government officials conducted extensive consultations to prepare a strategic plan for CINE in 1999. A diversified budget of approximately CAN$1 million was approved for CINE activities in research, response strategies to communities, community education, outreach, and communications.

A hallmark of CINE' work – one for which CINE is seen as a model internationally – has been good participatory techniques for activities conducted with communities. All community-based activities include agreement on the conduct of projects, local staff training and involvement, and return of results to communities before final reports are submitted.

In its work on POPs and other contaminants in traditional food and diets, CINE has insisted on including research to understand both the benefits and the risks of traditional food use. Benefits include health-giving properties in the form of both essential nutrients and cultural perceptions of wellness. CINE's reports present these benefits in contrast to the often poorly understood physical effects of contaminants such as POPs and heavy metals. Only by understanding all of the scientific and cultural issues can properly informed decisions be made about including, reducing, or eliminating certain traditional foods containing contaminants in the diet of Indigenous peoples.

The summary of CINE research on Arctic food use is described, with emphasis on the amount of traditional food consumed, the traditional food/market food dyad, and the benefits and risks of traditional food consumption.

OVERVIEW OF RESEARCH METHODS USED BY CINE FOR POPs RESEARCH

Community and laboratory studies form the basis of research activities reported in this chapter. Studies in the late 1980s with one community on Baffin Island (Nunavut) and two communities in the Sahtu (Northwest Territories) areas were followed by three major studies involving a total of forty-seven communities (figure 2.1). These covered a broad geographical area in Arctic Canada and a wide variety of Arctic traditional food species. Throughout, the intent was to understand community patterns of traditional food use and the ways that market food displaces traditional food in diets and to derive the exposure patterns to organochlorine and heavy metal contaminant risks and nutrient benefits for Arctic peoples. Data collection regarding food use consisted of twenty-four-hour dietary recall interviews, food frequency questionnaires, and socio-cultural questionnaires administered to individuals. Samples of food items as prepared in homes without added ingredients were analyzed at CINE laboratories for a spectrum of nutrients and organochlorine and heavy metal contaminants.

CINE needed to understand the total diet, throughout all seasons, to define traditional food-use patterns. Knowing the amount consumed was as important as knowing the contents of nutrients and contaminants in the traditional food. Knowledge of both the total dietary pattern and the cultural rationale for food use is necessary to establish the potential and need for creating informed educational messages to assist dietary change to improve health status.

It is well established internationally that when traditional lifestyles of Indigenous peoples give way to modern ways of living and industrial-based diets, chronic diseases increase. This is often signalled by population indicators of diets high in commercial fats and simple carbohydrates, reduced physical activity, and increased obesity.[5] Understanding patterns of food use by Arctic peoples has several potential benefits, including knowing the extent of contaminant

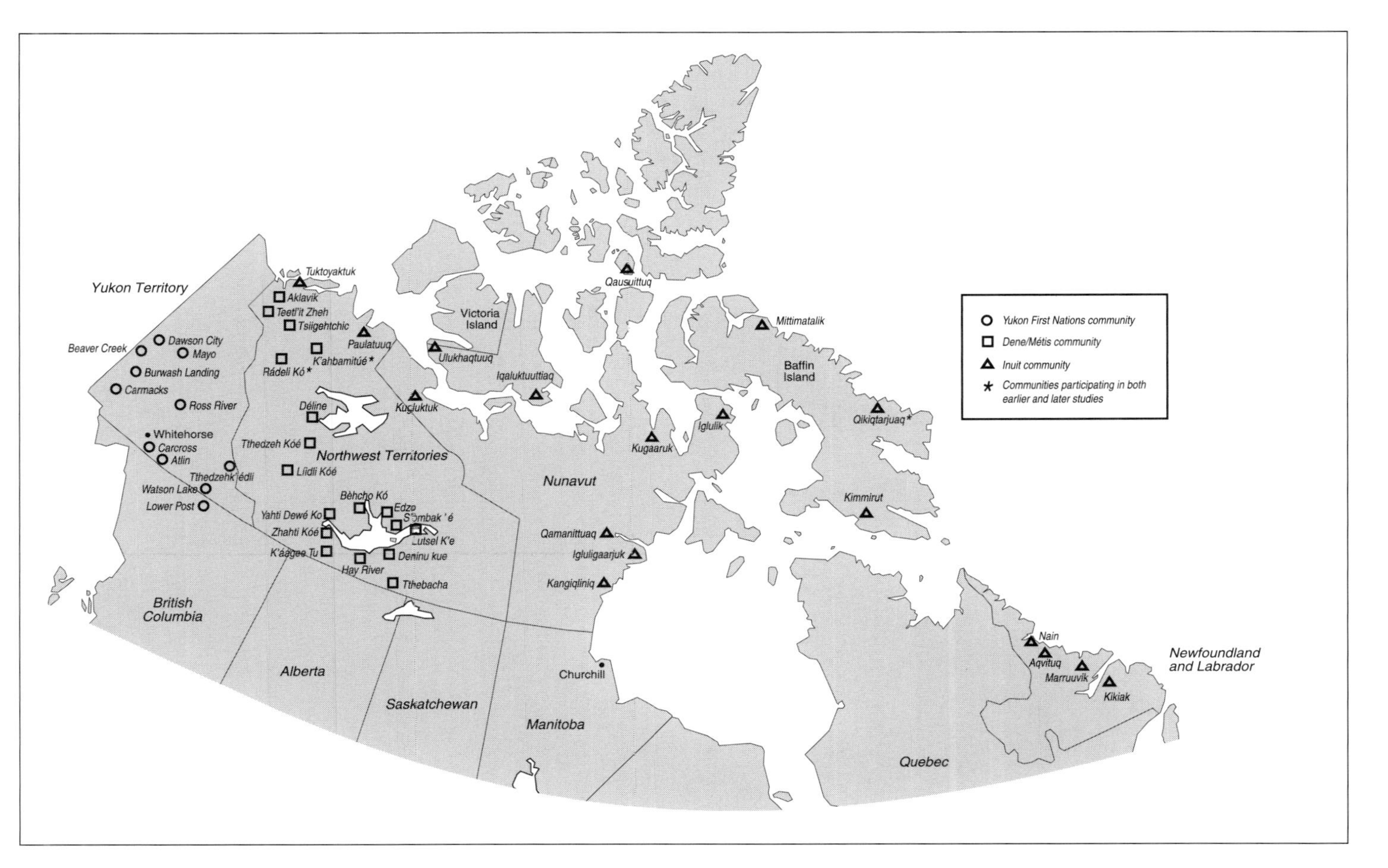

Figure 2.1 Sites of CINE Research in the Canadian Arctic

Table 2.1
Food Species in Three Cultural Areas of the Canadian Arctic

	Dene/Métis	*Yukon*	*Inuit*
Sea mammals	0	0	14
Land animals	17	16	14
Birds	16	26	70
Fish/Seafood	20	20	48
Plants	48	40	48
Total	101	102	194

exposure, knowing the extent of nutritional quality of the diet and specific food sources of nutrients and contaminants, and knowing what food items can be deleted or added to improve diet quality.

All studies began with community information sessions. For the larger studies, representatives from each community attended workshops in a centre established for their geographically defined region. The interview protocol was reviewed during these methodology workshops and the most representative communities in each region were selected. Interview teams included research managers and community-based interviewers trained in consistent interview protocol. Random samples of teen or adult individuals (Dene/Métis, Yukon First Nations, Inuit) were selected in the larger studies. (While all households and individuals had been included in the three communities (Baffin Island (1), Sahtu (2)) studied earlier, cost considerations for the expanded study dictated random sampling.)

OVERVIEW OF MAJOR FINDINGS

Figure 2.1 shows the community sites for this research, noting the three communities participating in both the earlier and the later studies. The work covered a broad geographical area and included three cultural groups: Yukon First Nations (eleven communities), Dene/Métis (eighteen communities), and Inuit (eighteen communities) in five regions. The numbers of species known to be consumed as traditional food, and still used, were clarified within community focus groups during the research and are shown in table 2.1. Several species were used commonly (e.g., caribou), but more than 300 independent species were reported. Only Inuit regularly consume sea mammal species.

The use of traditional food within communities varies by age and gender. Examples of this phenomenon from the three communities first studied are shown in figure 2.2, where the number of grams of dry weight traditional food/1,000 kilocalories of total diet is shown to vary significantly by age. Clearly, older men and women consume more total traditional food than do younger individuals, and men tend to consume more than do women.[6]

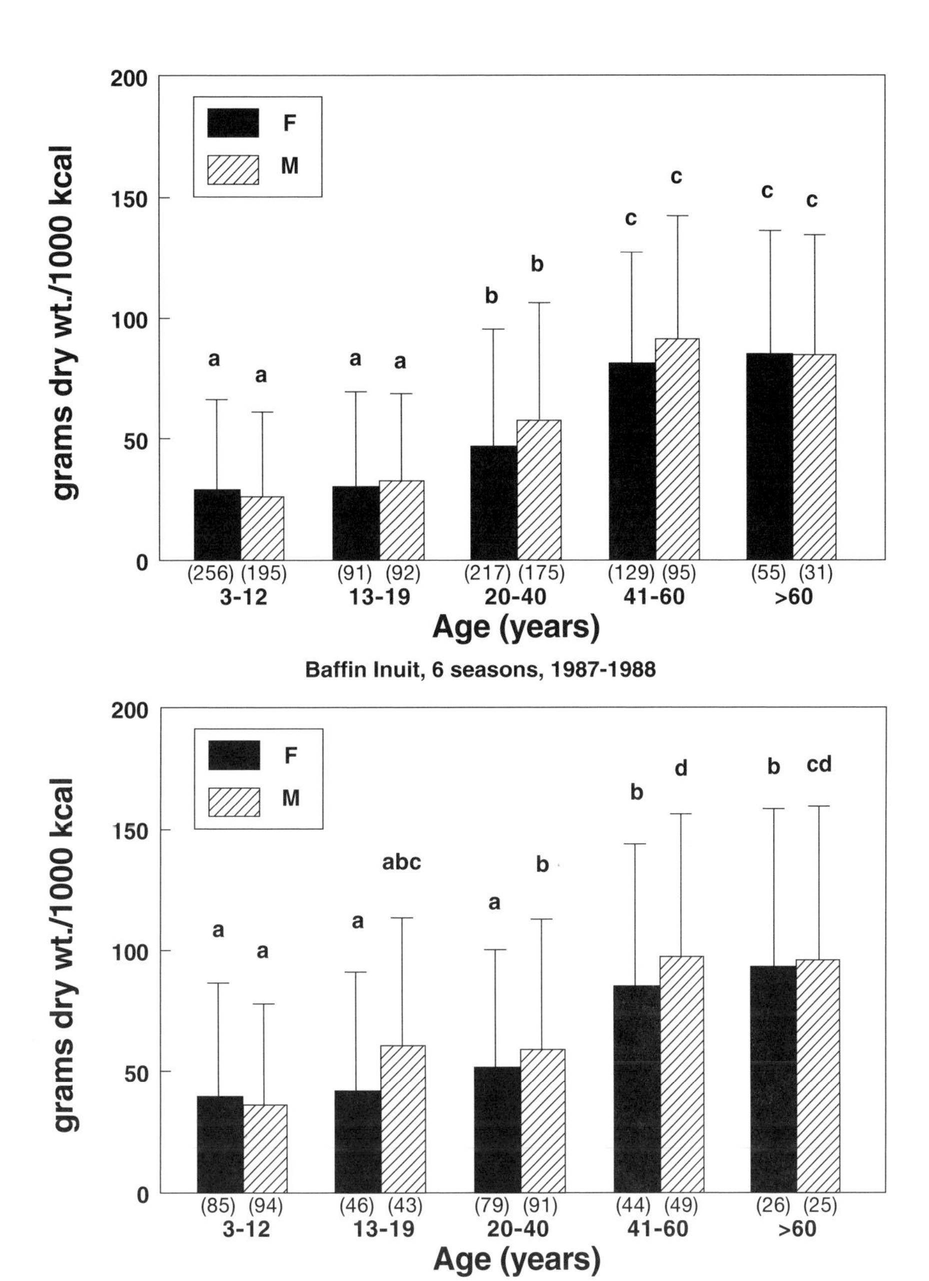

Figure 2.2 Seasonal Average Food Intake: Dry Weight/1,000 kcal. Bars indicated by different lowercase letters are statistically different (Bonferroni *t* test, age groups; $p < 0.05$). Values in parentheses on x-axis are number of records, total participants. F, female; M, male.

Source: H.V. Kuhnlein, "Benefits and Risks of Traditional Food for Indigenous Peoples: Focus on Dietary Intakes of Arctic Men," *Canadian Journal of Physiology and Pharmacology* 73 (1995): 765-71. Reprinted with permission.

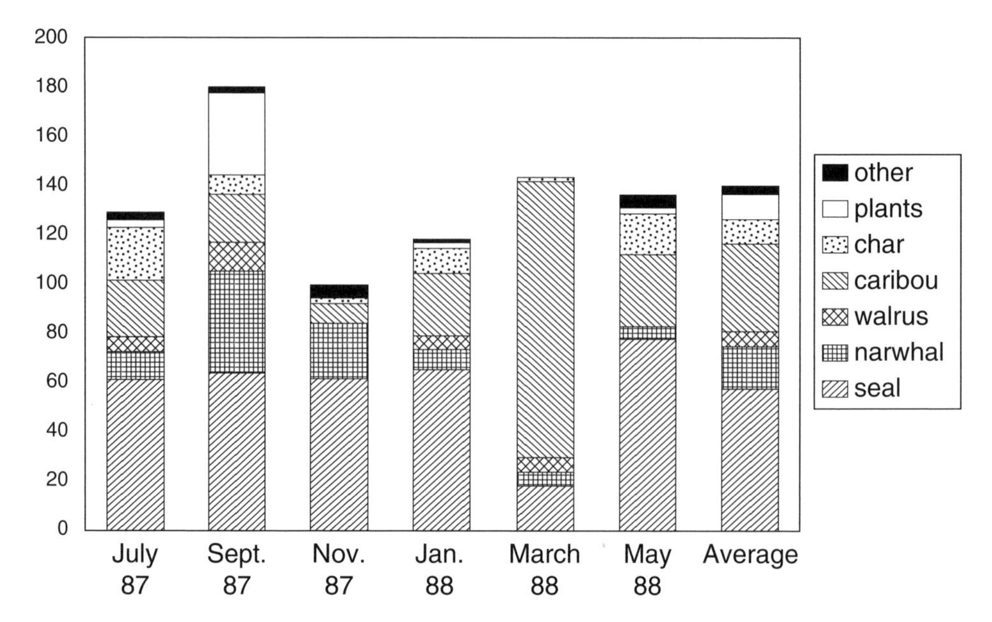

Figure 2.3 Total traditional food Consumed per Day Total Population (adjusted), Qikiqtarjuaq, Baffin Island, July 1987–May 1988. 1 unit on y-axis = 1,000 g.

Source: H.V. Kuhnlein, Benefits and Risks of Traditional Food for Indigenous Peoples: Focus on Dietary Intakes of Arctic Men, *Canadian Journal of Physiology and Pharmacology* 73 (1995): 765-71. Reprinted with permission.

The use of traditional food varies by season, which affects the amounts of market food used as well. The daily amount of traditional food consumed by one Baffin community through six seasons is shown in figure 2.3. Data show the peak season for traditional food use to be September, and minimum traditional food use is noted for November. The annual average consumption of traditional food for the entire community (450 individuals) is approximately 140 kg/day (309 lbs) for the time period the research was conducted. Recognizing that total dietary energy has contributions from both traditional and market food, figure 2.4 demonstrates how dietary intake data can discern the overall proportions of consumption of traditional and market food. For men in one Baffin community, the average market food contribution to energy was sixty-two per cent, while thirty-eight per cent came from traditional food species. In the Sahtu communities, men consumed similar amounts, with sixty-three per cent to sixty-nine per cent of energy from market food and the balance from traditional food species.[7]

Table 2.2 lists the twenty-five most frequently consumed market foods (by weight) in all Arctic areas. It is intuitively obvious that the nutrient quality of these food items is inferior to that of the traditional foods most frequently consumed. Table 2.3 lists the major contributors to dietary energy, protein, fat, vitamin A, iron, and zinc from traditional and market food reported in

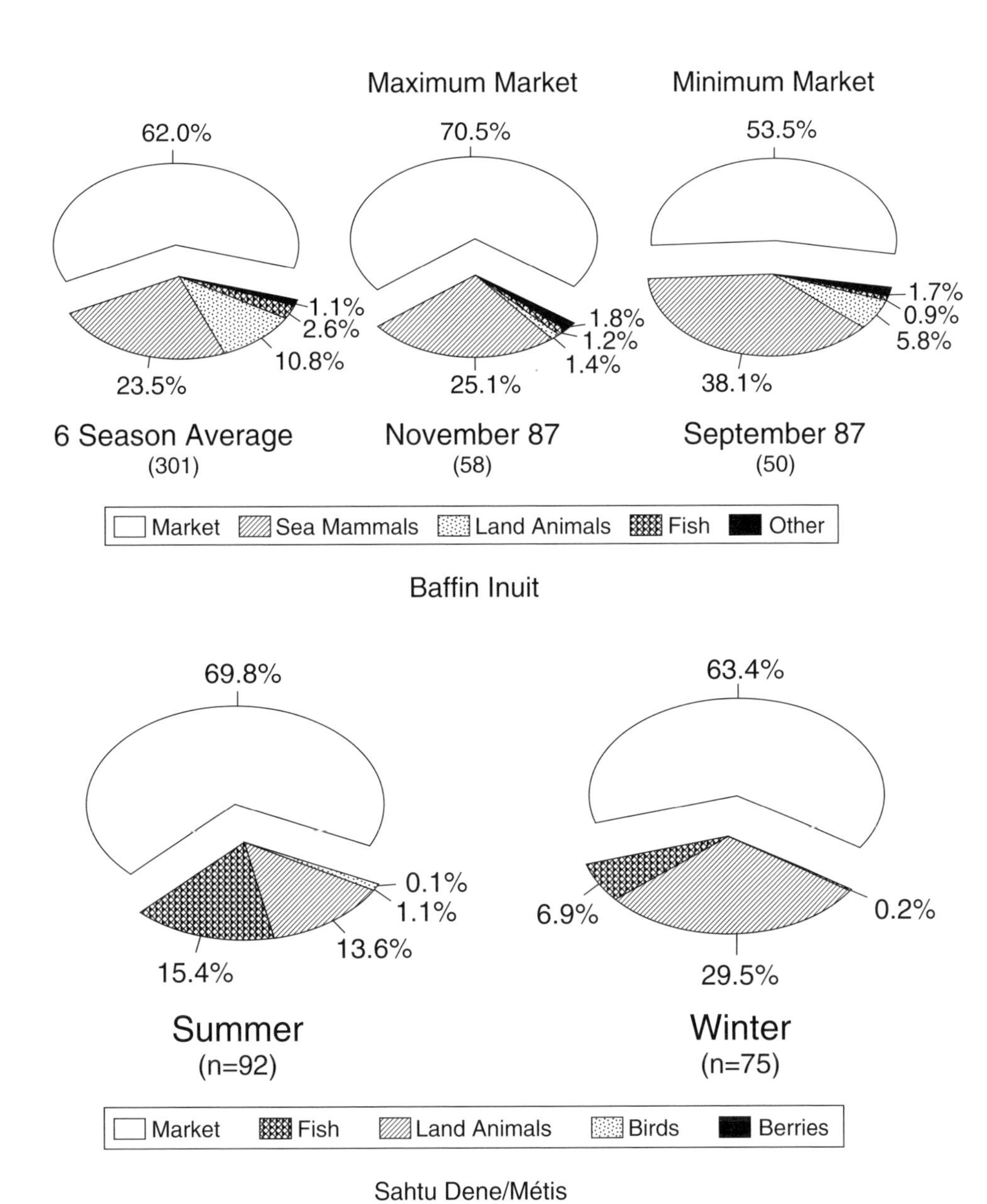

Figure 2.4 Use of Market and Traditional Food by Men (percentage of total energy intake). Values in parentheses are number of individual 24-h records. "Other" refers to birds, shellfish, and plants.

Source: H.V. Kuhnlein, "Benefits and Risks of Traditional Food for Indigenous Peoples: Focus on Dietary Intakes of Arctic Men," *Canadian Journal of Physiology and Pharmacology* 73 (1995): 765-71. Reprinted with permission.

Table 2.2
Most Commonly Consumed Market Food in the Arctic

Tea	Whitener, beverage
Sugar, granulated	Potatoes
Bread, white	Fat, vegetable, solid
Bannock/biscuits	Soup, dry mixes
Lard	Jelly/Jam
Drinks, crystals	Rice
Coffee, instant	Beef, ground
Milk, evaporated	Chicken, pieces fried
Corn flakes	Macaroni/cheese, mix
Cola-type beverages	Bacon
Butter	Oatmeal
Eggs	Frankfurters
Chips, salty snacks	

interviews with women in one Baffin community. Both traditional and market foods contribute to overall mean individual intakes of these nutrients, with traditional food making the major nutrient contributions.[8] Representing the same principle, figure 2.5 shows that Inuit women's intake of traditional food contributes significantly more of the nutrients represented (except calcium) than does the market food these women consume.[9]

Thus, in considering the contribution of traditional food to intake of organochlorine contaminants, one must understand the patterns and importance of traditional food to the overall dietary patterns and essential nutrient intake. Figure 2.6 shows the overall community average consumption of traditional food, as dietary energy contribution percentage, in the three major Arctic cultural areas. This varies from a low of six per cent in communities closer to commercial centres to a high of forty per cent in more remote communities.[10] Table 2.4 shows that the three cultural areas have similar overall mean individual adult intakes of carbohydrates and fats and that there are significant differences in intakes between days when traditional food is consumed and days when it is not. Significantly more carbohydrate, sucrose, fat, and saturated fat is taken in on days when only market food is consumed. The effect of increased dietary carbohydrate and fat from market food, with its implications for impending obesity and chronic disease, cannot be ignored. As use of traditional food is gradually declining among the younger generations and being replaced with low-nutrient-density market food while other lifestyle factors are also changing, health consequences are to be expected.[11]

It is also important to note the significance of the cultural values of using traditional food. Table 2.5 lists social and cultural values associated with harvest and use of traditional food by communities as reported in the three cultural

Table 2.3
Traditional and Market Food Making the Main Contribution to Intakes of Selected Nutrients for Inuit Women Residing on Baffin Island[a]

Food item	*Energy (Kcal)*	*Food item*	*Protein (g)*	*Food item*	*Fat (g)*
Bannock	279	Ringed seal meat	33	Ringed seal blubber	10
Ringed seal meat	189	Caribou meat	31	Bannock	8
Caribou meat	169	Narwhal mattak	7	Ringed seal broth	4
Table sugar	135	Bannock	6	Ringed seal meat	4
Ringed seal blubber	97	Arctic char flesh	5	Cookies	4
Cookies	94	Chicken meat	5	Ground beef	4
Soft drink	67	Walrus meat	4	Caribou meat	2
Narwhal mattak	50	Bearded seal meat boiled	3	Caribou fat	2
Ground beef	48	Narwhal meat	3	Vegetable shortening	2
Ringed seal broth	43	Bearded seal meat raw	3	Narwhal blubber	2
Food item	*Vit A (RE)*	*Food item*	*Iron (mg)*	*Food item*	*Zn (mg)*
Ringed seal liver	351	Ringed seal meat	23	Caribou meat	5
Ringed seal blubber	56	Caribou meat	6	Ringed seal meat boiled	4
Narwhal mattak	53	Walrus meat	3	Narwhal mattak raw	2
Narwhal blubber	26	Bearded seal meat	3	Ringed seal meat raw	1
Macaroni and cheese	21	Narwhal meat	3	Walrus meat	1
Carrots	21	Bannock	2	Bannock	1
Eggs	18	Ringed seal broth	2	Ground beef	1
Ringed seal meat	16	Ringed seal liver	1	Ringed seal pup meat	0.5
Vegetable/beef soup	16	Ringed seal pup meat	1	Narwhal mattak boiled	0.3
Arctic char flesh	12	Cornflakes	1	Chicken meat	0.3

Source: H.V. Kuhnlein, R. Soueida, and O. Receveur, "Dietary Nutrient Profiles of Canadian Baffin Island Differ by Food Source, Season, and Age," *Journal of the American Dietetic Association* 96 (1996): 155-62. Reprinted with permission.

[a]Main food contributors ranked by overall average of each season's mean intake (n=401, pregnant/lactating women not included).

areas. While it is important to evaluate exposure to POPs and other organochlorines, the cultural values of the foods containing these toxins must also be considered, especially when attempting to manage risk of POPs exposure.

Figure 2.7 illustrates how seven organochlorines in traditional food consumed by women in one Inuit and two Dene communities are expressed in mean and median daily intakes measured against current safe intake standards. This figure represents the range of average intakes by individuals for the communities; it does not necessarily represent annual average or usual individual intakes. Clearly, single-day intakes in the Baffin community often exceeded the safe intake guideline for chlordane, dieldrin, PCBs, and toxaphene, and consumption described in the dietary recall of some Sahtu women exceeded the guidelines for chlordane and toxaphene.[12] Figure 2.8 is an overall evaluation of

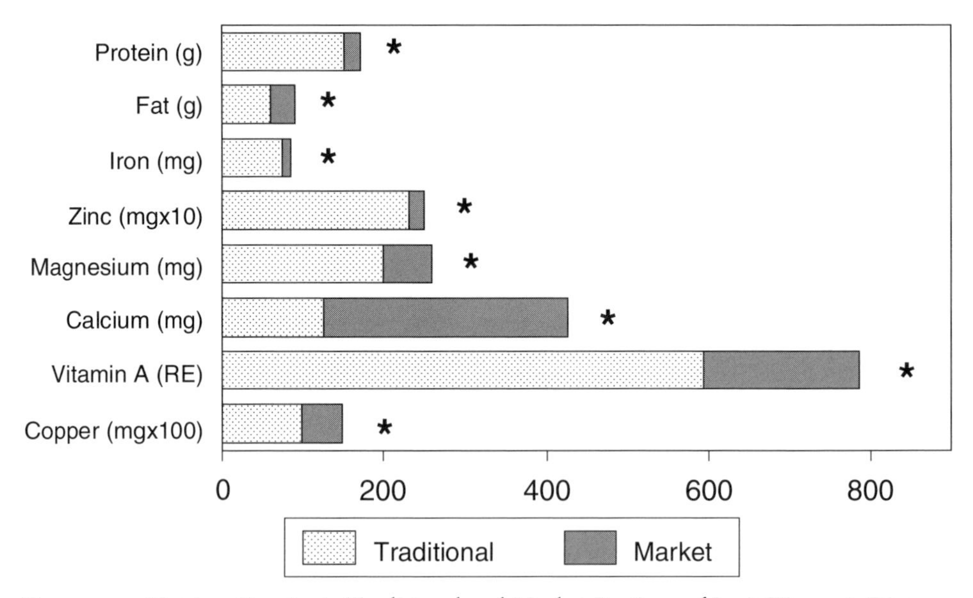

Figure 2.5 Nutrient Density in Traditional and Market Portions of Inuit Women's Diets (N=291). Values with an asterisk are significantly different ($p < 0.05$).

Source: H.V. Kuhnlein, O. Receveur, and H.M. Chan, Traditional Food Systems Research with Canadian Indigenous Peoples, *International Journal of Circumpolar Health* 60 (2001): 112-22. Reprinted with permission.

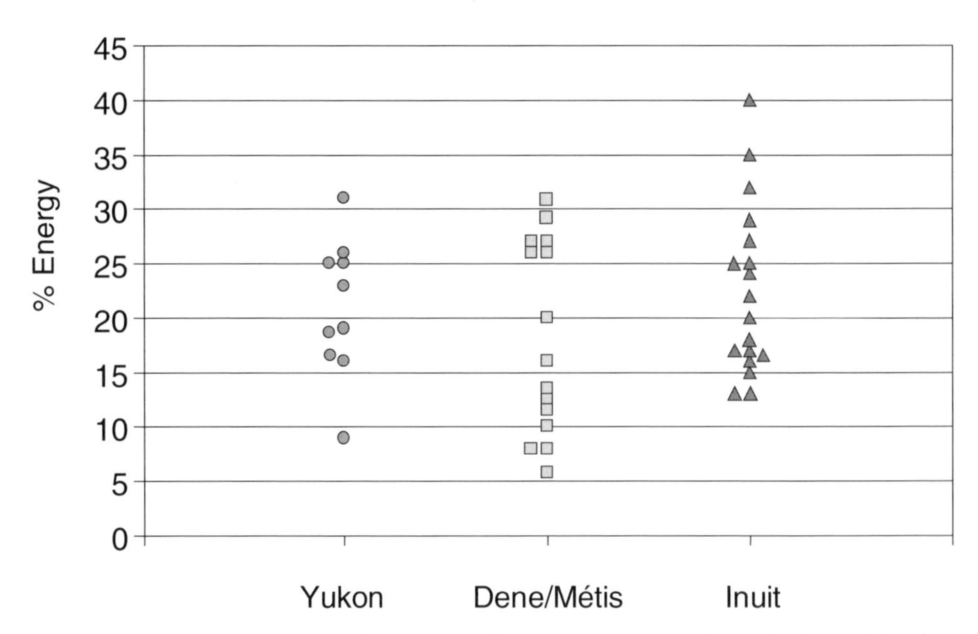

Figure 2.6 Average Percentage of Consumed Total Dietary Energy from Traditional Food in Yukon, Dene/Métis, and Inuit Communities

Source: H.V. Kuhnlein, O. Receveur, and H.M. Chan, Traditional Food Systems Research with Canadian Indigenous Peoples, *International Journal of Circumpolar Health* 60 (2001): 112-22. Reprinted with permission.

Table 2.4
Carbohydrate and Fat on Days with or without Traditional Food (TF) (least square means ± SEM)

% of energy		*With* TF	*Without* TF
Carbohydrate	Yukon	37–0.6 (n=413)	42–0.7* (n=389)
	Dene/Métis	38–1 (n=662)	44–1* (n=350)
	Inuit	37–0.5 (n=1092)	49–0.6* (n=783)
Sucrose	Yukon	9–0.3	12–0.4*
	Dene/Métis	9–1	13–1*
	Inuit	13–0.4	18–0.5*
Fat	Yukon	30–0.6	40–0.5*
	Dene/Métis	31–1	37–1*
	Inuit	30–0.4	35–0.5*
Saturated Fat	Yukon	11–0.3	14–0.2*
	Dene/Métis	11–0.3	13–0.3*
	Inuit	9–0.1	12–0.2*

Source: Canadian Arctic Contaminants Assessment Report II, Human Health (Ottawa: INAC, in Press)
* significant $p < 0.05$ (adjusted for season, site, gender, age)

Table 2.5
Important Socio-Cultural Benefits of Harvesting and Using Traditional Food

Contributes to physical fitness and good health	Brings respect from others
Is a favourite outdoor recreation activity	Builds one's pride and confidence
Provides people with healthy food	Provides education on natural environment
Keeps people "in tune with" nature	Contributes to children's education
Favours sharing in the community	Provides skills in survival
Saves money	Provides skills in food preparation at home
Is an essential part of the culture here	Is an opportunity to teach spirituality
Is an occasion for adults to display responsibility for their children	Is an opportunity to learn patience and other personality qualities
Is one way to practice spirituality	Is a way to strengthen the culture
Contributes to humility	

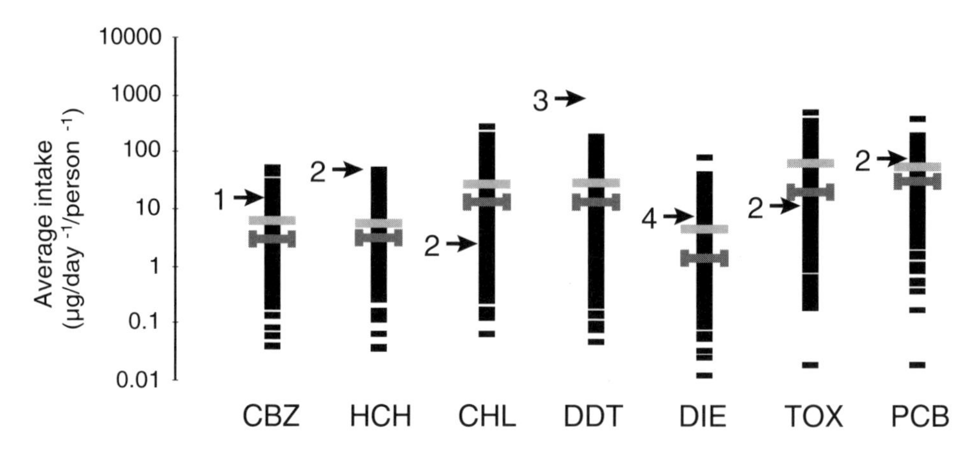

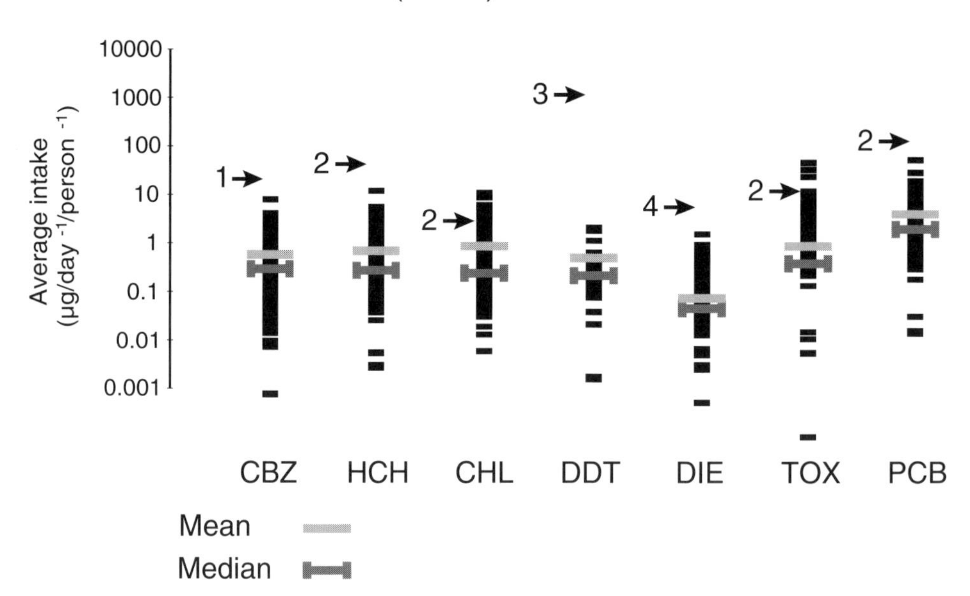

Figure 2.7 Distribution of Organochlorine Exposure from Traditional Food Consumed by Arctic Indigenous Women. The group mean and median intakes are presented in comparison to reference values: (1) Provisional Tolerable Daily Intake (PTDI), Health Canada; (2) Tolerable Daily Intake (TDI), Health Canada; (3) Acceptable Daily Intake (ADI), World Health Organization; (4) Tolerable Daily Intake (TDI), World Health Organization.

Source: Kuhnlein et al., "Arctic Indigenous Women Consume Greater than Acceptable Levels of Organochlorines," *Journal of Nutrition* 125 (1995): 2501-10. Reprinted with permission.

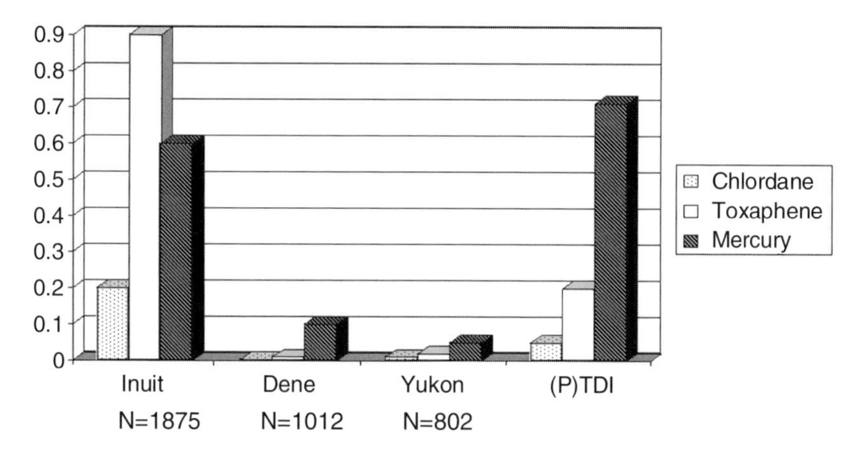

Figure 2.8 Mean Intakes of Chlordane, Toxaphene, and Mercury in Northern Canada (µg/kg/d)

Source: H.V. Kuhnlein, O. Receveur, and H.M. Chan, Traditional Food Systems Research with Canadian Indigenous Peoples, *International Journal of Circumpolar Health* 60 (2001): 112-22. Reprinted with permission.

the three larger studies for intakes of chlordane, toxaphene, and mercury. The average intake (expressed as µg/kg/day) for Inuit exceeded the PTDI (provisional tolerable daily intake) for chlordane and toxaphene, but not for mercury; the Dene and Yukon mean intakes did not exceed the PTDIs for these three contaminants.[13]

While it is known that sea mammal fats contain the highest amounts of organochlorines, it is important to understand the contribution of these fats to overall intake. Table 2.6 demonstrates how dietary intakes in the Baffin region presented exposure for chlordane, PCBs, and toxaphene. While caribou, ringed seal meat, and Arctic char were the major food items by weight, the three top contributors of chlordane, PCBs, and toxaphene were narwhal blubber, walrus blubber, and beluga blubber. For mercury, the top three contributors were caribou meat, ringed seal meat, and narwhal mattak (not shown). Thus, when considering how to revise the diet to offer protection against both organochlorine and heavy metal contaminants, several traditional food species must be carefully considered.

Note that food items containing the highest levels of contaminants are not always those making the greatest percentage contribution to exposure. Eating a large amount of a food with a low contaminant level may contribute more to total contaminant intake than does eating a small amount of food with a high contaminant level. Moreover, it is understood that these animal and fish food sources contain multiple contaminants as well as multiple essential nutrients. When multiple contaminants are contained in many, if not all, species in a traditional food system for one culture, the complexities of understanding

Table 2.6
Sources of Organochlorines in the Baffin Region (Contribution %)

Species	*Part*	*Weight*	CHL	PCB	TOX
Caribou	Flesh	38.2	0.9	1.3	0.1
Ringed seal	Flesh	18.7	0.8	2.4	8.9
Arctic char	Flesh	15.6	2.2	1.5	3
Narwhal	Muktuk	5	1.8	7	0.1
Walrus	Flesh	3.2	1.7	0.6	0.1
Ringed seal	Broth	2.9	0.2	1.1	0.1
Polar bear	Flesh	2.8	1.5	3.1	0.1
Narwhal	Blubber	1.9	37.9	44.5	35.6
Ptarmigan	Flesh	1.3	0	0	0
Beluga	Muktuk	1.2	1.7	1.7	0.9
Walrus	Blubber	1.2	34.9	22.2	43.1
Beluga	Blubber	0.4	11.1	8.5	6.3
Ringed seal	Blubber	0.3	1.9	1.3	0.3
Polar Bear	Fat	0.1	2.3	1.6	0.5
Total (%)		92.8	98.9	96.8	99.1

Source: Canadian Arctic Contaminants Assessment Report II, Human Health (Ottawa: INAC, in press).

dietary exposure and overall risk are formidable. Understanding dietary intake in detail is needed to determine the best way to modify the diet (which species, how much?) to reduce intake of certain or all contaminants, while maintaining maximum possible nutrient and cultural benefits that traditional food provides.

Discussion, Conclusions, and What's Ahead

These Arctic dietary studies have demonstrated the following:

a) variation in traditional food and market food contributions to total diet;
b) age and gender differences in food use;
c) seasonal differences in food use;
d) cultural and geographical differences in traditional food use;
e) higher intakes of carbohydrate and fat when traditional food is not consumed;
f) higher overall nutrient density of traditional food in comparison to market food;
g) cultural values associated with traditional food harvest and use;
h) cultural and geographical differences in contaminant intakes; and
i) influences of contaminant contents of food and dietary pattern on overall contaminant intakes by individuals and as a population.

Table 2.7 offers an overview of the identified risks and benefits of traditional food harvest and use by Arctic Indigenous peoples. Weighing the largely uncertain risks for chronic low-dose exposure to contaminants with the known probability of benefits is a complex process. Since the sciences of nutrition and toxicology have different paradigms of assessment, the probabilities of risk and benefit are on different scales. The exposure levels for Arctic populations are generally below the lowest observable effect levels (LOELs) found in occupational and accidental exposure incidences; however, they may be higher than public health guideline levels that incorporate safety factors ranging from several-fold to two orders of magnitude below LOELs. Therefore, the individual's risk for health effects, while unknown, is certainly low. Further, toxicological assessments include safety factors, making the individual's probability of risk uncertain. In contrast, nutrient benefits of traditional food use are known with certainty, food costs are easily measured, socio-cultural benefits from using traditional food are readily attained, and physical activity during hunting and fishing is obvious. Intuitively, physical activity and good diet help prevent chronic disease; research generally leads to this conclusion.[14] In considering these benefits, one must also recognize that there are serious nutritional, chronic-disease, socio-cultural, and economic effects (risks) of not using traditional food. If the use of traditional food is reduced because of potential but largely unproven health effects from low levels of contaminants in seasonal food, the alternative is to increase consumption in the Arctic of available market food which, at this time, is recognized as poor quality.[15]

Weighing the various benefits of using traditional food against risk of contaminants cuts across the disciplines of toxicology, nutrition, environmental policy, and public health practice. Each discipline has unique perspectives. If risk from a food source is not life-threatening, decisions are usually left with the consumer; this is particularly so for wildlife food, which is not under food industry or government regulation (i.e., wherein an item is removed from the marketplace when it exceeds certain residue limits). While consumer advisories based on residue limits and projected intakes can be issued, they are not widely respected, and consumption of the food continues when people intuitively recognize the significant benefits. In fact, restrictive fish consumption advisories have been seriously questioned because of the overriding nutritional benefits in contrast to uncertain risks.[16]

Good sense is obviously needed on a case-by-case basis for traditional food use. Continued advice to breastfeed infants is an example of how a traditional food with multiple known benefits to both infant and mother is still advised, even though contaminants are recorded in the milk of lactating women worldwide. Northern Indigenous women, who consume a wide variety of traditional food and whose milk is known to contain organochlorines, are still advised that breast milk is the best food for infants.[17]

Table 2.7
Identified Risks and Benefits of Traditional Food Use by Arctic Indigenous Peoples[a]

Risks (unknown or low probability)	*Benefits (known with certainty)*
Effects of low-level chronic exposure from: - heavy metals (Hg, Pb, d, As) - organochlorines (aldrin, chlordane Cl-benzenes, DDT, dieldrin, dioxin HCH, PAH, PCB, toxaphene) - radionuclides (^{137} Cs, ^{134} Cs, ^{90} Sr, ^{131} Io, ^{210} Po, ^{226} Ra, ^{210} Pb)	High nutrient density - protein, iron, zinc - vitamins A, D, E - fatty acids - other nutrients
Possible effects from simultaneous multiple contaminants: - neuro-behavioural - developmental - immune system - kidney damage - cancer	Prevention of chronic disease - obesity - diabetes - cardiovascular - other
Synergistic effects with drugs, alcohol, other products	Lower food costs
Accidents during hunting/fishing (known probability)	Physical activity in harvest
	Sociocultural values - cultural identity - preferred taste, flavours, etc. - favours community sharing - contributes to children's education - contributes the healthiest food - shows responsibility for others - participation in nature conservation - other

Source: H.V. Kuhnlein and H.M. Chan, "Environment and Contaminants in Traditional Food Systems of Northern Indigenous Peoples," *Annual Review of Nutrition* 20 (2000): 595-626. Reprinted with permission.

There are many political issues related to use of traditional food, contaminants in these foods, and multiple benefits from their use. Political and industrial will is needed to support research into understanding how the contaminants now in global environments are present in traditional food systems of Indigenous peoples, with special attention to effects of contaminant mixtures in high-nutrient foods.

Despite its considerable cost, appropriate education for Indigenous peoples should be provided to emphasize the reasoned considerations of risks and benefits of using traditional food and the consequences of adopting a diet of poor-quality market food.[18] While it is very frustrating and uncomfortable for Indigenous peoples to have different expert opinions on risks of consuming their traditional food, the best solution is to demystify the science so that the concepts of benefit and risk evaluations are easily understood.

Finally, it must become abundantly clear that the only reasonable and permanent solution to protecting health from organochlorines and other contaminants in the traditional food systems of Indigenous peoples is to intensify efforts in all possible global political arenas to stop emissions of persistent pollutants into global environments.

Acknowledgments

We thank all who have contributed to this research over many years. We acknowledge Rula Soueida for her assistance with the manuscript.

NOTES

1 H.V. Kuhnlein, "Nutritional and Toxicological Components of Inuit Diets in Broughton Island, Northwest Territories," Report to Health Canada, Yellowknife, 1989; D. Kinloch, H.V. Kuhnlein, and D.C.G. Muir, "Inuit Foods and Diet: A Preliminary Assessment of Benefits and Risks," *Science of the Total Environment* 122 (1992): 247–78.

2 H.V. Kuhnlein, "Dietary Evaluation of Food, Nutrients, and Contaminants in Fort Good Hope and Colville Lake, Northwest Territories," Report to Health Canada, Yellowknife, 1991.

3 O. Receveur et al., "Variance in Food Use in Dene/Métis Communities," Report to Arctic Environmental Strategy (Ottawa: INAC 1996); O. Receveur et al., "Yukon First Nations' Assessment of Dietary Benefit/Risk," Report to Arctic Environmental Strategy (Ottawa: INAC 1998); H.V. Kuhnlein et al., "Assessment of Dietary Benefit/Risk in Inuit Communities," Report to Northern Contaminants Program (Ottawa: INAC 2000).

4 H.V. Kuhnlein, "CINE Strategic Plan, Centre for Indigenous Peoples' Nutrition and Environment" (Montreal: McGill University 1999).

5 P. Bjerregard and T.K. Young, *The Circumpolar Inuit – Health of Population in Transition* (Copenhagen: Munksgaard 1998); B.M. Popkin, "The Nutrition Transition in Low Income Countries: An Emerging Crisis," *Nutrition Reviews* 52, no. 9: 285–98.

6 H.V. Kuhnlein, "Benefits and Risks of Traditional Food for Indigenous Peoples: Focus on Dietary Intakes of Arctic Men," *Canadian Journal of Physiology and Pharmacology* 73 (1995): 765–71.

7 H.V. Kuhnlein, "Benefits and Risks of Traditional Food for Indigenous Peoples."

8 H.V. Kuhnlein, R. Soueida, and O. Receveur, "Dietary Nutrient Profiles of Canadian Baffin Island Inuit Differ by Food Source, Season, and Age," *Journal of the American Dietetic Association* 96 (1996): 155–62.

9 H.V. Kuhnlein, "Dietary Nutrient Profiles"; H.V. Kuhnlein, O. Receveur, and H.M. Chan, "Traditional Food Systems Research with Canadian Indigenous Peoples," *International Journal of Circumpolar Health* 60 (2001): 112–22.

10 Kuhnlein, "Traditional Food Systems Research."

11 H.V. Kuhnlein and O. Receveur, "Dietary Change and Traditional Food Systems of Indigenous Peoples," *Annual Review of Nutrition* 16 (1996): 417–42; Bjerregard, *The Circumpolar Inuit*; O. Receveur, M. Boulay, and H.V. Kuhnlein, "Decreasing Traditional Food Use Affects Diet Quality for Adult Dene/Métis in 16 Communities of the Canadian Northwest Territories," *Journal of Nutrition* 127, no. 11 (1997): 2179–86.

12 H.V. Kuhnlein et al., "Arctic Indigenous Women Consume Greater Than Acceptable Levels of Organochlorines," *Journal of Nutrition* 125, no. 10 (1995): 2501–10.

13 Kuhnlein, "Traditional Food Systems Research."

14 Popkin, "The Nutrition Transition"; Bjerregard, *The Circumpolar Inuit*; Kuhnlein, "Dietary Change and Traditional Food Systems."

15 H.V. Kuhnlein and H.M. Chan, "Environment and Contaminants in Traditional Food Systems of Northern Indigenous Peoples," *Annual Review of Nutrition* 20 (2000): 595–626.

16 G.M. Egeland and J.P. Middaugh, "Balancing Fish Consumption Benefits with Mercury Exposure," *Science* 278 (1997): 1904–05.

17 E. Dewailly et al., "Polychlorinated Biphenyls (PCBs) and Dichlorodiphenyl Dichloroethylene (DDE) in Breast Milk of Women Living in the Province of Quebec, Canada," *American Journal of Public Health* 86 (1996): 1241–6; J. Van Oostdam et al., "Human Health Implications of Environmental Contaminants in Arctic Canada: A Review," *Science of the Total Environment* 230, nos 1–3 (1999): 1–82.

18 J. Jensen, K. Adare, and R.G. Shearer, *Canadian Arctic Contaminants Assessment Report* (Ottawa: INAC 1997).

3
Canadian Research and POPs: The Northern Contaminants Program

RUSSEL SHEARER AND SIU-LING HAN

CONTAMINANTS IN THE CANADIAN ARCTIC: REASON FOR CONCERN

Canada's North comprises three territories with ninety-three communities, most of them home to small populations of First Nations, Métis, or Inuit people. The total population numbers fewer than 100,000 and most are spread thinly across more than 3,775,000 square kilometres (1,458,000 sq. mi.).[1] The population in the North is young, with forty-five per cent under the age of twenty-five, and about half is Aboriginal. Throughout thc North, cultural identities, including stewardship of the land, remain strong and traditional harvesting and arts and crafts are important dimensions of the economy.

Scientists had been monitoring contaminants in the Canadian Arctic since the 1970s, but by the late 1980s some disturbing findings heightened concern over this issue:

- Increased evidence that traditional/country foods contained contaminants raised questions about resulting contaminant levels in northern Aboriginal peoples. Thus, in 1985–87 PCBs were measured in the blood of Inuit from the community of Broughton Island, Nunavut, known to have a relatively high per capita intake of traditional/country foods. Results showed that blood PCBs exceeded the tolerable levels set by Health Canada in sixty-three per cent of the women and men under fifteen years of age, in thirty-nine per cent of the women aged fifteen to forty-four, in six per cent of the men fifteen years and older, and in twenty-nine per cent of women forty-five years and older. This study also revealed elevated organic mercury levels in some individuals.[2]

- Most alarming was the finding that PCB levels in the milk of Inuit women from the east coast of Hudson Bay in Nunavik (northern Quebec) were

> approximately five times higher than those in women in southern Canada. The elevated levels in these Inuit were attributed to their position at the top of the Arctic marine food chain, as consumers of marine mammal tissues.[3]

Levels of persistent organic pollutants (POPs) in Canadian Inuit populations are among the highest observed in the world. In the Arctic, the major concern is long-term chronic exposures as animals and people are exposed to low levels over their entire lifetimes. Because very little work had been done in the Arctic on contaminants – particularly those of anthropogenic origin – before the mid-1980s, there were inadequate data to define the scope, magnitude, and likely duration of the problem.

POPs enter the Arctic through long-range transport on air and water currents, with the atmosphere being the primary pathway. Many of the pesticides and industrial chemicals are no longer used in Canada, and have, in many cases, been banned or restricted in most of the developed world. They are still in use in many developing nations and continue to revolatilize from formerly treated soil in regions where use has been banned.[4]

Arctic ecosystems are particularly vulnerable to POPs. Their characteristics, including low water solubility, ability to remain for long periods of time in the atmosphere in a gas phase or on particles, and high resistance to biological or chemical degradation, favour their long-range transport. Through numerous research programs, POPs have been detected throughout the Arctic ecosystem, including in air, surface seawater, suspended sediments, snow, fish, marine mammals, seabirds, caribou, and other terrestrial animals and plants.[5] Many of these chemicals are highly lipophilic and persistent; they accumulate in lipid-rich tissues of Arctic animals, reaching highest concentrations in the long-lived animals at the top of the food chain. Unexpectedly high levels have been found in some Arctic biota, especially fish and marine mammals, which are significant components of the diets of many Arctic residents. Chemical contamination of these traditional/country foods provides a critical path of contaminant transfer to human consumers, particularly northern Aboriginal peoples. Ninety-one per cent of Aboriginal households consume traditionally harvested meat and fish, and twenty-two per cent have reported that all their meat and fish are obtained through harvest.[6]

There are limited data directly linking POPs to adverse effects in wildlife. Evidence suggests that potential effects include reproductive and immune disorders and cancer.[7] In instances of human accidental or chronic exposure to high levels of contaminants, toxicological effects detected include reproductive disorders, immune dysfunction, developmental abnormalities, and reduced learning abilities. The extent to which these and other effects result from the levels being found in Arctic biota remains to be determined and is presently the subject of major investigation.[8]

The greatest concern is for exposure of the fetus and infant to the mother's lifetime concentrations of contaminants, which can be transferred to offspring during pregnancy and through nursing. The unborn child may be at risk of subtle effects related to learning ability, memory, and resistance to infection.[9]

CANADA'S RESPONSE: THE NORTHERN CONTAMINANTS PROGRAM

The early years

In 1985, an ad hoc committee of federal and territorial researchers, coordinated by Indian and Northern Affairs Canada (INAC), began to collaborate on the issue. The committee first conducted a baseline literature review and determined that there was a definite need to assess the extent of the contamination of fish and wildlife in Canada's North and to determine the implications for the health of northerners, particularly Aboriginal peoples. The main conclusion was that contaminants in the North are serious and widespread and that it was highly unlikely that small quantities of PCBs found at military sites were the cause.

Subsequently, a co-operative program (the Northern Contaminants Program (NCP)) was designed around an integrated ecosystem approach comprising monitoring, research, and evaluation. The main priority was to assess the extent of contamination of traditional/country foods used by northern people. The committee prepared a benchmark report summarizing the current state of knowledge on the subject and the gaps that remained, which was published in 1992 as a special issue of the journal *Science of the Total Environment*. Once the results of these initial studies were publicly released, the communications and education aspects of the program consisted largely of "fire-fighting" in response to inflammatory media articles. Results of the media coverage were often tragic: some people ceased to eat country foods and, in one isolated case, a mother stopped breastfeeding an infant, which resulted in malnutrition.

In December 1989, the committee convened a workshop to develop a long-term inter-agency research and monitoring strategy to deal with the issue. The strategy was designed in collaboration with Aboriginal organizations, four federal departments, and the (then) two territorial governments and encompassed the governments' responsibilities for delivering advice to northerners concerning health aspects of their traditional/country food diets and for pursuing contaminant sources and emission controls in the international forum. INAC chaired the inter-agency committees at both the policy and technical levels, ensuring coordination of the initiative.

Phase I of the NCP

Phase I of the NCP began in 1991 and ended in 1997 with funding made available under the federal government's Green Plan/Arctic Environmental Strategy (AES). In response to results of the co-operative studies undertaken in the mid- to late 1980s, the NCP embarked upon a comprehensive and coordinated approach to addressing the issue of Arctic contamination. The program assessed the risk to northern ecosystems and human health from the long-range transport of persistent contaminants to the Arctic. Thus, the NCP set an overall goal "To reduce, and wherever possible eliminate, contaminants in traditional/ country foods" and a work plan that would allow an understanding of the source, geographic extent, scope, magnitude, and duration of the problem and would identify and support whatever actions were necessary to move towards solutions. A formal INAC-led committee structure of responsible federal, territorial, and Aboriginal organizations was established to manage the NCP. The strategic research plan and priorities for the NCP were developed based on a multi-disciplinary ecosystem approach focusing on five main categories: identify contaminant sources and their transport to the Arctic; assess contaminant levels in fish and wildlife; assess effects of contaminants on the health of northern ecosystems, including human health; provide timely health advice to northern people; and establish international controls through agreements and co-operation with other countries and the circumpolar community.

Four federal government departments were involved in the NCP: Environment Canada (EC), which conducted research on atmospheric transport, wildlife, and the ambient environment; Health Canada, which advised on human health and assessed the risks of traditional foods; Fisheries and Oceans Canada (F&O), which conducted research on fish and marine mammals; and INAC, which coordinated the program, emphasizing communications and education, international control actions, and Aboriginal partnerships. The two territorial governments at the time also participated, particularly the departments of Renewable Resources and Health. Five northern Aboriginal organizations were partners, together representing all the Aboriginal peoples in the North: the Council of Yukon First Nations, The Dene Nation, the Métis Nation–N.W.T., the Inuit Tapirisat of Canada (now the Inuit Tapiriit Kanatami), and the Inuit Circumpolar Conference. Universities, research institutes, communities, and individuals provided research, advice, and information. In the last few years, as evidence of the risks to human health became evident, the Inuit of northern Quebec and Labrador were included in the program and Santé Québec (the Quebec Department of Health) became a participant.

The inclusion of northern Aboriginal organizations, made possible by AES funding, was an important facet of the NCP-Phase I. The role of the Aboriginal organizations in NCP decision-making was critical in orienting the program to the direct human implications of an issue that, until then, had been seen primarily from a scientific perspective. Through the leadership of the Aboriginal organizations, the NCP began to revise not only the actual projects to meet the needs of northern communities but also the manner in which the projects were carried out.

The strong partnerships in management of the program and the collaborative working relationships in its delivery resulted in development of an evolving infrastructure and capacity for addressing contaminants issues in the North. The improved capacity increased not only communities' abilities to address contaminant issues themselves but also NCP scientists' awareness of the needs of northerners and their openness to approaches to research that such awareness often entails. The completion of six years' work under the AES/NCP provided an excellent foundation, both scientifically and structurally, for moving into a second phase of the contaminants program in Canada's North.

From a scientific perspective, a critical achievement of the NCP-Phase I was the substantiation of the theory that contamination of the Canadian Arctic is due to sources outside the Arctic. The extensive spatial monitoring of wildlife, air, precipitation, and water completed over the six-year period allowed Canada to identify the contaminants that are occurring in the North, namely POPs, and analysis of these data provided the evidence to establish that the sources were external and that the main transport pathway was atmospheric. Much of the world's scientific data demonstrating contaminant movement to the North and the behaviour of chemicals in and among various components of the environment have come from NCP researchers.

NCP-Phase I generated several key findings that have led to calls for action:

- Contaminants have been detected in all components of the northern food chain.

- Animals high in the food chain and high in fat, such as marine mammals, have the highest levels of POPs.

- Sources of POPs and some metals are distant, and the contaminants are transported from the industrial and agricultural areas of the world to Canada's North by air and water currents.

- Inuit women have levels of PCBs and other POPs in their milk and blood that are five times higher than those of women in southern Canada and among the highest levels in the world.

- In parts of the North, levels of PCBs in mothers' blood are at, or exceed, levels that have been associated, in studies in the Great Lakes region, with neurobehavioural effects on children.

The strength of the evidence generated by the NCP has been such that, in only six years, Canada greatly accelerated its domestic response to persistent toxic substances through initiatives such as the federal Toxic Substances Management Policy. More importantly, Canada took an active leadership role in getting the issue of long-range transport of persistent toxic substances onto the international agenda. So compelling was the evidence generated by the NCP that POPs, which were not being addressed by the global community when the program began, are now the subject of legal controls signed in 1998 for the northern hemisphere (CLRTAP POPs Protocol) and are part of a legally binding global agreement (the Stockholm Convention).

Within Canada, the evidence generated over this period allowed us to identify the species carrying the highest contaminant loads and those that are very low in contaminants. We now know something about contaminant exposure of humans through both diet and body-burden data. There are now some data that tell us which contaminants people are most exposed to, what the main sources of human exposure are in different parts of the North, and what this means for the actual levels in people in various regions of the North.

Additionally, Canada assessed the importance of traditional/country food diets, which was crucial to support its arguments for international controls and plans for management of risk. For example, NCP data quantified the nutritional and dietary importance of traditionally harvested foods to northerners, providing context for the concerns engendered by contamination of the food chain but also heightening awareness of northerners' dependence on these foods and therefore of the critical and immediate need for action to reduce contaminant levels in those foods.

The need to generate relevant information and communicate it in a way that best allows northerners to make informed decisions about contaminants, nutrition, and consumption of traditional foods has emerged as an underlying principle of NCP policy and activity. Intensive deliberations conducted in 1993 laid the foundations for the NCP's evolving strategy on communications, community involvement, and capacity-building. The northern Aboriginal organizations, as the most credible sources of information on contaminants for Aboriginal peoples, play the lead role in the development of the NCP's communications strategies, which encompass diverse initiatives such as the establishment of guidelines making consultation and collaboration by scientists with communities a condition of project funding, the technical and financial support of communities to develop their own contaminants research projects, the creation of a contaminants curriculum for integration into northern schools, and the identification of regional contaminants expertise for communities.

Key reports and an international declaration for action

In June 1997, INAC and EC released the *Canadian Arctic Contaminants Assessment Report* (*CACAR*) and an accompanying "Highlights Report." Summarizing the leading-edge Canadian research, the *CACAR* provided evidence that the major sources of contaminants are from outside the Arctic, that the major pathway is the atmosphere, and that contaminants pose a threat to the health of northern Aboriginal peoples. The overall conclusion of the *CACAR* is: "Contaminants in the food chain are not thought to pose a direct threat to the health of adult humans. Contaminant levels in traditional/country food are low enough that a single serving, or even many servings, will not make someone sick. However, lifetime stores of contaminants in people may be at a level where the unborn child may be at risk of subtle effects related to learning ability, memory, and resistance to infection."[10]

Also in June 1997, the eight Arctic countries released the *State of the Arctic Environment Report* (*SOAER*), which summarizes all of the scientific data on contaminants generated by these nations under the Arctic Monitoring and Assessment Programme (AMAP), the circumpolar environmental monitoring activity of the Arctic Council, which began in 1991.

Canada's input to the AMAP, coordinated through the NCP, was significant and is detailed in the *CACAR*. The circumpolar data reaffirm Canadian findings of significant accumulation of contaminants in northern food chains, high exposure of Arctic human populations to environmental contaminants, and long-range transport of contaminants from source regions. The major conclusion of the AMAP's *SOAER* was that certain Arctic populations are among the most exposed populations in the world to certain contaminants and that some of these contaminants are carried to the Arctic via long-range transport and accumulate in animals that are used as traditional foods.[11]

On 13 June 1997, responding to the AMAP report, ministers of the eight circumpolar countries, including Canada, declared in the Alta Declaration on Environmental Protection and Sustainable Development: "**We welcome** with appreciation the AMAP reports and commit to take their findings and recommendations into consideration in our policies and programmes. **We agree** to increase our efforts to limit and reduce emissions of contaminants into the environment and to promote international co-operation in order to address the serious pollution risks reported by the AMAP. **We will** draw the attention of the global community to the content of the AMAP reports in all relevant international fora, particularly at the forthcoming Special Session of the General Assembly, and **we will** make a determined effort to secure support for international action which will reduce Arctic contamination."[12]

On 25 June 1997, the prime minister of Norway, speaking on behalf of all Arctic countries, delivered the speech to the UN General Assembly in New York and hosted a press conference and round table to draw international attention to the issue.

Use of Program Information in Key Management/Policy Initiatives

The key objectives of the NCP are:

1 To provide information that assists Northerners in making informed decisions about their food use.

2 To reduce and, where possible, eliminate contaminants in northern traditionally-harvested foods.

These objectives lead to decisions on food consumption and on contaminant controls. The NCP addresses these objectives through the following sub-programs: Monitoring the Health of Arctic Peoples and Ecosystems and the Effectiveness of International Controls, Human Health, Education and Communications, and International Policy. The following schematic illustrates the use of program information in key management/policy initiatives.

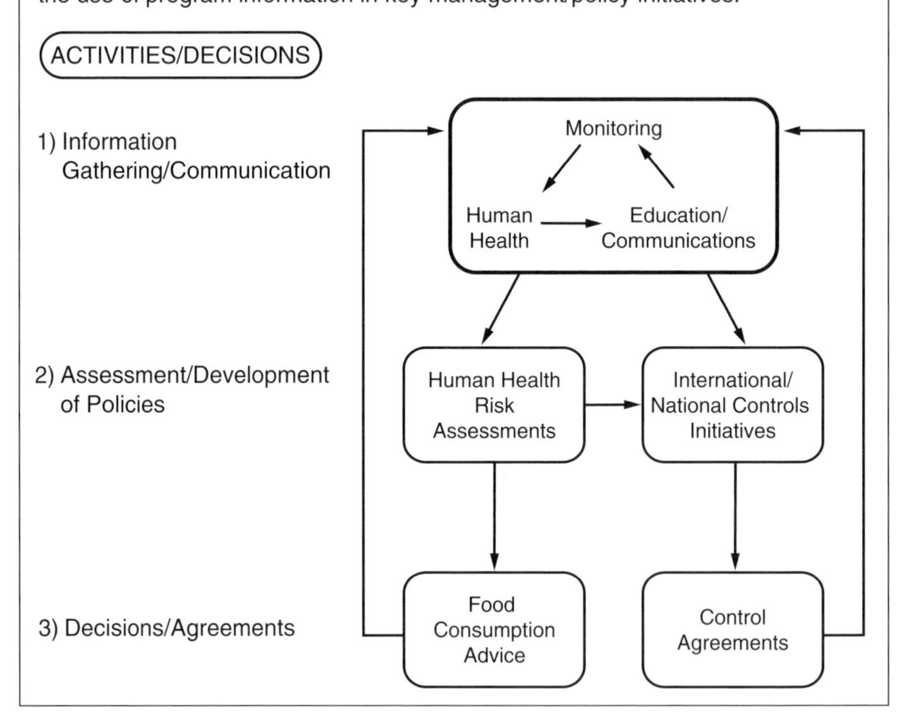

Figure 3.1

The Arctic ministers also agreed in the Alta Declaration to continue the AMAP to monitor and report on levels of contaminants in all relevant components of the Arctic environment and "to work vigorously for the early completion and implementation of a protocol" on POPs being negotiated under the United Nations Economic Commission for Europe's (UNECE) Convention on Long-range Transboundary Air Pollution (LRTAP).

Since then, three international agreements on contaminants of concern to the Arctic have been signed. The UNECE CLRTAP POPs and Heavy Metals protocols have been ratified by ten and twelve countries, respectively. In the case of both protocols, sixteen ratifications are necessary for entry-into-force. Canada signed and ratified both protocols in 1998. In May 2001, 111 countries signed the global POPs agreement, known as the Stockholm Convention. As of September 2002, twenty-one countries had ratified; fifty ratifications are necessary to have the convention enter into force. Canada signed the Stockholm Convention and was the first country to ratify it on 23 May 2001.

Consultations in the planning of Phase II of the NCP

Results of the *CACAR* provided a starting point for the extensive consultations conducted from 1996 to 1998 throughout the North with Aboriginal organizations, northern communities, contaminants researchers, program managers, and other stakeholders. The intent of the consultations was to find the common elements between northern community concerns and priorities regarding contaminants and the scientific needs critical for addressing the issue of contamination in Canada's Arctic. A series of workshops was held throughout 1996/97/98, including the "Ideas" Workshop in Yellowknife, N.W.T., in November 1996 and the Contaminants Results Workshop at the Institute of Ocean Sciences in Sidney, British Columbia, in January 1997. Building on the draft conclusions of the *CACAR*, these workshops provided guidance on the future priorities of NCP-Phase II. Priorities for research and monitoring under Phase II are based on an understanding of the species that are most relevant for human contaminant exposure and of geographic locations and populations that are most at risk.

Considerable effort was made to ensure that results of the NCP are used in carrying out human health risk assessments and in substantiating the need for national and international controls for the contaminants of concern as outlined in figure 3.1, *Use of Program Information in Key Management/Policy Initiatives*.

Phase II of the NCP

In 1998, the NCP began a second five-year phase addressing immediate health and safety needs of northerners concerning contaminants in traditional/country foods. The objective was "to reduce and wherever possible eliminate contaminants in traditionally harvested foods, while providing information that assists informed decision-making by individuals and communities in their food use." The five main components of the NCP-Phase II were a) human health research, to determine the risks to humans from consuming traditional/country foods and impacts on the developing fetus through contaminant exposure from the mother's diet; b) monitoring of the health of Arctic peoples and ecosystems and the effectiveness of international controls, to collect physical and biological data necessary to support human health risk assessments and international controls; c) education and communications, to provide northerners with the information needed to make informed decisions on their food use; d) international policy, to control the input of contaminants to the Arctic by coordinating international monitoring programs and taking action in the international negotiating sessions leading to regional and global agreements; and e) Aboriginal partnerships, to ensure appropriate communications and participation occur with northern communities. Each of these five components is interconnected and contributes to addressing immediate health and safety issues associated with contaminants in traditional/country foods.

The NCP nominally addresses concerns that are exclusively territorial but encompasses health issues relevant to all marine mammal consumers. Research in other northern areas such as Nunavik (northern Quebec) and Labrador indicate that the impact of contaminants on the health and safety of humans and ecosystems extends beyond the territorial boundaries. As a result, the NCP-Phase II officially extended its geographic scope to include the Inuit regions of Nunavik and Labrador.

Management Structure and Innovative Partnerships. Contamination of the Arctic food chain is a multi-jurisdictional issue of concern to federal, territorial, and Aboriginal governments, but it can be addressed only through co-operation from the international community on a global scale. Investigation of the concerns must yield information that meets internationally acceptable scientific standards and responds to the needs expressed by communities and by individual consumers of traditional/country foods.

A program that addresses such scientifically and politically complex issues requires well-developed management, planning, and implementation structures and strategies. The multi-disciplinary nature of the NCP has enabled it to develop such structures and strategies. The program derives much of its strength from the partnership approach that forms the basis of its management process. This approach encompasses representatives who address the key areas of Arctic contaminants research based on an ecosystem approach, northern community concerns, needs and priorities, and the international and domestic agendas for the control of toxic substances.

NCP-Phase II is directed by a management committee chaired by INAC, with a make-up similar to that of the Phase I management committee, including representatives from the five northern Aboriginal organizations (Council of Yukon First Nations, Dene Nation, Métis Nation-N.W.T. (up to its dissolution in 2000), Inuit Tapiriit Kanatami, and Inuit Circumpolar Conference Canada); Yukon, Northwest Territories, and Nunavut territorial governments; Nunavik Nutrition and Health Committee; and four federal departments (Environment, Fisheries and Oceans, Health, and Indian and Northern Affairs). The management committee is responsible for establishing NCP policy and research priorities and for final decisions on the allocation of funds. Three territorial contaminants committees in the Yukon, Northwest Territories, and Nunavut (formally established in 2000) and the Nunavik Nutrition and Health Committee support this national management committee.

In 1998, the management committee redesigned the NCP-Phase II. The two main initiatives undertaken were the development of blueprints representing the long-term vision and strategic direction for NCP-Phase II and the implementation of a more open and transparent proposal-review process. This new management structure is designed to ensure that the NCP remains scientifically defensible and socio-culturally aware while achieving progress towards its broad policy objectives.

Blueprints were developed for each of the four main NCP subprograms: Human Health; Monitoring the Health of Arctic People and Ecosystems and the Effectiveness of International Controls; International Policy; and Education and Communications. The blueprints, which are evolving documents reviewed at least annually, are used to provide the necessary guidance to project proponents in the development of proposals as well as to peer reviewers, review teams, and the management committee in the evaluation of proposals.

Under a revamped proposal-review process, the NCP Technical Committee, established under Phase I to provide a technical review of proposals, was replaced with an external peer-review process facilitated by review teams. The process includes both a scientific review by external peer reviewers, facilitated by technical review teams (the Human Health and Abiotic/Biotic Monitoring review teams) and a social/cultural review facilitated by a review team, the three Territorial Contaminants Committees (TCCs), the Nunavik Nutrition and Health Committee (NNHC), and the Labrador Inuit Association (LIA). The management committee considers the recommendations of both the technical and the socio-cultural reviews in making final funding decisions. Proposals submitted under the Education and Communications subprogram are evaluated by peer reviewers facilitated by a review team and the three TCCs and the NNHC. All peer reviewers, review teams, TCCs, and the NNHC use evaluation criteria and the blueprints to review and rate proposals.

As a condition for funding, written approval of consultation from the appropriate northern community authority or national-level Aboriginal organization is required for all projects involving field work or analyses of samples collected in the North. The Social/Cultural Review Team ensures that each proposal has attached consultation forms signed by the applicable regional or national authority.

Under Phase II, great strides have been made in developing a realistic and practical community consultation process that must be followed to obtain funding from the NCP. The process, which specifies requirements for all project proposals, was revised in 2001 to improve community consultation and clarify and streamline procedures.

All project proposals to be considered for funding under the NCP must include a signed Statement of Consultation form describing the specific details of the consultation that has occurred or is intended to take place. When any portion of the project – in any year of the project – is conducted within Nunavut, the N.W.T., the Yukon, Labrador, or Nunavik (including archived samples previously collected from one or more of these regions), consultation with the appropriate Aboriginal organization or regional/community bodies must be initiated during the development of the proposal and continue through the duration of the project. For proposed projects that have no northern component at any stage, proponents are asked for a brief statement explaining why no northern consultation is considered necessary. Any project proposal submitted to the NCP without a completed Statement of Consultation form is returned to the project proponent and not considered for funding.

All project proponents are also required to read and abide by the NCP Guidelines for Responsible Research, which provide direction to project leaders and scientists for planning communications and developing research agreements with communities.

ABORIGINAL PARTNERSHIPS

Through partnerships, the northern Aboriginal organizations have been able to develop their own internal capacity to work on contaminants issues and other important environmental issues with their constituents. The Aboriginal partners play several roles within the NCP. By participating in the NCP management structures, they provide advice and represent northern Aboriginal interests to INAC and the other NCP government partners, contributing input on communications as well as on research priorities. For instance, the Aboriginal partners have been able to work closely with many of the scientific partners involved in Arctic research, advising them on issues such as methods for improving community involvement, appropriate contacts in regions and communities, and specific communications problems related to their work.

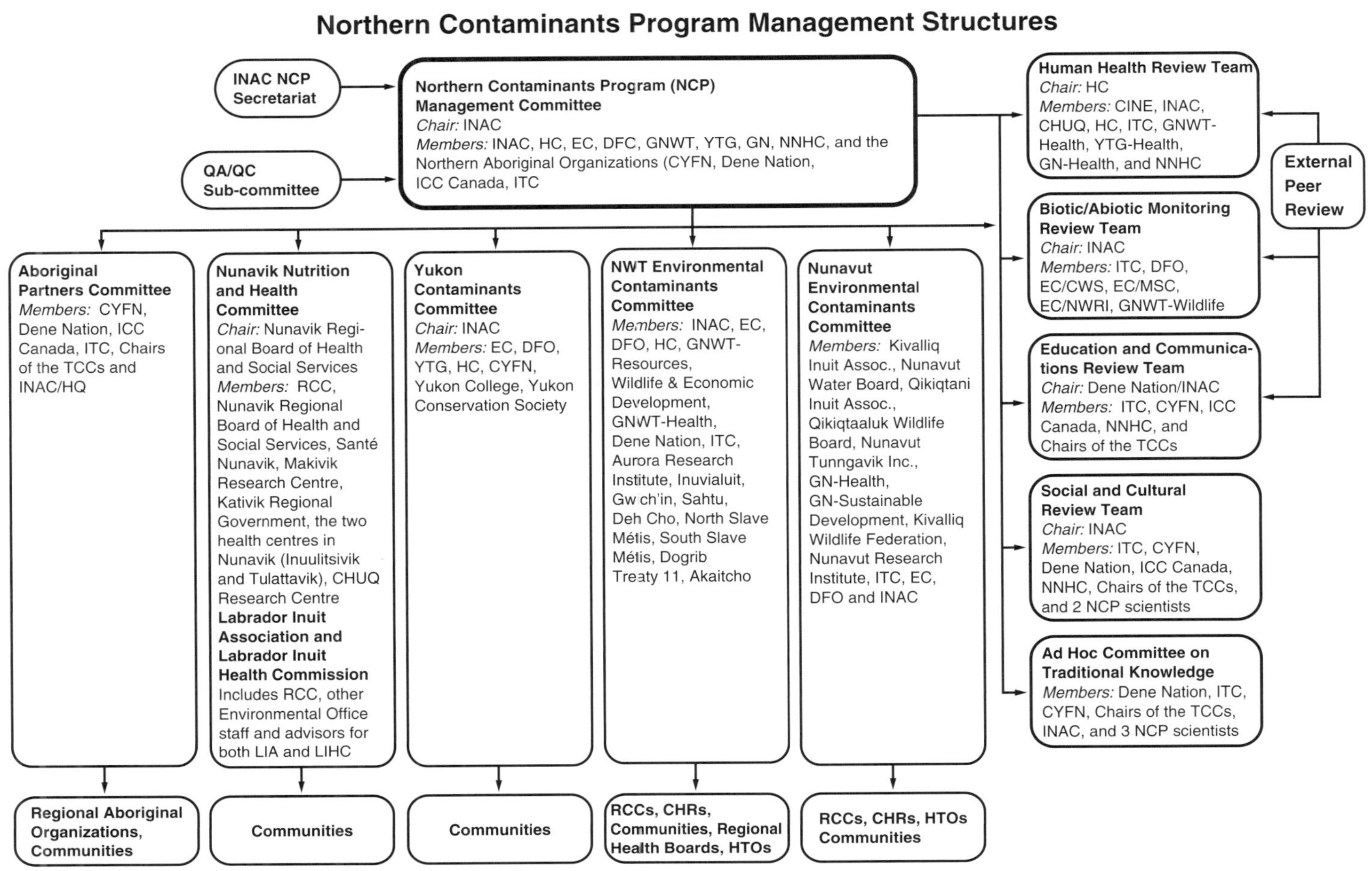

Northern Contaminants Program Management Structures
INAC NCP Secretariat
QA/QC Sub-committee
Northern Contaminants Program (NCP) Management Committee
Chair: INAC
Members: INAC, HC, EC, DFC, GNWT, YTG, GN, NNHC, and the Northern Aboriginal Organizations (CYFN, Dene Nation, ICC Canada, ITC
Human Health Review Team
Chair: HC
Members: CINE, INAC, CHUQ, HC, ITC, GNWT-Health, YTG-Health, GN-Health, and NNHC
Biotic/Abiotic Monitoring Review Team
Chair: INAC
Members: ITC, DFO, EC/CWS, EC/MSC, EC/NWRI, GNWT-Wildlife
Education and Communications Review Team
Chair: Dene Nation/INAC
Members: ITC, CYFN, ICC Canada, NNHC, and Chairs of the TCCs
Social and Cultural Review Team
Chair: INAC
Members: ITC, CYFN, Dene Nation, ICC Canada, NNHC, Chairs of the TCCs, and 2 NCP scientists
Ad Hoc Committee on Traditional Knowledge
Members: Dene Nation, ITC, CYFN, Chairs of the TCCs, INAC, and 3 NCP scientists
External Peer Review
Aboriginal Partners Committee
Members: CYFN, Dene Nation, ICC Canada, ITC, Chairs of the TCCs and INAC/HQ
Nunavik Nutrition and Health Committee
Chair: Nunavik Regional Board of Health and Social Services
Members: RCC, Nunavik Regional Board of Health and Social Services, Santé Nunavik, Makivik Research Centre, Kativik Regional Government, the two health centres in Nunavik (Inuulitsivik and Tulattavik), CHUQ Research Centre
Labrador Inuit Association and Labrador Inuit Health Commission
Includes RCC, other Environmental Office staff and advisors for both LIA and LIHC
Yukon Contaminants Committee
Chair: INAC
Members: EC, DFO, YTG, HC, CYFN, Yukon College, Yukon Conservation Society
NWT Environmental Contaminants Committee
Members: INAC, EC, DFO, HC, GNWT-Resources, Wildlife & Economic Development, GNWT-Health, Dene Nation, ITC, Aurora Research Institute, Inuvialuit, Gwich'in, Sahtu, Deh Cho, North Slave Métis, South Slave Métis, Dogrib Treaty 11, Akaitcho
Nunavut Environmental Contaminants Committee
Members: Kivalliq Inuit Assoc., Nunavut Water Board, Qikiqtani Inuit Assoc., Qikiqtaaluk Wildlife Board, Nunavut Tunngavik Inc., GN-Health, GN-Sustainable Development, Kivalliq Wildlife Federation, Nunavut Research Institute, ITC, EC, DFO and INAC
Regional Aboriginal Organizations, Communities
Communities
Communities
RCCs, CHRs, Communities, Regional Health Boards, HTOs
RCCs, CHRs, HTOs Communities

Figure 3.2

As a result, the NCP is better able to respond to the needs and wishes of northerners when designing and delivering programs and projects.

The Aboriginal partners have successfully improved two-way communications links and systems with their Aboriginal constituents, enabling better representation of their priorities within the NCP, and, equally important, promoting better use of NCP information and opportunities for funding within the regions and communities. As a result, the capacity to address contaminants issues at the community level has in recent years increased significantly, which is demonstrated by the community-driven project proposals that have been approved for funding. Such projects contribute to technical as well as administrative capacity at the community level and include specific scientific contaminants studies and communication projects by local environment committees.

The Aboriginal organizations have also been largely responsible for the formal and informal integration of traditional knowledge (TK) into the NCP. The Dene Nations' Elders/Scientists Retreats and the Council of Yukon First Nations' extensive documentation and project work are good examples. The efforts of the Aboriginal organizations have resulted in the development of a document entitled *TK for Dummies* used by scientists and community workers alike.

Finally, the capacity developed within the Aboriginal organizations has permitted them to participate nationally and internationally to ensure their positions are considered in policy development, including in arguments for control of contaminants. The Aboriginal partners participate actively in the Arctic Council and played an important role in pushing the UNECE Executive Body to move forward with a POPs protocol to the CLRTAP. Northern Aboriginal organizations worked closely with the federal government through a coalition called the Canadian Arctic Indigenous Peoples Against POPs (CAIPAP) during the UNEP POPs negotiations that led to the Stockholm Convention. They feel that their involvement is a model for their participation in other health and environmental issues and there have clearly been significant gains, aided in great part by the NCP, in their capacity to address these types of issues.

Within Canada, the increased capacity of Aboriginal organizations for addressing the contaminants issue has enabled them to have a stronger voice in national consultations.

LOOKING AHEAD

During the course of the NCP, significant action to address persistent and bioaccumulative contaminants has occurred on both the international and the domestic fronts. It can be argued that much of the international action since 1989 has been driven by NCP results and NCP participants. Not only have the data generated by the NCP been the keystone of calls for international action on long-range, persistent, bioaccumulative, and toxic substances, but Canada's leadership role in the CLRTAP POPs Protocol and the Stockholm Convention negotiations has largely been precipitated and facilitated by the NCP. It has been a remarkable achievement not only to obtain world controls on these substances of concern to northern Aboriginal peoples, particularly since international controls are generally viewed as impossible to achieve, but also to have successfully made Canada a world leader in this area. Nationally, NCP data have strengthened, accelerated, and substantiated domestic policies, such as the federal government's Toxic Substances Management Policy, and legislation on toxic substances in the Canadian Environmental Protection Act.

The NCP has also been highly innovative from a policy and implementation perspective. Through an evolving process, the NCP set new standards for community and Aboriginal peoples participation in scientific programs. The partnerships forming the management structure of the NCP are unique and have proven highly productive and mutually beneficial. Through this structure, the NCP was given the capacity to address human health and Arctic ecosystem contamination using approaches that ensure scientific integrity and relevance while responding to community concerns and priorities.

Results of NCP-Phase II will be published in winter 2002/2003 in the *Canadian Arctic Contaminants Assessment Report II* (*CACAR II*). *CACAR II* will not be simply a publication of past results; this substantive report will be the critical component of a long-term strategy for future work in the North and will be used as the foundation for consultation with northerners, particularly Aboriginal peoples, for determining the specific focus of such work. It will provide the rationale and strengthen the case for future funding.

The *CACAR II* data to be published include important baseline information for the development of a global monitoring program for persistent organic pollutants, which will be part of the Stockholm Convention, and for the UNEP global assessment for mercury to be released in 2003.

The NCP also represents Canada's main contribution to the Arctic Monitoring Assessment Programme under the Arctic Council. Much of the data in *CACAR II* will also be published in the AMAP Arctic Assessment report and in the State of the Arctic Environment report, which will be released at the Arctic Council ministerial meeting in October 2002 in Finland.

The NCP's objective, "to reduce and wherever possible eliminate contaminants in traditionally harvested foods, while providing information that assists informed decision-making by individuals and communities in their food use," is as pertinent today as it was when the program commenced, if not more so. Given the heightened awareness of elevated contaminant levels in northern populations, the achievement of this objective is far from complete. The focus of a new NCP will likely continue to be on protecting human health, monitoring the new international controls, directing education and communications initiatives, and addressing emerging priorities such as mercury contamination in the North.

CHRONOLOGY OF EVENTS/MILESTONES IN THE EVOLUTION OF THE NCP

Date	*Event*
Early 1970s	Scientists begin measuring long-range-transported atmospheric contaminants in Canada's North.
1985	• Distant Early Warning (DEW) line site PCB clean-up begins. • INAC establishes inter-agency working group on contaminants in Native diets and conducts a baseline literature review. Conclusion: Contaminants in the North are serious and widespread and it is highly unlikely that small quantities of PCBs found at DEW line sites are the cause. A "co-operative program" is put in place with monitoring, research, and evaluation elements. The first priority is to assess the extent of contamination of local food sources used by northern people. By 1989 the scope widens to include organochlorines, metals, and radionuclides.
1989	• Results of study showing elevated PCBs and mercury among residents of Broughton Island is delivered to the community. • Results of a study showing elevated PCBs, other POPs, and mercury in Inuit residents of Nunavik is published. • Scientific Evaluation Meeting (Ottawa) summarizes the present state of knowledge concerning contaminants in northern Canada. Approximately fifty scientists are in attendance representing a broad spectrum of interests, including representatives from two Aboriginal organizations and the other seven circumpolar nations. An integrated ecosystem approach is adopted for assessment of the issue. • Strategy meeting (Toronto) to set a long-term northern contaminants research and monitoring plan, including representatives of federal, provincial, and territorial governments, universities, and northern Aboriginal organizations, leads to the development of the 5-Year Strategic Action Plan on Northern Contaminants.
1991	• NCP established under the Green Plan's Arctic Environmental Strategy (1991–1997). In addition to a Technical Committee, a Science Managers Committee, which includes five Aboriginal Partners, is struck to oversee the program's policy and funding issues. • Arctic Monitoring and Assessment Programme (AMAP) is established under the circumpolar Arctic Environmental Protection Strategy (AEPS). Canada and the seven other Arctic countries sign the Declaration on Protection of the Arctic Environment in Rovaniemi, Finland and adopt the AEPS. • A Task Force on POPs, led by Canada and Sweden, is established under the United Nations Economic Commission for Europe's Convention on Long-range Transboundary Air Pollution (UNECE CLRTAP). • First annual NCP Results Workshop.

CHRONOLOGY OF EVENTS (*Continued*)

1992
- Scientists publish a benchmark study on the state of knowledge of contaminants in the Canadian Arctic in a special issue of the international research journal *Science of the Total Environment.*
- Yukon Contaminants Committee is established to deal with the broader issue of contaminants related to the health advisory issued as a result of elevated levels of toxaphene and PCBs in lake trout and burbot from Lake Laberge.
- Rio "Earth Summit" (World Summit on Sustainable Development).

1994
- Guidelines for Responsible Research, developed by the Aboriginal Partners, are implemented.
- UNECE CLRTAP Task Force Report presented to the Convention and a Preparatory Working Group is established, led by Canada, to prepare a draft POPs protocol. The report is also used to justify to UNEP Governing Council the need for global POPs controls.

1995
- UNEP Governing Council makes decision to conduct a global POPs assessment.
- Vancouver, Canada: Canada and the Republic of the Philippines co-host an international experts meeting on POPs.

1996
- Ottawa, Canada: Arctic Council is established by the eight Arctic countries.
- Iqaluit, Canada: Canadian Polar Commission holds conference – "For Generations to Come: Contaminants, the Environment, and Human Health in the Arctic."
- Yellowknife, Canada: NCP Consultation and Strategy Meeting is held – The "Ideas" Workshop.

1997
- *Canadian Arctic Contaminants Assessment Report* (CACAR) and *Highlights Report* are released.
- AMAP's *A State of the Arctic Environment Report* is released.
- Tromsø, Norway: First AMAP Symposium on Pollution in the Arctic Environment.

1998
- NWT Environmental Contaminants Committee (NWT–ECC) is established.
- NCP Phase II begins, with a focus on immediate human health and safety issues.
- NCP blueprints are developed and implemented.
- *AMAP Assessment Report: Arctic Pollution Issues* is released.
- UNECE CLRTAP – POPs and Heavy Metals protocols signed by thirty-six countries including Canada. Canada ratifies both protocols this same year.
- Regional Contaminant Coordinator positions established within the NCP.

CHRONOLOGY OF EVENTS (*Continued*)

	• NCP Management Committee revamps NCP proposal review process to include external peer reviewers and technical and social/cultural review teams. A strategy meeting the following year results in further modifications, including the establishment of the education and communications review team.
1999	• Communications tours begin.
2000	• Nunavut Environmental Contaminants Committee (NECC) is formally established.
2001	• Stockholm Convention on POPs signed by 111 countries including Canada. Canada is the first country to ratify the convention. • Education and communications strategy meeting leads to replacing "written consent" requirements with a process for consulting with communities and Aboriginal organizations that is a prerequisite for project funding.
2002	• NCP contributes data and information to the UNEP global mercury assessment to be completed by 2003. • World Summit on Sustainable Development in Johannesburg. • AMAP's *Arctic Pollution 2002* is released. • Inari, Finland: Second AMAP International Symposium on Environmental Pollution of the Arctic.
2002/03	• *CACAR II* and *Highlights Report* are released.
2003	• Ottawa, Canada: NCP Symposium on Contaminants in the Canadian Arctic.

NOTES

1 L.A. Barrie et al., "Arctic Contaminants: Sources, Occurrence and Pathways," *Science of the Total Environment* 122, nos 1–2 (1992): 1–74.

2 D. Kinloch, H.V. Kuhnlein, and D.C.G. Muir, "Inuit Foods and Diet: A Preliminary Assessment of Benefits and Risks," *Science of the Total Environment* 122, nos 1–2 (1992): 247–78.

3 E. Dewailly et al., "High Levels of PCBs in Breast Milk of Inuit Women from Arctic Quebec," *Bulletin of Environmental Contamination and Toxicology* 43, no. 1 (1989): 641–6.

4 Barrie, "Arctic Contaminants."

5 Barrie, "Arctic Contaminants"; R.W. Macdonald et al., "Contaminants in the Canadian Arctic: 5 Years of Progress in Understanding Sources, Occurrence and Pathways," *Science of the Total Environment* 254, nos 2–3 (2000) 93–234; D.C.G. Muir et al., "Arctic Marine Ecosystem Contamination," *Science of the Total Environment* 122, nos 1–2 (1992): 75–134; W.L. Lockhart et al., "Presence and Impli-

cations of Chemical Contaminants in the Freshwaters of the Canadian Arctic," *Science of the Total Environment* 122, nos 1–2 (1992): 165–243; D.J. Thomas et al., "Arctic Terrestrial Ecosystem Contamination," *Science of the Total Environment* 122, nos 1–2 (1992): 135–64; D.C.G. Muir et al., "Spatial and Temporal Trends and Effects of Contaminants in the Canadian Arctic Marine Ecosystem: A Review," *Science of the Total Environment* 230, nos 1–3 (1999): 1–82.

6 Government of the Northwest Territories, "Northwest Territories Renewable Resource Harvester Survey, Winter 1990," GNWT Bureau of Statistics.

7 Muir, "Arctic Marine Ecosystem"; Kinloch, "Inuit Foods and Diet"; R.J. Norstrom and D.C.G. Muir, "Chlorinated Hydrocarbon Contaminants in Arctic Marine Mammals," *Science of the Total Environment* 154 (1994): 107–28; Muir, "Spatial and Temporal Trends"; J. Van Oostdam et al., "Human Health Implications of Environmental Contaminants in Arctic Canada," *Science of the Total Environment* 230, nos 1–3 (1999): 1–82.

8 Lockhart, "Presence and Implications"; Muir, "Spatial and Temporal Trends."

9 Kinloch, "Inuit Foods and Diet"; Van Oostdam, "Human Health Implications."

10 S.-L. Han and K. Adare, *Highlights of the Canadian Arctic Contaminants Assessment Report: A Community Reference Manual,* (Ottawa: Indian and Northern Affairs Canada 1997), p. iv.

11 AMAP, *Arctic Pollution Issues: A State of the Arctic Environment Report* (Oslo, Norway: Arctic Monitoring and Assessment Programme 1997), p. 172.

12 Alta Declaration, 2. See www.grida.no/prog/polar/aeps/alta.htm for complete text.

4

Circumpolar Perspectives on Persistent Organic Pollutants: The Arctic Monitoring and Assessment Programme

LARS-OTTO REIERSEN, SIMON WILSON, AND VITALY KIMSTACH

BACKGROUND

In a speech in Murmansk in October 1987, Mikhail Gorbachev, then president of the Union of Soviet Socialist Republics (U.S.S.R.), called for international co-operation to address pollution in the northern part of the U.S.S.R. and the Arctic. This initiative was one of the first signals that the cold war was coming to an end and one of the first steps in a process that ultimately led to the establishment of the Arctic Council.

A January 1989 Finnish initiative to promote international co-operation in the Arctic led to a consultative meeting on the protection of the Arctic environment in September 1989 in Rovaniemi, Finland. Participants discussed possibilities for environmental co-operation among the Arctic states and approved a follow-up work plan. Among other things, Canada agreed to prepare a background report on persistent organic pollutants (POPs) in the Arctic, and Norway and the U.S.S.R. agreed to review existing national and international monitoring programs operating in the Arctic and to present proposals for future action. Between September 1989 and June 1991, experts met several times to develop these reports.

The Canadian report included new and surprising results from a study documenting that Inuit women living in northern Quebec had levels of PCBs in their breast milk that were five times (or more) higher than those of women living farther south in the same province. It concluded that there were insufficient data on sources, sinks, and pathways and on spatial and temporal trends of POPs in the Arctic and little knowledge about their possible biological effects in the region and recommended further collection of data on POPs in the Arctic environment, including their effects on ecosystems and humans.[1]

The report from Norway and the U.S.S.R. proposed putting in place an Arctic monitoring program setting POPs, heavy metals, and radionuclides as priorities and assessing their levels in humans and the environment. It also recommended that the program address, as a lower priority, pollution problems associated with petroleum hydrocarbons and acidifying substances; include biological and human health effects studies; and make efforts to identify sources. The report included recommendations for all important parts of an international monitoring program, e.g., parameters, media, stations and sampling, methodology, quality assurance (QA) and quality control (QC), and administration.[2]

On the basis of these preparatory documents, ministers of the eight Arctic countries[3] adopted the Arctic Environmental Protection Strategy (AEPS)[4] at a ministerial meeting in Rovaniemi in June 1991. As a part of this strategy, the Arctic Monitoring and Assessment Programme (AMAP) was established, initially as a task force and in 1993 as a formal AEPS Working Group. Under the AEPS, AMAP was requested "to monitor the levels of pollutants and to assess their effects in the Arctic environment."

At the 1991 Rovaniemi meeting, ministers also agreed to support the United Nations Economic Commission for Europe (UNECE) in its work in this field. A UNECE task force was examining the potential for a protocol to the Convention on Long-range Transboundary Air Pollution (CLRTAP) to control the entry of POPs into the environment. In November, delegates from the eight Arctic countries brought their concerns about POPs in the Arctic to the attention of the executive body for the CLRTAP. The mandate of the UNECE task force was subsequently modified to provide, by 1994, the basis for elements of a possible CLRTAP protocol on POPs.

AMAP'S MONITORING PROGRAM

Between 1990 and 1992, experts from the eight Arctic countries, along with representatives from Arctic Indigenous peoples' organizations (the Nordic Saami Council, Inuit Circumpolar Conference (ICC), and the Russian Association of Indigenous Peoples of the North (RAIPON)), planned the first detailed AMAP monitoring program.[5] It was designed to follow contaminants from their sources, via transport pathways, into all components of the Arctic environment and its ecosystems and ultimately into top predators, including humans.

The monitoring program for evaluating human exposure gave priority to the analysis of contaminants in maternal and cord blood.[6] As the program identified newborns and infants as a critical group for possible health effects of contaminants, cord-blood analyses were included to document the exposure of the fetus to contaminants during pregnancy. By taking samples from both Indigenous and non-Indigenous populations, researchers could see if there were differences in their exposure to contaminants and possibly trace any differences to lifestyle or diet.

Table 4.1
Part of the Original AMAP Monitoring Program for POPs

Program	*Media*	*Persistent Organics*								
		Planar CB	PCB	DDT/ DDE/ DDD	HCH	HCB	Chlordane	Toxaphene	Dioxins/ Furans	Mirex
Atmospheric	air	–	E	E	E	R	E	E	R	R
Terrestrial	soil	ES	ES	ES	ES	ES	ES	ES	ES	ES
	mammals	ES	ES	ES	ES	ES	ES	ES	ES	ES
Freshwater	sediment	R	E	E	E	E	E	R	R	R
	water	R	ES	ES	ES	ES	ES	ES	R	R
	biota	R	E	E	E	E	E	E	R	R
Marine	sediment	R	E	E	E	E	R	R	R	–
	biota	R	E	E	E	E	R	R	R	–
Human	blood	E	E	E	–	E	E	–	E	E

E-essential; ES-essential in sub-regions; R-recommended

Table 4.2
Part of the Original AMAP Monitoring Program for Human Health

Media	*Priority*
Human blood, tissue, and placenta	E
Human hair, urine, and breast milk	R

E-essential; R-recommended

A significant contribution to the work – to ensure comparable data on contaminants in humans around the circumpolar Arctic – came from a Canadian laboratory that offered to analyze fifty human samples from all eight Arctic countries. These samples were analyzed for levels of POPs such as DDTs, PCBs, and chlordanes. By the time the AMAP report was produced in 1997, all Arctic countries except the United States had joined in this part of the human health program. Other contaminants found in samples of human placenta and blood (e.g., mercury, lead, cadmium, and radionuclides) were analyzed by national laboratories and reported to the human health assessment group.

Key to implementing AMAP was the decision to proceed step-by-step to allow the eight Arctic countries to adapt or develop their national programs based on AMAP's recommendations and according to their own priorities and financial and scientific resources. Part of the general strategy of AMAP was to build upon ongoing national and international research and monitoring activities. AMAP recognized that research (in addition to national monitoring work) would provide much of the information necessary to assess levels of contaminants and their effects in the Arctic, where logistical constraints limit monitoring activities. (In many other regional monitoring programs, research is largely ignored.) To address QA/QC issues, AMAP encouraged all participating laboratories to join appropriate international QA/QC programs and, where relevant, adopt existing international recommendations for methodology and parameters to be analyzed. By doing this, AMAP was able to compare the levels of contaminants observed in the Arctic with levels published from other monitoring and research programs being conducted at lower latitudes.

At the beginning of the 1990s, few international marine monitoring programs included POPs other than PCBs and DDTs, and monitoring of biological effects was only just being introduced. Atmospheric programs focused mostly on acidifying components and heavy metals (e.g., the Co-operative Programme for Monitoring and Evaluation of the Long-range Transmission of Air Pollutants in Europe (EMEP) under the CLRTAP). AMAP was possibly the first international monitoring effort to design and implement a single program to cover all major ecological systems (atmospheric, marine, freshwater, and terrestrial) and humans and at the same time fully integrate its monitoring and assessment activities.

With ongoing Arctic monitoring and research programs that included POPs, Canada, Norway, and Sweden were well placed to contribute to AMAP. In such cases, the focus was on harmonizing the existing national work with AMAP. Where countries did not already have a comprehensive Arctic monitoring program, new initiatives were required. Denmark/Greenland, for example, had been conducting ad hoc environmental monitoring work linked mainly to areas near mine sites for several years. A new long-term monitoring program for Greenland was specifically established according to AMAP's recommendations, with initial field operation and sampling of human blood starting in 1994.

During 1993–94, the AMAP assessment strategy was developed[7] and, in 1994, AMAP established eight assessment experts groups, each responsible for drafting different sections of the report: POPs, heavy metals, radioactivity, oil and petroleum hydrocarbons, acidification, climate and UV, contaminant pathways, and human health. To equitably share the costs associated with the AMAP assessment, each of the eight Arctic countries accepted the responsibility to lead/co-lead one or more of the assessments. For example, Sweden and Canada became lead countries for the POPs assessment, nominating and supporting the work of Dr Cynthia de Wit (Sweden) and Dr Derek Muir (Canada) as lead authors in this work.

An Assessment Steering Group (ASG) was established, composed of all lead authors, representatives from Arctic Indigenous peoples' organizations, and the AMAP board (representing the AMAP working group and secretariat) to ensure coordination among the different chapters in dealing with cross-disciplinary issues. For example, to ensure that treatment of POPs was consistent in both the POPs and the human health chapters, it was agreed that the POPs chapter would follow the contaminants from their sources, via pathways, and finally describe levels and effects in biota, whereas the human health chapter would cover all information dealing with levels of POPs in human media and possible effects on human health.

AMAP's assessments were based to a large extent on information and results from recent (largely unpublished) monitoring and research work. As a part of the assessment, two tools were established to support the experts in their work.

An AMAP Project Directory (AMAP PD) was developed to register information on relevant monitoring and research projects and programs within the Arctic and adjacent areas. More than 500 projects were described in the AMAP PD at the time of the first AMAP report in 1997, and a further 300 projects are registered in the PD that has been established to support the 2002 AMAP assessment.

A number of AMAP Thematic Data Centres (TDCs) have been established to

- provide access to data from recent monitoring and research activities conducted as part of the AMAP national implementation plans;

Table 4.3
ASG Members and Their Lead Tasks

	AMAP *Phase 1 (1991-97)*		AMAP *Phase 2 (1998-2002)*	
Task	*Name*	*Country*	*Name*	*Country*
Chair of ASG	Lars-Erik Liljelund	Sweden	a. Cynthia de Wit	Sweden
			b. Helgi Jensson	Iceland
POPs	Cynthia de Wit	Sweden	Cynthia de Wit	Sweden
	Derek Muir	Canada	Derek Muir	Canada
Pathways	Keith Pucket	Canada		
	Harald Loeng	Norway		
Pathways and Climate Change	–	–	Robbie Macdonald	Canada
			Keith Pucket	Canada
			Harald Loeng	Norway
Human Health	Jens C. Hansen	Denmark	Jens C. Hansen	Denmark
	Andrew Gilman	Canada	Andrew Gilman	Canada
Radioactivity	Per Strand	Norway	Per Strand	Norway
	Mikhail Balonov	Russia	Yuri Tsaturov	Russia
Heavy Metals	Rune Dietz	Denmark	Suzanne Marcy	USA
	Glen Shaw/ Jozef Pacyna	USA/ Norway		
	Valentin Koropalov	Russia		
Acidification	Juha Kämäri	Finland	Juha Kämäri	Finland
Oil	Andrew Robertson	USA	Hein Rune Skjoldal	Norway
Climate & UV	Elisabeth Weatherhead	USA	John Calder	USA
			Elisabeth Weatherhead	USA
Indigenous Peoples Representative	Jan Idar Solbakken	Saami Council	Jan Idar Solbakken	Saami Council
Chair of AMAP	David Stone	Canada	a. Hanne Petersen	Denmark
			b. Helgi Jensson	Iceland
Executive Secretary	Lars-Otto Reiersen	Norway	Lars-Otto Reiersen	Norway
Other	Helgi Jensson	Iceland		

- provide a means to ensure that data are treated in a consistent manner and undergo uniform statistical analysis, etc., including application of objective quality assurance procedures;
- begin the process of establishing a long-term archive of Arctic-relevant monitoring data, for use in future assessments (e.g., of temporal trends); and
- meet the terms of reference of the ministerial declarations, charging AMAP with establishing databases of sources, types, and levels of radionuclide contamination of the atmospheric, aquatic, and terrestrial environments of the Arctic and northern areas.

To date, six AMAP TDCs have been established:

- Atmospheric contaminants data, at the Norwegian Institute for Air Research (NILU), Kjeller, Norway
- Marine contaminants data, at the International Council for the Exploration of the Sea (ICES), Copenhagen, Denmark
- Freshwater contaminants data, at the University of Alaska Fairbanks (UAF), Fairbanks, United States
- Terrestrial contaminants data, at the University of Alaska Fairbanks (UAF), Fairbanks, United States
- Radioactivity data, including sources, levels, and trends, at the Norwegian Radiation Protection Authority (NRPA), Oslo, Norway
- Human health data, at the National Institute of Public Health (NIPH), Copenhagen, Denmark

AMAP TDCs are located at established centres with appropriate expertise and facilities for conducting the types of international data-handling work required. Some of these centres also conduct data-handling work for other international monitoring programs, enabling harmonized reporting of data to meet the needs of different regional programs.

Data are reported to the AMAP TDCs and made available from the TDCs to scientists engaged in AMAP assessments under strict conditions that protect the rights of data owners. These conditions are described in the AMAP Data Policy Document.

AEPS MINISTERIAL MEETING IN NUUK AND THE POPs INITIATIVE UNDER UNECE

Although the Arctic Monitoring and Assessment Programme was still being finalized and implemented during the 1990 to 1994 period, some Arctic countries already had national research and monitoring programs that covered some of AMAP's priority issues. Based on results from these ongoing activities, AMAP scientists and environmental administrators were able to present a preliminary report

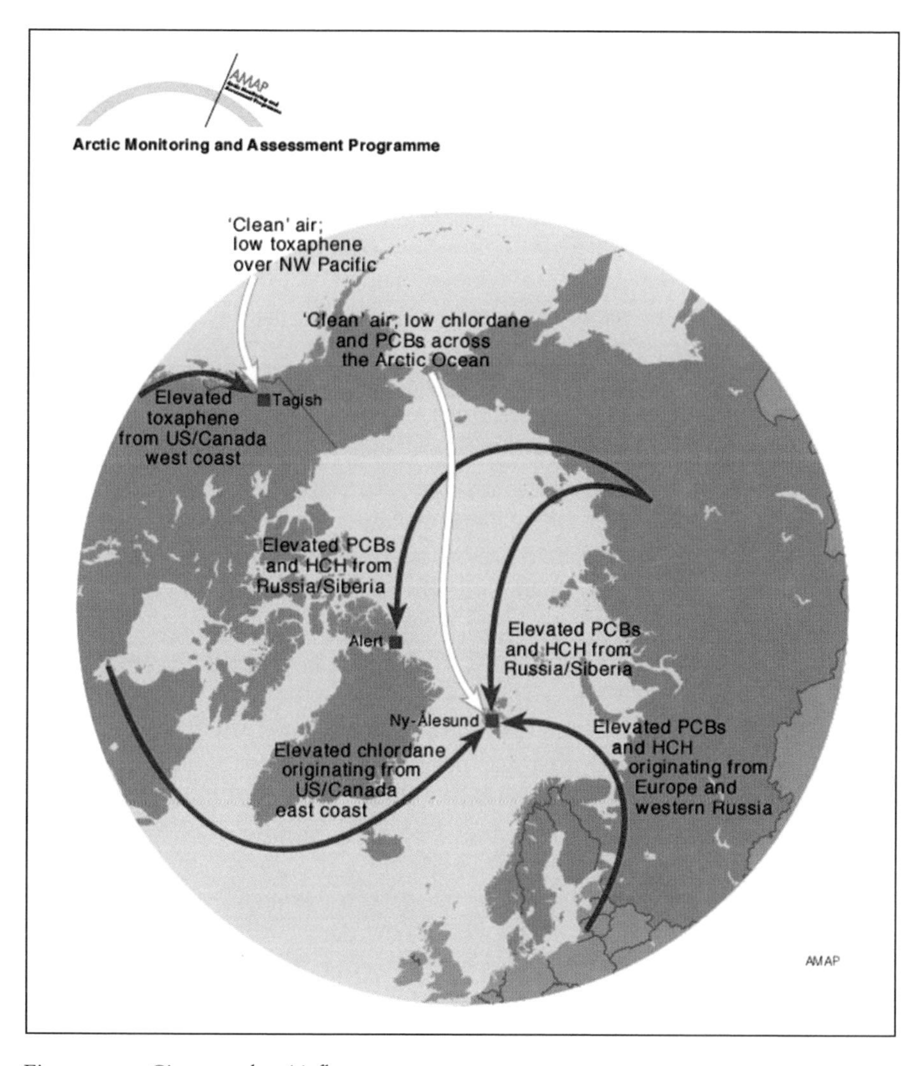

Figure 4.1 Circumpolar Airflows

Source: Arctic Pollution Issues: A State of the Arctic Environment Report (Oslo, Norway: Arctic Monitoring and Assessment Programme 1997), 79.

on pollution of the Arctic to the Second AEPS Ministerial Conference, in Nuuk, Greenland, in September 1993. The report provided further evidence that high levels of some POPs and heavy metals were observed in some areas of the Arctic; however, information was still lacking for extensive areas. Of particular concern were the indications that POPs were accumulating in top predators, including Arctic Indigenous peoples consuming from the marine food chain.[8] Although no significant local sources of POPs existed, at least two atmospheric stations measuring POPs in the Arctic at that time found air concentrations that indicated atmospheric transport of POPs that could be traced to sources at southern latitudes.

In its report to the AEPS ministerial meeting in 1993, AMAP recommended that "Given the increasing substantiation of reasons for concern related to persistent organics in the Arctic, the eight Arctic countries agree to support activities that will lead to the development of a protocol to control the emissions of these substances under the UNECE LRTAP Convention." In response to this, the ministers agreed "to support the development of appropriate protocols under LRTAP auspices, and to consult with non-ECE nations whose emissions and discharges may affect the Arctic, to achieve their participation in the protocols and to continue to take measures to reduce and/or control the use of a number of persistent organic pollutants."[9]

Under UNECE, the work to provide the basis for a protocol on POPs was proceeding, led by Canada and Sweden. After the Nuuk ministerial meeting, Canada and Norway took a further step in this initiative. A UNECE meeting held in February 1994 had the specific aim to draft a special report on the POPs issue to be presented to the UNECE executive body in November 1994.

Although the first AMAP assessment, covering the period 1991–97, was finalized only in January 1997 and formally released in June 1997, AMAP had exceptionally good links to the UNECE process. David Stone (Canada) was both chair of AMAP and co-chair of the UNECE Task Force on POPs. New information coming out of AMAP was fed into the UNECE process as soon as it became available. This close co-operation between AMAP and UNECE allowed AMAP's compelling new information on POPs in the Arctic to inform the process leading to the Protocol on Persistent Organic Pollutants under the CLRTAP that was opened for signature in Århus, Denmark, in June 1998.

THE 1997 AMAP ASSESSMENT REPORT

"In comparison with most other areas of the world, the Arctic remains a clean environment. However, the following conclusions illustrate that, for some pollutants, combinations of different factors give rise to concern in certain ecosystems and for some human populations. These circumstances sometimes occur on a local scale, but in some cases may be regional or circumpolar in extent."[10]

Sources and pathways of POPs

POPs of concern in the Arctic originate mainly in temperate and warmer areas of the world. Pesticides that are banned or restricted in many countries continue to be used extensively in agricultural and pest control applications in developing countries; the environmental implications of new substances being introduced as replacements for banned substances are not well known. Industrial sources in Europe, North America, and Asia also contaminate the Arctic.

Sources within the Arctic include military installations (PCBs from early warning radar sites), industrial sources (dioxins/furans from smelters and municipal and industrial waste disposal), and past use of pesticides for pest control. The extent of contamination from many of these sources and of possible continuing use of restricted substances in some areas is not well documented.

POPs reach the Arctic mainly via atmospheric (figure 4.1) and riverine pathways (figure 4.2). In the atmosphere, POPs travel in a series of "hops," depositing from the atmosphere under cold conditions and then revolatilizing when temperatures increase. The polar region forms cold traps for these compounds.

It is possible that transport of POPs by sea ice results in their release in marginal ice areas. For example, release of particulates following melting in the marginal ice areas in the Greenland and Barents seas may be an important mechanism for focusing contaminants from a wide area of the Arctic into these regions. The potential for damage to ecosystems is enhanced by the tendency of transport pathways to focus POPs in areas of high biological productivity.

Biomagnification in Arctic food chains increases the potential threat from POPs

The processes of bioaccumulation and biomagnification allow POPs to reach very high concentrations in top predators even when levels in air, soil, and water are low. Biomagnification is a process that occurs in food chains where animals consume other animals for food, thereby consuming all the contaminants accumulated by their prey. Since many POPs are not broken down or excreted, concentrations increase with each step from prey to predator. Bioaccumulation of POPs is particularly strong in Arctic marine food chains, leading to high levels of POPs in top predators such as seals, polar bears, and, ultimately, man. The role of fat reserves in allowing many Arctic animals to survive the cold climate and the importance of fat in the diet of top predators, including humans, further promote the biomagnification of lipid-soluble POPs.

Studies have shown that POPs can affect the immune system and lead to developmental, behavioural, and reproductive effects in a number of species, including birds, fish, and mammals. Current levels of POPs found in some species in some areas are approaching or even exceeding thresholds associated with these types of effects. The concern for Arctic ecosystems is that effects on individuals may ultimately lead to effects on the population or the community. Temporal data show that levels of some POPs in the Arctic environment are not yet decreasing even though the uses of POPs have been restricted for some time; in some cases, levels are still rising.

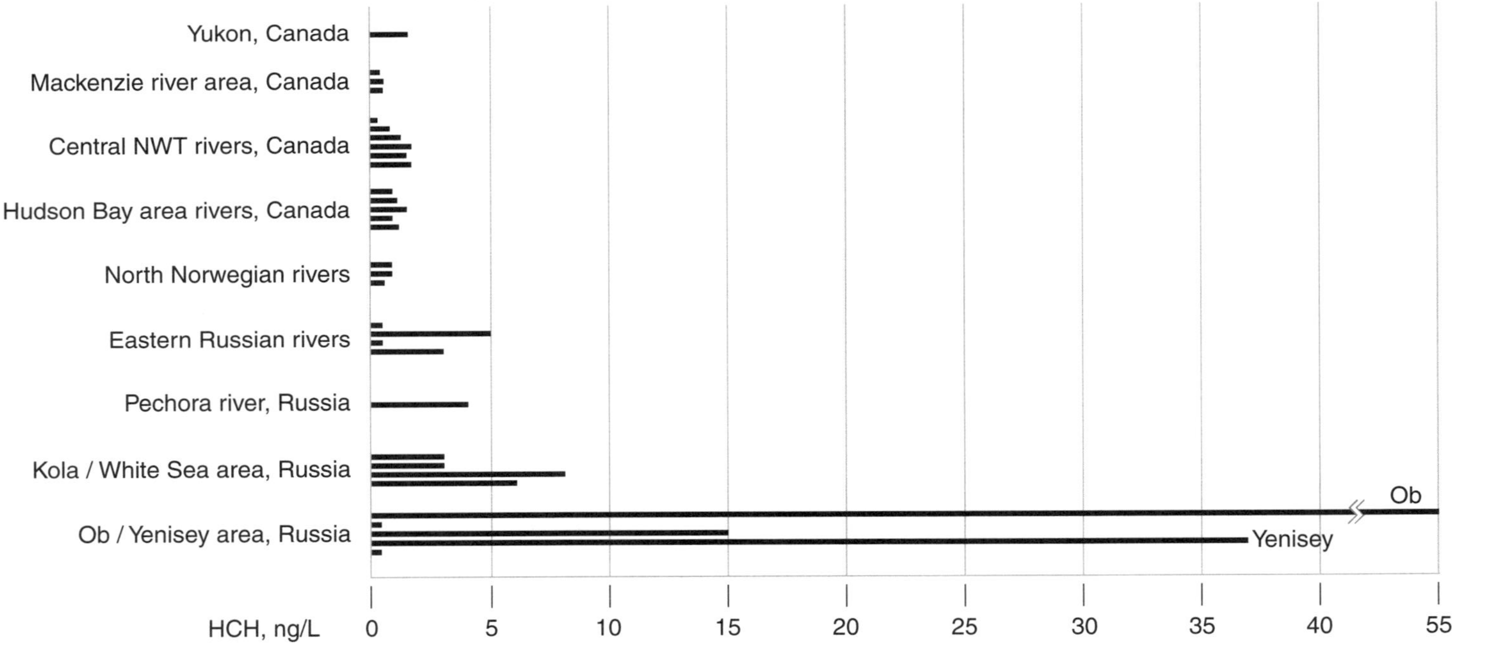

Figure 4.2 POPs Reaching the Arctic in Measured Quantities by Several Rivers
Data for river water and sediments indicate a substantial input to the Arctic of certain POPs by the large Russian rivers that flow northward to the Arctic Ocean (Ob, Yenisey, Lena). This may indicate recent or continuing use of banned or restricted substances, such as DDT; however, these data need further confirmation.

Source: "Report to the Ministers: Update on Issues of Concern to the Arctic Environment including Recommendations for Actions" (Oslo, Norway: Arctic Monitoring and Assessment Programme 1993), Report 93:4.

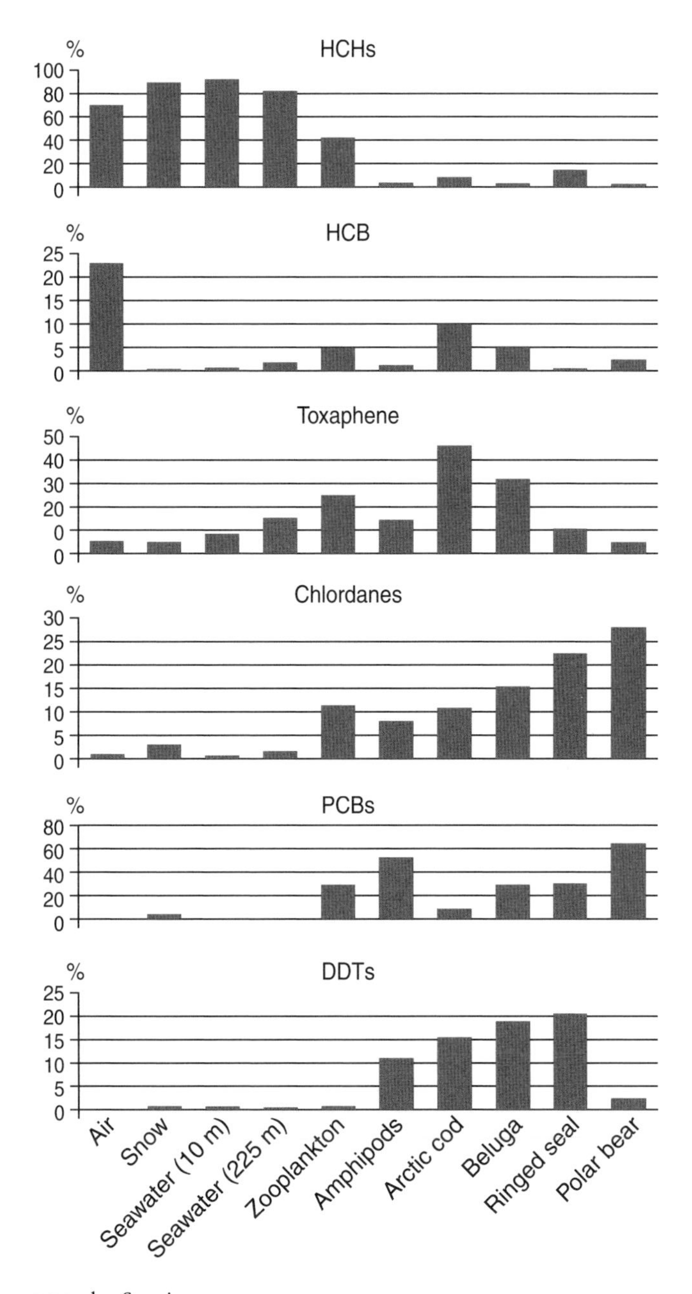

Figure 4.3 POPs by Species
POPs are found in all compartment of the Arctic environment. Figure 4.3 shows the distribution between selected environmental compartments or species of six major classes of organochlorine contaminants, illustrating the process of transfer between compartments and bioaccumulation in the marine mammal food chain.

Source: Arctic Pollution Issues: A State of the Arctic Environment Report (Oslo, Norway: Arctic Monitoring and Assessment Programme 1997).

Intake of POPs from traditional food consumed by Indigenous peoples in the Arctic

POPs accumulate in the fatty tissues of many species near the top of the Arctic food chain that are consumed by Indigenous peoples as part of their traditional diet. Indigenous coastal communities consume relatively high quantities of marine mammals; inland communities tend to consume more reindeer and wild meat. More objective dietary assessments are required, particularly in northern Russia, to better estimate the exposure risk to Indigenous communities from the consumption of traditional country foods.

Because the periods of fetal and early childhood development are those associated with the greatest vulnerability to toxic substances, contaminant intakes by pregnant women and children are of great concern.

Studies in Canada have shown that average intakes of contaminants are generally higher in the Baffin Inuit population as a result of a diet including large amounts of marine mammals and fish, while the Sahtu Dene/Métis peoples, who consume mainly caribou and fish, have lower intakes. Baffin Inuit exhibit one-day intakes that exceed tolerable daily intake (TDI) for chlordane and toxaphene by nearly tenfold. It is possible to gauge the relative, but not the absolute, risk by comparing estimated one-day intake with currently held TDI. Exceeding TDI is of particular concern if the individual exceeds the limit for a significant period of his/her life. Existing TDI guidelines need to be improved for Arctic residents based upon studies of interactions among individual contaminants present in Arctic foods and between contaminants and nutrients.

Corresponding levels of POPs in human blood and breast-milk samples

Primarily as a result of dietary factors, certain populations of Inuit in Canada and Greenland seem to be particularly exposed to some POPs. Because many POPs can pass the placental barrier, POPs levels as measured in the blood of pregnant women are reflected in their offspring; mothers can also pass POPs to their newborns through breast milk. In some Canadian Inuit women, breast-milk concentrations of PCBs and DDT are several times greater than the levels found in southern Canadian women. The effects of this exposure to infants are not fully understood.

Levels of POPs in Arctic Indigenous peoples have been investigated in an international pan-Arctic study conducted by AMAP. There are gaps in the available data for Alaska and much of Siberia; however, from results obtained so far, the highest levels of a number of POPs occur in peoples living in Greenland and the Canadian Arctic and, for DDT, in populations in northern Russia. It is important to remember that the social and cultural well-being and the physical health of many Arctic Indigenous peoples depend on the harvest and consumption of traditional country foods. In the majority of cases, the benefits of these traditional diets outweigh the disadvantages associated with contaminants; however, greater knowledge is required to advise Indigenous peoples appropriately on the security of their food.

Advice concerning human exposure

Several groups of people in the Arctic are highly exposed to POPs derived from long-range transport or local sources that accumulate in animals used as traditional foods. However, weighing the well-known benefits of breast milk and traditional food against the suspected but not yet fully understood effects of POPs, it is recommended that consumption of traditional food continue, with recognition that there is a need for dietary advice to Arctic peoples so they can make informed choices concerning the foods they eat. Breastfeeding should continue to be promoted.

POPs levels respond to controls

Environmental levels of a number of POPs decreased in the 1970s and the early 1980s following the introduction of bans and restrictions; however, concentrations do not appear to have changed significantly over the last fifteen to twenty years.

DECISIONS AT THE 1997 AEPS MINISTERIAL MEETING

Based on their consideration of the first AMAP assessment report at the Fourth AEPS Ministerial Conference in June 1997 in Alta, Norway, ministers committed "to take [the AMAP] findings and recommendations into consideration in our policies and programmes." They further agreed "to increase efforts to limit and reduce emissions of contaminants into the environment and to promote international co-operation in order to address the serious pollution risks reported by AMAP" and to "draw the attention of the global community to the content of the AMAP reports in all relevant international fora" and "make a determined effort to secure support for international action which will reduce Arctic contamination."[11]

At the June ministerial meeting, Norway proposed that an action plan be initiated to ensure a follow-up to the AMAP findings. A preparatory meeting was held in Copenhagen in November 1997 to discuss possible projects that might be part of such an action plan. In these discussions, POPs, heavy metals, and radionuclides were identified as the key issues for attention. At this time, it also became apparent that Russia most probably would not sign the UNECE Protocol on POPs that was under final preparation, and therefore a special project on PCBs in Russia was proposed with the aim of assisting Russia to develop a national plan of action to phase out the use of PCBs and handle PCB-contaminated waste appropriately. By establishing such a national plan, Russia would be in a better position to sign and meet the requirements of the protocol. The PCB project proposal was further developed by the AMAP secretariat, in close co-operation with Russian authorities, and presented to the Arctic Council in September 1998.

The first phase of the Russian PCB project was the preparation of an inventory of PCB production, use, and storage in the Russian Federation, including the territories of the former Soviet Union (FSU). Phase two called for a feasibility study of possible pilot projects for handling PCB-contaminated wastes, and phase three, the full implementation of selected pilot projects.

THE 1998 MINISTERIAL MEETING IN IQALUIT

At the First Arctic Council Ministerial Meeting, held in Iqaluit, Canada, in September 1998, AMAP presented the *AMAP Assessment Report: Arctic Pollution Issues,* which was the scientific documentation behind its 1997 *Arctic Pollution Issues: A State of the Arctic Environment Report.*[12]

Based on this detailed documentation and recommendations presented by AMAP, the ministers reaffirmed their agreement "to work vigorously for the early ratification and implementation of the Protocols on the elimination or reduction of discharges, emissions and losses of Persistent Organic Pollutants (POPs) ... under the framework of the United Nations Economic Commission for Europe Convention on Long-Range Transboundary Air Pollution,"[13] to "encourage other states to do the same, with the aim to bring the Protocols into force as early as possible" and to "fully support regional cooperation to facilitate the delivery of the measures that are needed to meet the obligations of the Protocols. ..."[14]

The ministers instructed their senior Arctic officials (SAOs) to continue to develop an overall plan of action, now known as the Arctic Council Action Plan (ACAP). The ACAP was to complement existing legal arrangements and include actions of a wide scope on pollution prevention and remediation measures and on the identification and implementation of specific co-operation projects.

The AMAP secretariat was asked to administer the first phase of the PCB project in Russia, with financial support from all eight Arctic countries and the Netherlands. The result of phase one – the PCB inventory – was presented at the second Arctic Council ministerial meeting, in Barrow, Alaska, in October 2000. The report documented that PCBs had been produced in Russia up until 1993 and that significant quantities of PCBs were still in use in transformers and condensers. The FSU was the second largest PCB producer in the world,

Table 4.4 The Overall Mass Balance for the Production and Use of PCBs in Russia

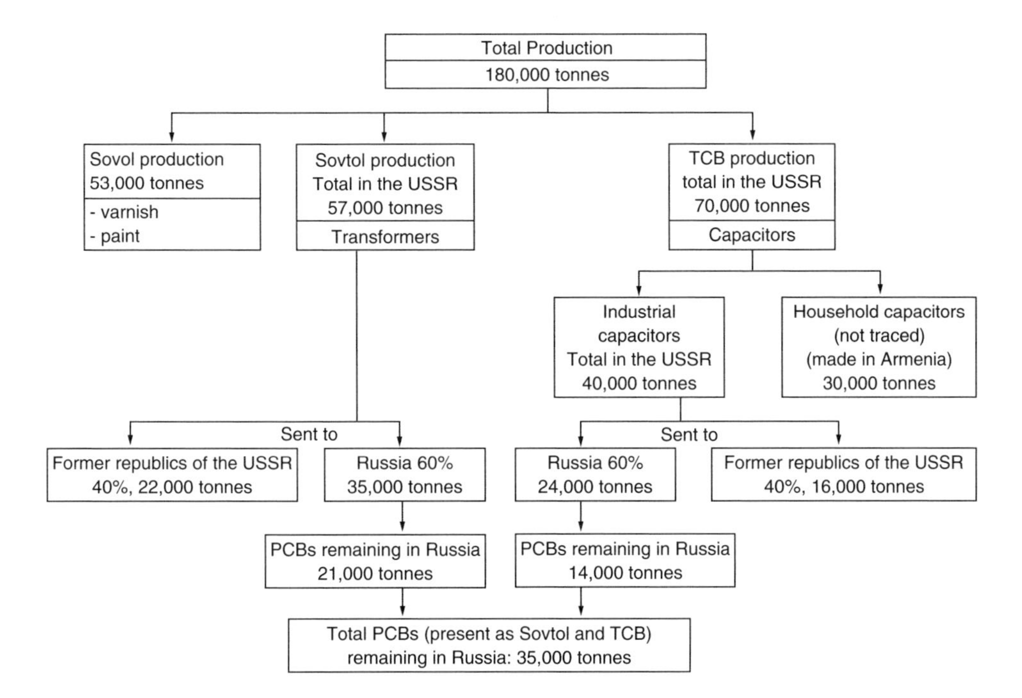

after the United States (table 4.4). At the Barrow ministerial meeting, the AMAP secretariat was asked to continue its involvement in the project and to administer phase two of the PCB project – the feasibility study. The report from this study is to be presented at the Arctic Council ministerial meeting in Finland, in October 2002. It is anticipated that this will lead to the initiation of phase three of the project, under which selected pilot projects will be implemented. Phase three will require close co-operation between Russian companies, international financial institutions, and the Arctic countries.

In June 1998 the UNECE CLRTAP Protocol on Persistent Organic Pollutants, together with a similar protocol on heavy metals, was signed in Århus, Denmark, by all Arctic countries except Russia. The adoption of the UNECE agreement on POPs paved the way for an extended international agreement on POPs – this time at the global, rather than regional, level. Negotiations were begun, under the auspices of UNEP, to work towards such an agreement.

The detailed scientific information from AMAP was introduced into the UNEP process, and, as with the CLRTAP process, the Arctic information proved very effective and compelling justification of the need for international action on POPs.

The Stockholm Convention on Persistent Organic Pollutants made it much easier for the Arctic countries to implement several projects on POPs in Russia that had been initiated under ACAP. It also opened possibilities for financial support from the Global Environment Facility (GEF) to assist Russia with financial support to develop a national plan of action to phase out use of POPs. Perhaps the most significant outcome of Russia signing the Stockholm Convention is that the Arctic countries can now take a unified stand in demonstrating to the rest of the world their concern about POPs and their commitment to action to reduce emissions of POPs.

THE 2000 MINISTERIAL MEETING IN BARROW

At the second Arctic Council Ministerial Meeting, in Barrow, in October 2000, the Arctic Council Action Plan (ACAP) was endorsed and an ACAP Steering Committee established under the leadership of Norway. Projects on dioxins and furans and obsolete pesticides in the Russian Federation were adopted to complement the ongoing PCB project, with AMAP once more requested to coordinate implementation of the PCB project's second phase. An ACAP project on mercury was also initiated.

THE 2002 AMAP ASSESSMENT

While the full report of the second AMAP assessment is to be delivered to ministers at the third Arctic Council ministerial meeting, in October 2002, AMAP presented updated information in Barrow, summarizing preliminary findings from its 1998–2002 assessment period.[15]

The evidence that POPs are widespread in the Arctic has been confirmed by new results from geographical areas not covered under the first AMAP assessments. Evidence that POPs not only accumulate in but also affect Arctic wildlife is also increasing.

Some of the highest levels of POPs such as PCBs are found around Svalbard and Franz Josef Land (figure 4.4), and recent results indicate that polar bears with high PCB levels suffer from impaired defence against infections. High PCB levels may also be affecting the survival of cubs. Effects of persistent organic pollutants have been documented in other Arctic species as well, including the northern fur seal, glaucous gull, peregrine falcon, and dog whelk.

Although PCBs and other organic pollutants have been regulated for several decades in Arctic countries, the levels in the environment are mostly a legacy of past emissions. PCBs from old uses and equipment are still spreading in the environment. Also, there is evidence that some of the regulated persistent pesticides are still being used or may be leaking from inadequate storage/disposal sites. Consequently, although levels in some areas (generally close to sources) are decreasing, it may take many years for levels in areas remote from sources to decline.

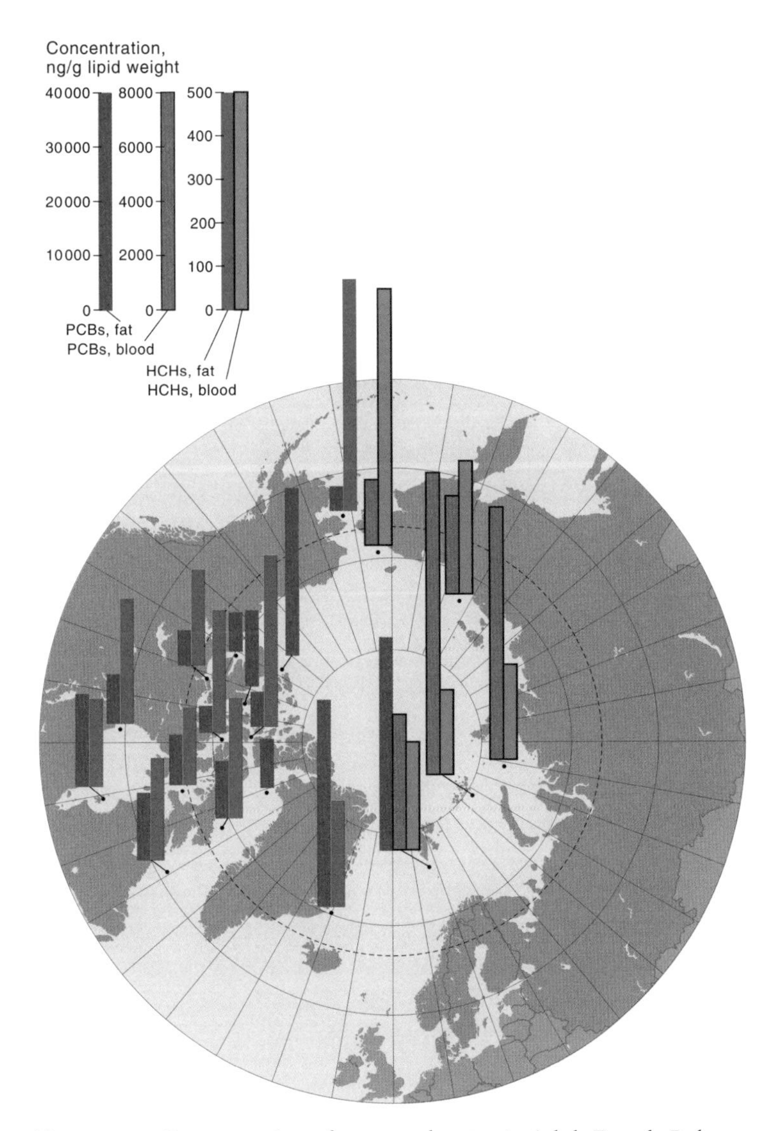

Figure 4.4 Concentration of PCBs and HCHs in Adult Female Polar Bears. Data from AMAP phase 1 (fat) and recent studies (blood). At Svalbard, PCB levels measured in fat are approximately five times higher than in blood from the same animals. If this relationship is the same in other areas, high fat PCB concentrations can be expected in bears in the Russian Arctic.

Source: Arctic Pollution 2002 (Oslo, Norway: Arctic Monitoring and Assessment Programme 2002), p. 22.

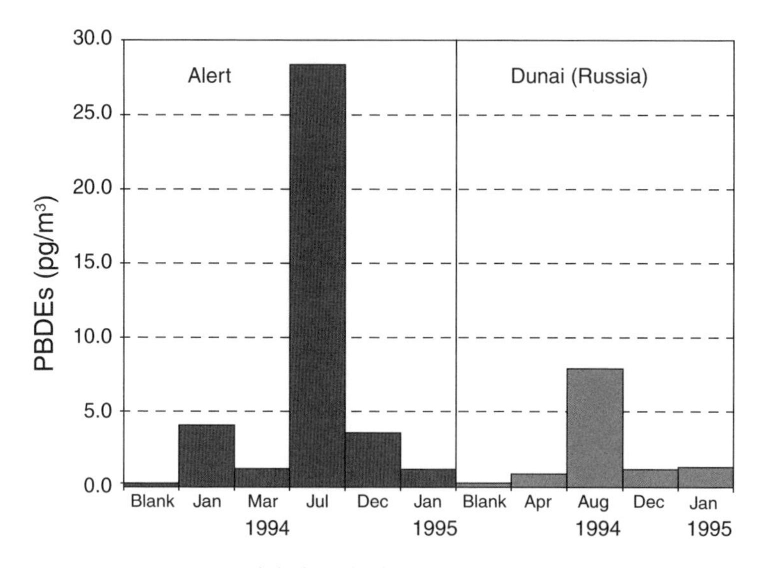

Figure 4.5 Brominated diphenyl ethers in air samples from northern Ellesmere Island (Alert, Canada) and the Lena River delta (Dunai, Russia) in 1994–95

Source: Arctic Council Secretariat, "The First Ministerial Meeting of the Arctic Council, Iqaluit. The Iqaluit Declaration," 1998.

Data also show that new persistent organic pollutants are arriving in the Arctic, some of which are currently being produced in large quantities. Examples include polybrominated diphenyl ethers (PBDEs) and polychlorinated naphthalenes (PCNs). These compounds are used as flame retardants and chlorinated paraffins and in cutting oils and compounds such as perfluorooctane sulfonate (PFOS), a surfactant and flame suppressant. PCNs are chemically similar to PCBs, and some PCNs have toxic properties similar to those of chlorinated dioxins, furans, and dioxin-like PCBs.

"NEW" POPs IN AIR

One of the knowledge gaps identified in the first AMAP assessment concerns arriving POPs, including currently used pesticides and other chemicals being introduced as replacements for banned substances. Archived air sample extracts (1994–95) from Alert, Canada, were analyzed for PBDEs and PCNs. PBDEs were detected at low pg/m^3 levels in pooled monthly air samples; levels observed in Alert were higher than those in Dunai, Russia, possibly attributable to the higher usage of these compounds in North America than in Russia. Peak concentrations of PBDEs were observed during July and August, indicating enhanced volatilization and transport from mid-latitude source regions during the summer months (figure 4.5).

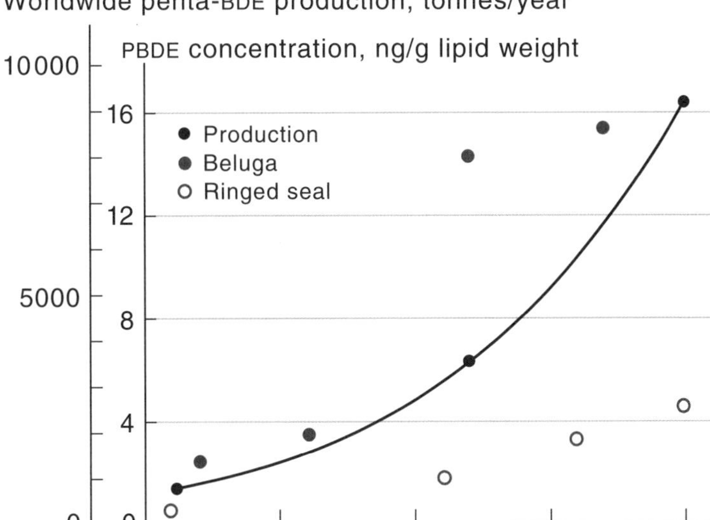

Figure 4.6 Comparison of Temporal Trends of PBDEs in Ringed Seal and Beluga in the Canadian Arctic with Estimated Global Production of Penta-BDE Over the Same Period

Source: Arctic Pollution 2002 (Oslo, Norway: Arctic Monitoring and Assessment Programme 2002), p. 25.

PCNs were also identified in concentrations ranging from 0.01 to 0.99 pg/m^3 at Alert in 1994/95 and from 0.03 to 2.33 pg/m^3 at Dunai. PCN concentrations at high Arctic sites are lower than those observed at lower Arctic latitudes over the eastern Arctic Ocean, Norwegian Sea, and Barents Sea.

"NEW" POPS IN BIOTA

Studies into temporal trends of PCBs have also documented trends in brominated and chlorinated diphenyl ethers and chlorinated paraffins in beluga blubber samples from the southeast Baffin Island stock. No significant decline in PCBs, toxaphene, or DDT was observed over the period 1982–96; however, a significant increase was observed in short-chain chlorinated paraffins and brominated diphenyl ethers from 1982 to 1996, while chlorinated diphenyl ethers declined significantly. Figure 4.6 illustrates the increase in concentration of the brominated diphenyl ether 2,2'4,4' TeBDE in beluga blubber.

PCNs have also been detected in marine mammals and birds; short-chain chlorinated paraffins, which can travel over long distances, have been detected in Arctic sediment and biota; and PFOS has been detected in polar bears and seals in Arctic North America and on Svalbard.

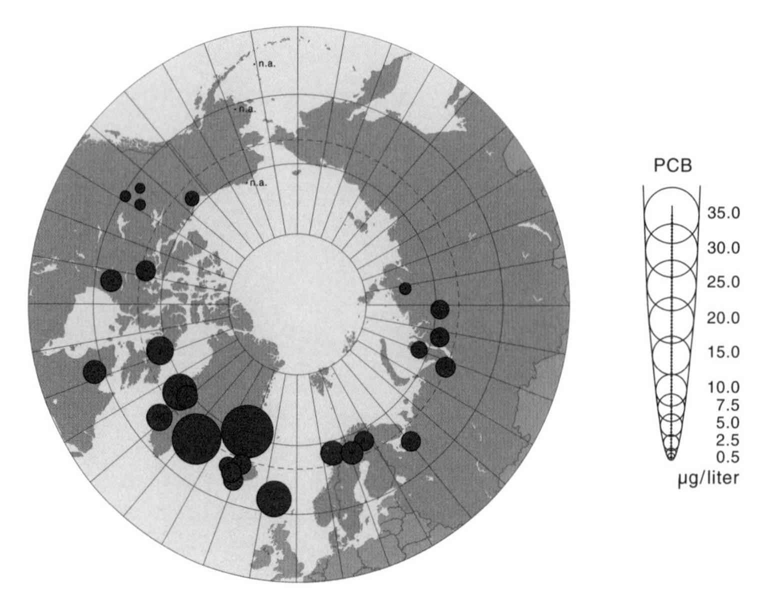

Figure 4.7 PCB Concentrations in Human Blood

Source: Arctic Pollution 2002 (Oslo, Norway: Arctic Monitoring and Assessment Programme 2002), p. 93.

The use of brominated flame retardants has increased dramatically in the past decade and annual worldwide production now exceeds 200,000 tonnes (1 metric tonne = .98 ton). Most are used in the industrial areas of the northern hemisphere – potential source regions to the Arctic. In areas outside the Arctic, PBDEs have been found in human breast milk and in the tissues of several animal species. PBDE levels in North America are increasing, while in Europe levels in biota increased until the mid-1980s and in humans until the late 1990s. In the Arctic, at present, PBDE levels are much lower than levels of "legacy" POPs such as PCBs.

The second AMAP assessment contains much new information on biological effects of POPs and better documentation of the levels in humans throughout the circumpolar region. PCBs are a special concern because of evidence that a number of Indigenous populations have blood levels that exceed nationally derived guidelines (levels of concern and action levels).

AMAP findings also continue to highlight mercury as a problem in the Arctic. Ministers in Barrow noted "with concern that releases of mercury have harmful effects on human health and may damage ecosystems of environmental and economic importance, including in the Arctic" and called "upon the United Nations Environment Programme to initiate a global assessment of mercury that could form the basis for appropriate international action in which the Arctic States would participate actively."[16]

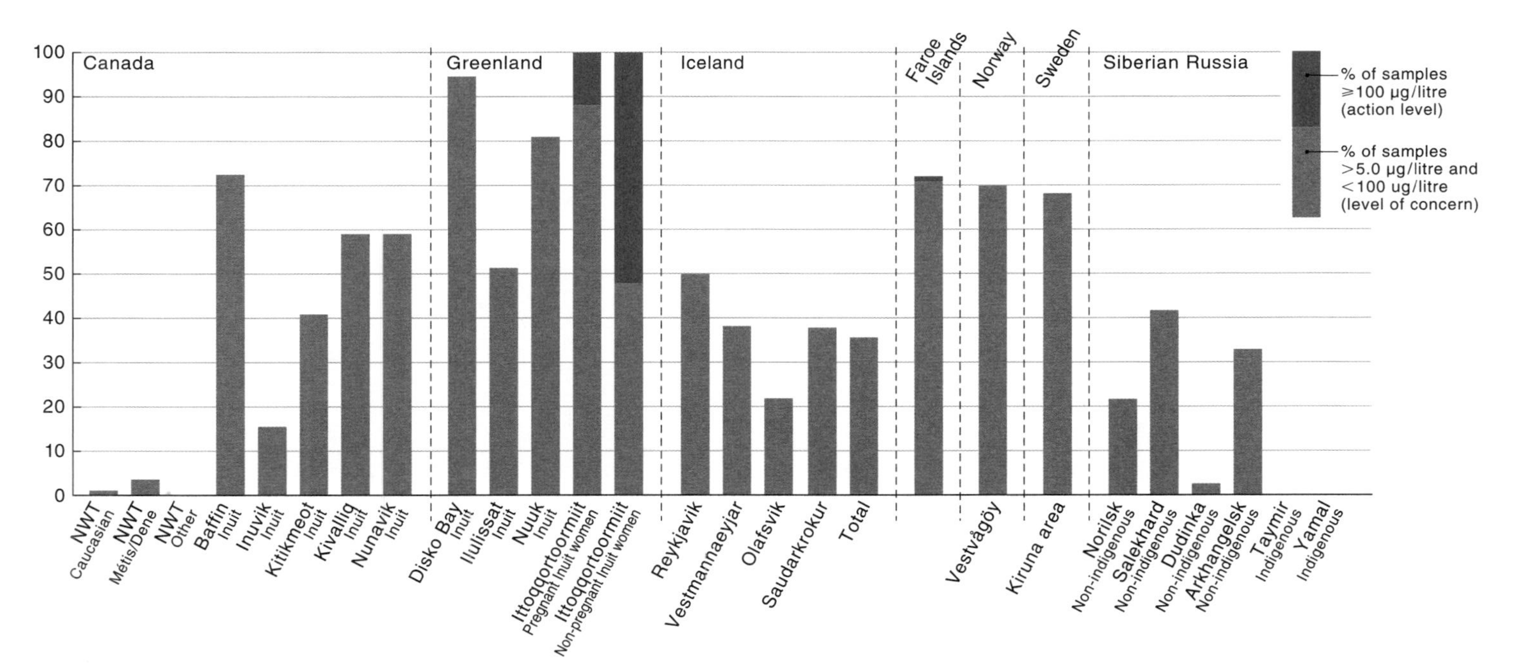

Figure 4.8 PCBs in Blood, Measured as Aroclor Exceedance

Source: Arctic Pollution 2002 (Oslo, Norway: Arctic Monitoring and Assessment Programme 2002), p. 89.

Partly in response to the Arctic concerns, UNEP initiated a global study on the health and environmental impacts of mercury. This study, which will be reported later in 2002, may ultimately lead to development of a global agreement on mercury, extending the UNECE heavy metals protocol in the same way that the Stockholm Convention extended the UNECE POPs protocol. AMAP has already begun compiling new information on mercury in the Arctic.

ARCTIC INDIGENOUS PEOPLES AND AMAP

Since its inception, AMAP has had a strong focus on the contaminants exposure of Arctic Indigenous populations (due to high exposures related to traditional diets) and non-Indigenous northern populations. This has been reflected in the design of the monitoring and assessment program, in the AMAP findings, and in the involvement of representatives of the Indigenous peoples' organizations in the preparation and drafting of the AMAP assessments. Representatives of Arctic Indigenous peoples have been active at all levels in the AMAP process.

In the AMAP assessment presented in 1997 there was a lack of environmental and human health data from Alaska and from large parts of the vast Arctic areas of northern Russia, particularly eastern Siberia. Consequently, the United States initiated work in Alaska to acquire important data on the exposure of Indigenous peoples that might be related to lifestyle and food. In northern Russia, a special project initiated by RAIPON, the relevant Russian authorities, and AMAP specifically focuses on Indigenous peoples' food security and health. The project concentrates on four areas across the north of Russia, from the Kola Peninsula in the west to Chukotka in the east. Financing from the GEF, the eight Arctic countries, the Nordic Council of Ministers, UNEP, World Meteorological Organization, World Wildlife Fund, and the Salamander Foundation in Canada supports collection and analysis of thousands of samples of water, sediments, fish, birds, terrestrial animals, seals, and whales, greatly improving our knowledge about contaminants in northern Russia. In addition, blood samples have been taken from pregnant women, and surveys of diet and lifestyle have been conducted. Preliminary results of this work indicate that Indigenous peoples of northern Russia are exposed to increased levels of POPs and mercury, although the routes of exposure may be different from those seen in Arctic Canada and Greenland.

So far, the results of the AMAP studies have demonstrated that some Indigenous peoples of the Arctic, especially those living mainly from the marine food web and consuming large quantities of marine mammals, are among the most POPs-exposed people on Earth. The Indigenous people whose traditional diet includes mainly reindeer meat have been among the most exposed people to radioactivity from global nuclear weapons tests and accumulation of radionuclides in the lichen–reindeer–human food chain.

That people inhabiting and harvesting traditional food from some of the cleanest areas on Earth receive some of the highest contaminant exposures found anywhere – even though POPs are not widely used there – is a matter of great concern. Chemicals of little or no benefit to these people are brought to them by atmospheric transport, ocean currents, and rivers, accumulate in their foods, and result in exposures that have the potential to damage their health. At the same time, many of these people are dependent on the traditional foods that they can hunt or collect locally for their everyday diet. In many areas of the Arctic, alternative foods are either not available or not affordable. Traditional foods are a vital part of the cultural identity of many northerners. This is the Arctic dilemma, and one that calls for international action. Actions taken by the Arctic Council can address regional problems through programs like ACAP, but many other problems, including transport of POPs from areas as far away as the tropics or the industrial areas in the mid-latitudes, require international actions such as the Århus protocols and the Stockholm Convention.

THE FUTURE

Climate change is a concern for Arctic populations. What will be the effects on the natural environment, precipitation, ocean currents, wind patterns, sea ice, wildlife, and their way of life? These questions are linked to others: how will such changes influence the physical transport of contaminants by winds and oceans, and how might possible changes in ecosystems affect contaminant bioaccumulation and human exposure? The effects over the short term and the long term are not yet clear – the only certainty is that changes will occur.

AMAP, the Arctic Council Working Group on Conservation of Arctic Fauna and Flora (CAFF), and the International Arctic Science Committee (IASC) have been asked by ministers to assess the effects of climate change in the Arctic. Work is under way and the Arctic Climate Impact Assessment (ACIA) is to be presented to the Arctic Council in autumn 2004.

The climate scenarios to be addressed in the ACIA have been selected. Figure 4.9 shows the change in air temperature that will occur in 100 years based on the results of five different models. One prediction shows an increase of five degrees Celsius in central parts of the Arctic Ocean and two to three degrees around the margins of the Arctic Ocean. This would have a dramatic effect on the natural environment and the lives of Indigenous peoples. More precipitation may lead to more deposition of airborne contaminants in the North; greater global erosion and flooding would introduce more contaminants to the Arctic via marine pathways. On the other hand, an increase in temperature might weaken the mechanisms that cause POPs to be transported to the Arctic and increase rates of degradation of some of the contaminants, making them less persistent.

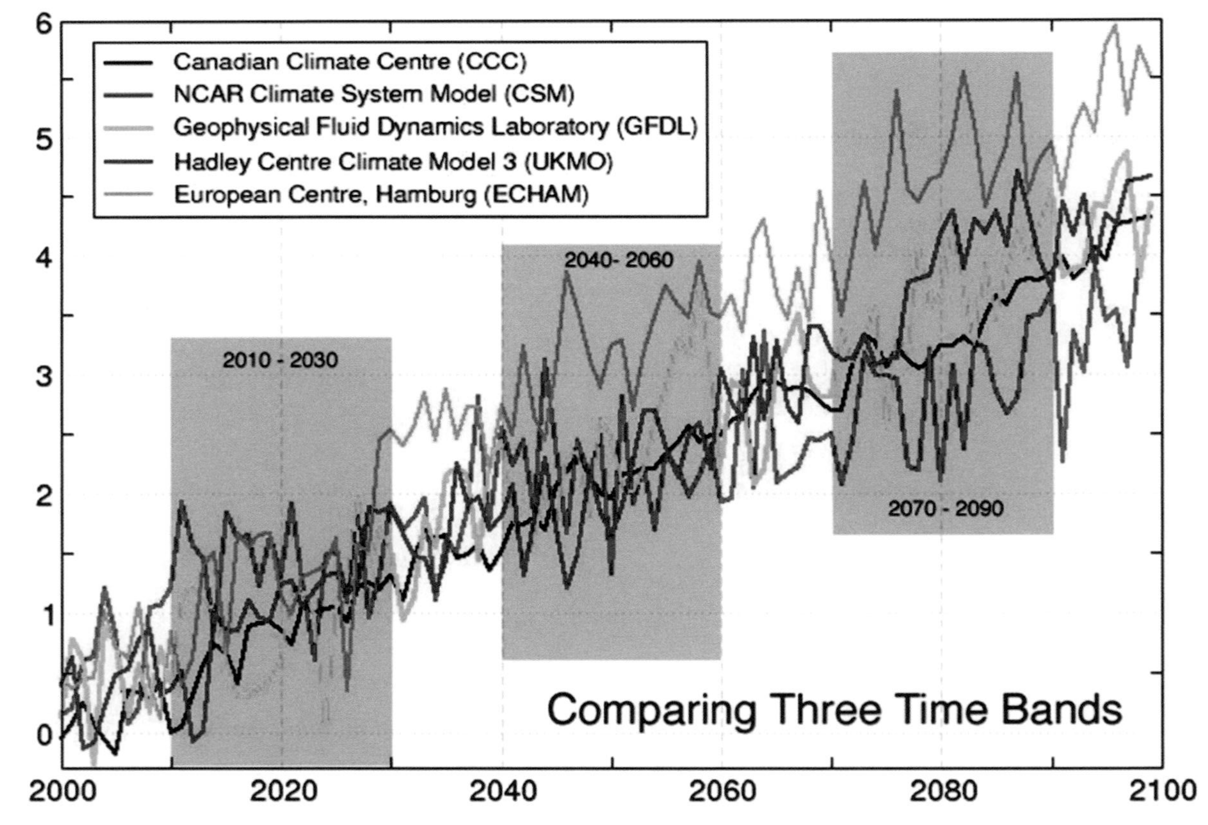

Figure 4.9 The Change in Air Temperature That Will Occur in 100 years according to Results from Five Differei Models

Source: AMAP, in preparation.

FOLLOWING INTERNATIONAL ACTIONS: MITIGATING AND MONITORING

The Stockholm Convention may require additional monitoring to document the effectiveness of the convention and to confirm that countries take the measures they have agreed to. A global long-term monitoring network for POPs covered by the convention is one component; a monitoring program for POPs not (yet) covered by the convention is another. The Arctic component of monitoring networks will be important, as the Arctic seems to act as a sink for a number of POPs. AMAP, through its network of stations and TDCs, would be able to contribute to these new goals. AMAP has already been invited to participate in discussions on monitoring linked to the Stockholm Convention. The international agreements made so far achieve nothing unless they are fully and effectively implemented.

NOTES

1 J. Jensen, "Report on Organochlorines," in *The State of the Arctic Environment Reports* (Rovaniemi, Finland: Arctic Centre, University of Lapland 1991), 335–84.
2 L-O. Reiersen, *State of the Arctic Environment: Updated Draft Proposal for Arctic Monitoring and Assessment Programme* (Norway: State Pollution Control Authority 1991).
3 The eight countries are Canada, Denmark/Greenland, Finland, Iceland, Norway, Russia, Sweden, and the United States.
4 Arctic Environmental Protection Strategy (Rovaniemi: AEPS 1991).
5 AMAP, "The Monitoring Programme for the Arctic Monitoring and Assessment Programme, AMAP" (Oslo, Norway: Arctic Monitoring and Assessment Programme 1993), Report 93: 3.
6 Arctic Environmental Protection Strategy.
7 AMAP, "Guidelines for the AMAP Assessment" (Oslo, Norway: Arctic Monitoring and Assessment Programme 1995), Report 95: 1.
8 AMAP, "Report to Ministers: Update on Issues of Concern to the Arctic Environment, including Recommendations for Actions" (Oslo, Norway: Arctic Monitoring and Assessment Programme 1993), Report 93: 4.
9 "The Arctic Environment: Second Ministerial Conference. The Nuuk Ministerial Declaration on Environment and Development in the Arctic" (Copenhagen: Ministry of Foreign Affairs 1993).
10 AMAP, *Arctic Pollution Issues: A State of the Arctic Environment Report* (Oslo, Norway: Arctic Monitoring and Assessment Programme 1997), p. vii. (Available in English, Danish, Greenlandic, Norwegian, Saami, and Russian languages.)
11 Arctic Council Secretariat, "The Fourth Ministerial Meeting under the Arctic Environmental Protection Strategy, Alta. The Alta Declaration on the Arctic

Environmental Protection Strategy & Senior Arctic Affairs Officials (SAAOs) Report to Ministers," 1997, p. 2.

12 AMAP, 1997.

13 Arctic Council Secretariat, "The First Ministerial Meeting of the Arctic Council, Iqaluit. The Iqaluit Declaration," 1998, para. 18.

14 Arctic Council Secretariat, "Report of Senior Arctic Officials to the Arctic Council, Iqaluit, 1998."

15 AMAP, "AMAP Report on Issues of Concern: Updated Information on Human Health, Persistent Organic Pollutants, Radioactivity, and Mercury in the Arctic" (Oslo, Norway: AMAP 2000), Report 2000: 4.

16 Arctic Council Secretariat, "The Barrow Declaration on the Occasion of the Second Ministerial Meeting of the Arctic Council," 2000, para. 13.

5

The Deposition of Airborne Dioxin Emitted by North American Sources on Ecologically Vulnerable Receptors in Nunavut

BARRY COMMONER, PAUL WOODS BARTLETT, KIMBERLY COUCHOT, AND HOLGER EISL

INTRODUCTION

The picture that most North Americans have of the Arctic – a pristine, snowy wilderness, sparsely peopled and unpolluted – is, unfortunately, unreal. Although there are few sources of pollution in the region itself, it receives emissions from sources far to the south that are transported mostly by the prevailing air currents. The North American Commission for Environmental Cooperation (CEC) engaged the Center for the Biology of Natural Systems (CBNS) to model the rates of deposition, in the new Canadian territory of Nunavut, of airborne dioxins (polychlorinated dibenzo-p-dioxins (PCDDs) and polychlorinated dibenzofurans (PCDFs)) emitted by identifiable North American sources of these pollutants.[1]

The study was a response to evidence that Nunavut is especially vulnerable to the long-range air transport of dioxin. There are no significant sources of dioxin in Nunavut or within 500 kilometres (311 miles) of its boundaries, yet dioxin concentrations in Inuit mothers' milk are twice the levels observed in southern Quebec.[2] This is apparently a result of the elevated dioxin levels in the Indigenous diet – traditional foods such as caribou, fish, and marine mammals. Ninety-eight per cent of human exposure to dioxin is through ingesting animal foods, especially those rich in fat. In temperate climates, dioxin enters the food chain from the air; taken up through pasture and other leafy food crops by animals, it appears in milk and beef, which in the United States account for about two-thirds of diet-mediated exposure to dioxin.[3] In the Arctic, dioxin enters the major terrestrial (caribou) food chain chiefly through lichens, mosses, and shrubs and the marine (seal, walrus) food chain chiefly through algae. Since these avenues of entry cannot be protected from airborne pollutants, remedial measures must be directed at the sources of dioxin emissions. It is necessary also to estimate the degree to which each source contributes because relatively few of the numerous sources account for most of the airborne dioxin that is deposited on receptors.

METHODOLOGY

We used an air transport model to estimate the amounts of dioxin deposited on the terrestrial and marine areas of eight sites in Nunavut from each of 44,091 sources in Canada, the United States, and Mexico. The HYSPLIT (Hybrid Single-particle Lagrangian Integrated Trajectory) air transport model was originally developed by the National Oceanic and Atmospheric Administration (NOAA) to track the movement of inorganic radionuclides. For the Nunavut project, we fine-tuned the HYSPLIT-4 model for dioxin to simulate dioxin deposition conditions in the Arctic environment of Nunavut.[4] The model is designed to operate with NOAA's global meteorological archive, which incorporates global meteorological data for a three-dimensional grid, 190 x 190 kilometres (118 miles x 118 miles) horizontally with fourteen vertical atmospheric layers (up to 12,000 metres, or 13,123 yards), tabulated at six-hour intervals. The model's study period, 1 July 1996 through 30 June 1997, incorporated the NOAA global weather data for that period.

The model computes the transport, degradation, and deposition of dioxin emitted by a source at a given geographical location (designated by latitude and longitude) by estimating the atmospheric behaviour of the emitted material as if divided into discrete "puffs." The model tracks each puff, recording its location and dioxin content at one-hour intervals until – due to diffusion and degradation – the airborne dioxin concentration reaches negligible levels (figure 5.1). For this purpose, the model includes algorithms that calculate the rates of degradation of dioxin congeners (largely through reaction with OH radicals) as affected by solar flux, cloud cover, and season as well as rates of deposition (en route and at designated receptors) in wet and dry conditions.

Since the model estimates the amount of dioxin emitted from a single source that is eventually deposited on an individual receptor, determining the total of amounts deposited on each receptor requires computer runs from each of the numerous known sources. Although 44,091 sources of airborne dioxin were identified for this study, such a comprehensive approach was not feasible; a computer run averaging 6.5 hours from each source would be required. Instead, the model used 105 standard (hypothetical) source points suitably distributed in Mexico, the United States, and Canada. For each standard point, these computer runs estimated the fraction of emitted dioxin deposited at each of the receptors (i.e., the Air Transport Coefficient, ATC). A separate program ("Transport Coefficient," TRANSCO, developed earlier by Dr Mark Cohen,

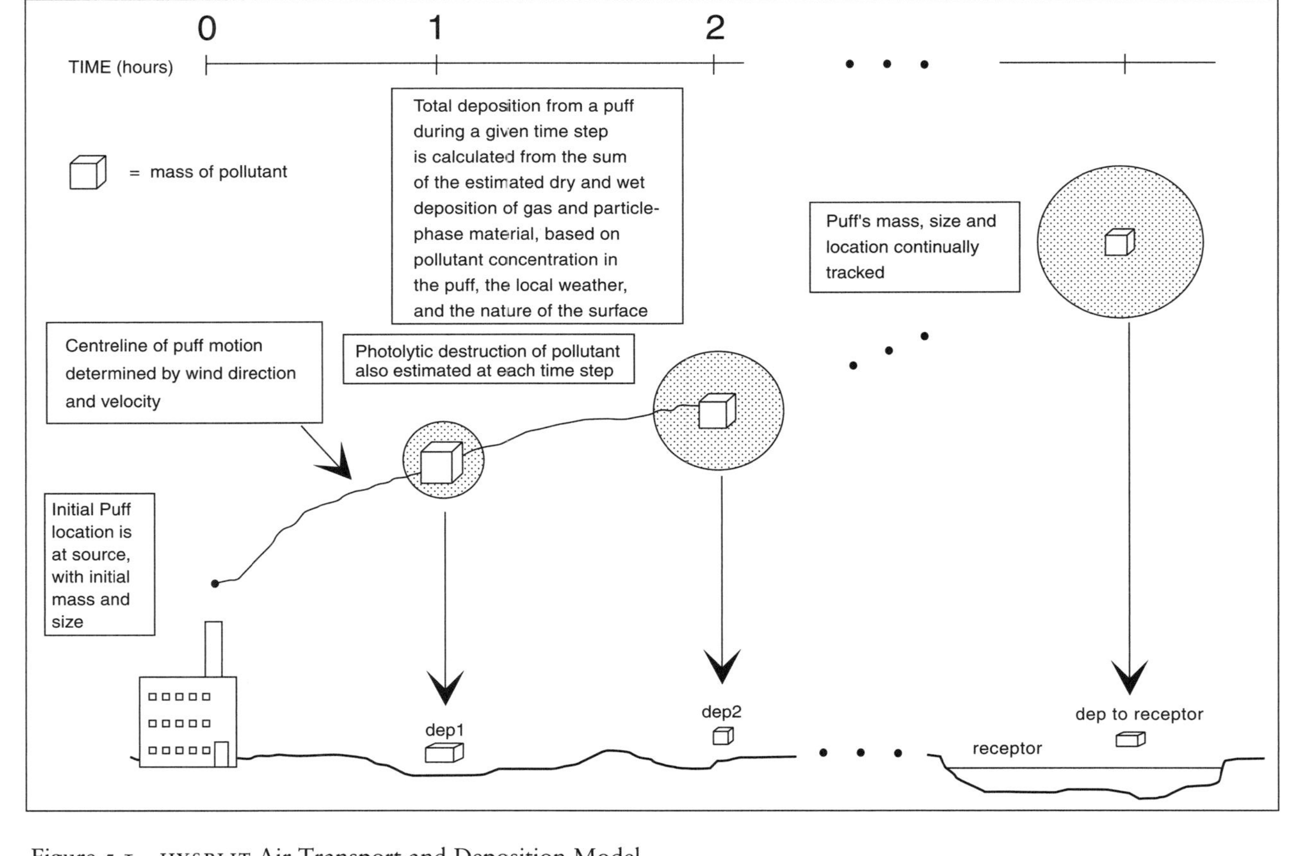

Figure 5.1 HYSPLIT Air Transport and Deposition Model
A simplified diagram of the operation of the air transport model. The total annual dioxin emission from the source is arbitrarily divided into uniform "puffs" that are released continually at a set interval (one hour, in this example). The size (shaded circles) of the puff increases with time due to diffusion and dispersion; the mass of the puff (open cube) decreases with time due to deposition and degradation.

then at CBNS) includes a spatial interpolation program, which estimates the ATC values for each of the numerous *actual* sources from the values of the four nearest standard points and their geographic orientation relative to the receptor. The actual source's estimated ATC value multiplied by its annual rate of dioxin emission yields the amount of the emitted dioxin that is deposited at each of the receptors. In practice, the model was initially run separately for each of four representative PCDD and PCDF congeners. Then a program (also included in TRANSCO) used these results to interpolate the ATC values for the remaining seventeen toxic PCDD/PCDF congeners and eight homologue groups. By means of these interpolations, the model, together with the TRANSCO program, enabled us to estimate the amounts of each of the toxic congeners and homologue groups deposited on each of the receptors from each of the 44,091 sources over the one-year study period.

The HYSPLIT-3 air transport model has been evaluated by comparison with month-long dioxin measurements taken by CBNS at three sites in Vermont and Connecticut in August/September 1996.[5] Dr Mark Cohen at NOAA recently evaluated the improved HYSPLIT-4 dioxin model by comparing modelled estimates with measured deposition in Siskiwit Lake, Isle Royale, Lake Superior.

RECEPTORS

The following criteria were used to select appropriate receptor areas: a) that they serve as avenues of entry of airborne dioxin into the Inuit food chains; b) that they are representative of the three major ecozones in Nunavut: Southern Arctic, Arctic Cordillera, and Northern Arctic; c) that they cover the three administrative Nunavut regions: Kitikmeot, Kivalliq (Keewatin), and Qikiqtaaluk (Baffin); and d) that they are representative of the geographic extent of Nunavut.

The chief land-based food chain that supports the Inuit diet is lichen–caribou. Lichens absorb nutrients largely from airborne materials and are a major source of food for caribou, an important part of the Inuit diet. The chief marine-based food chain is algae–crustacean–fish (chiefly cod)–seal–polar bear/whale; fish and these marine animals are major components of the Inuit diet. Algae, the avenue of entry, chiefly occur under land-attached ice, which generally extends to about fifty kilometres (31 miles) from the coast.[6]

Based on these considerations, we define the receptors that conduct airborne dioxin into the Inuit diet as 1) land areas in which caribou are plentiful and are regularly hunted by Inuit, and 2) marine areas within about fifty kilometres of the coastline, in which seals, polar bears, and whales are plentiful and are hunted by Inuit. Trout and Arctic char are also important in the Inuit diet. Since the freshwater streams and lakes in which they occur are abundantly distributed in the land areas, estimates of deposition on these receptors will be representative of the level of dioxin contamination in these freshwater fish.

Nunavut is described in considerable detail in the *Nunavut Atlas*, which includes detailed maps for each of the fifty-seven sectors of the territory.[7] One set of maps describes the occurrence of land-based and marine wildlife, and another set outlines the areas in which the Inuit hunt, fish, and trap wildlife. Based on this information, we selected eight such sectors as receptors; they are identified by the names of nearby Inuit communities (figure 5.2). Together, these sectors are representative of most of the area of Nunavut. They range from latitude 56° (Sanikiluaq) to latitude 74° (Arctic Bay) and from longitude W60° (Broughton Island) to longitude W120° (Ikaluktutiak). In each sector we identified a land-based receptor area and a marine (land-attached ice) receptor area of 10,000 to 20,000 square kilometres (3,860 to 7,720 square miles) in size.

DIOXIN SOURCE INVENTORIES

For the air transport model, the geographic location (latitude and longitude) and the annual rate of dioxin emission for each known source are essential. Relatively few facilities have actual dioxin emission measurements available; for others, emission is estimated from capacity or throughput data and an appropriate emission factor. Estimates are derived from the available measurements of dioxin emission from typical sources and, ideally, take into account the influence of the facility's operational characteristics and type of air pollution control system. Typically the emission factor is expressed as the annual amount of dioxin emitted per ton (1 ton = 1.02 metric tonnes) of the facility's throughput (e.g., fuel burned). Because of the considerable uncertainty involved in this procedure, the emission factors, and hence the emission rates, are estimated as high and low values, which generally may range by up to two orders of magnitude. This same range affects the modelled estimates of dioxin deposition. The data reported below are based on mid-range values. Mid-range emission values are characteristic of source type as a class but may be less accurate for certain individual sources, especially untested facilities.

Certain sources cannot be localized because they are mobile (e.g., diesel trucks) or are too numerous to be individually identified (e.g., backyard trash burners). In these cases, emission estimates are based on the estimated number of sources in an appropriate unit area, and their collective location is represented by the area centroid. Examples of the dioxin emissions from such area-based sources are total emissions of diesel trucks operating in a given U.S. county or emissions from the estimated amount of backyard trash burning in a Mexican municipality. The sources that make up the dioxin emission inventories for Canada, the United States, and Mexico include a total of twenty-three classes of facilities. The three inventories are summarized in figure 5.3.

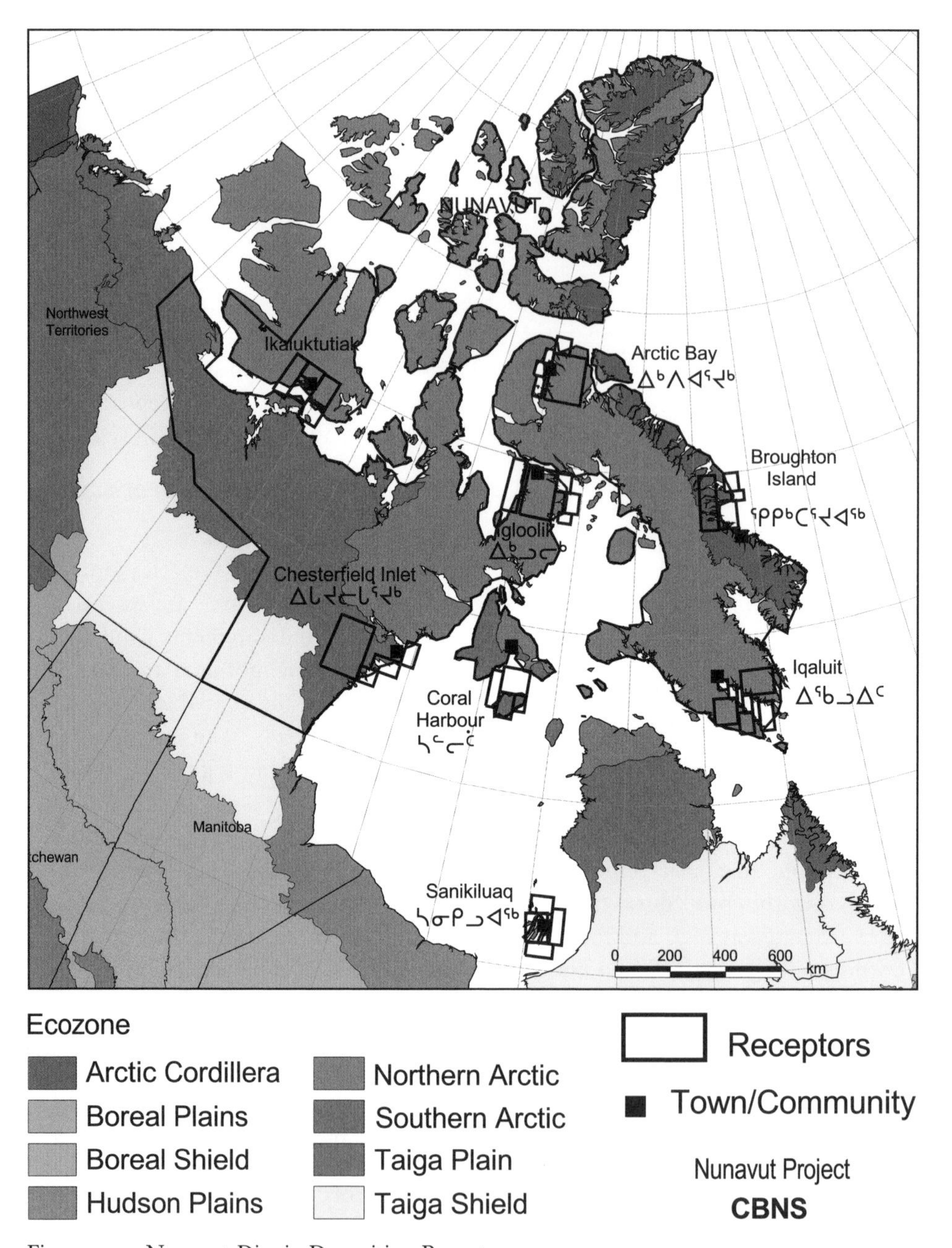

Figure 5.2 Nunavut Dioxin Deposition Receptors
The eight receptor sites are identified by the nearby communities (black squares). At each site, the land and water receptors consist of the total area of one or more rectangles located on land or water.

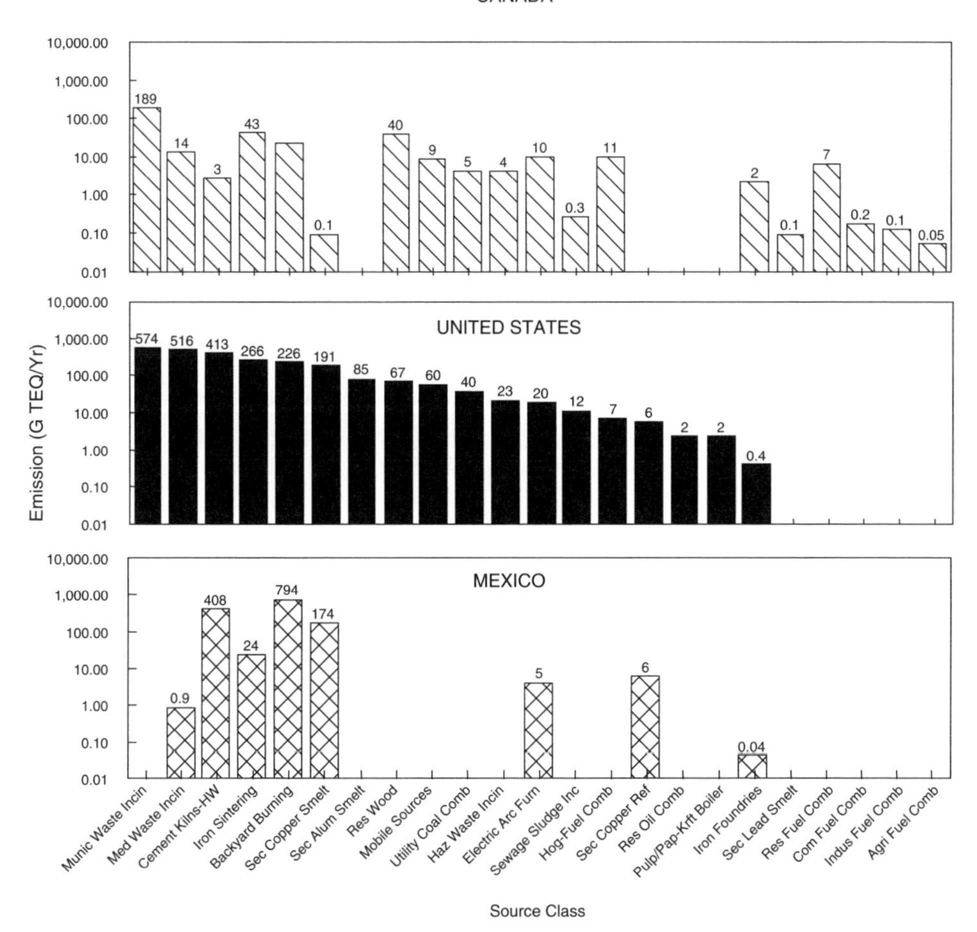

Figure 5.3 The North American Dioxin Source Emission Inventories
The dioxin source inventory for Canada is based on one prepared for 1997 by Environment Canada, except for the inventory for backyard burning, which was prepared by CBNS. The inventory for the United States is based on one prepared for 1995 by the U.S. EPA, updated to 1996/97 by CBNS, with added inventories for backyard burning, iron sintering plants, and several metallurgical processes. The inventory for Mexico was prepared by CBNS based on data provided by the Instituto Nacional de Ecología (INE) and commercial sources.

Environment Canada has published an inventory of dioxin sources as of 1997 that was made available to us in electronic form.[8] The inventory includes estimated dioxin emission rates of 16,729 sources in nineteen classes, representing 1,493 point sources and 15,236 area sources. A few of these source classes overlapped with source class categories in the U.S. inventory, and adjustments were made to avoid duplication. In addition, since Environment Canada had not yet completed an analysis of backyard waste burning, we prepared such an analysis for inclusion in the Canadian inventory. Backyard waste burning is the practice, common in rural areas of Canada and the United States (i.e., areas not served by centralized waste collection), of disposing of household waste in a burner generally made of a suitably aerated steel drum. The amount of dioxin emitted annually from backyard waste burners in Canada was estimated from demographic data (which distinguished between rural and urban populations on the basis of population density), data on *per capita* production of residential waste, and an emission factor developed in a recent U.S. Environmental Protection Agency (EPA) study of the combustion of household waste in barrel burners.[9] The outcome of this analysis is an estimate of the amount of dioxin emitted annually from the backyard burners in each Canadian postal zone.

The U.S. dioxin emission inventory is largely based on that produced by the U.S. EPA as of 1995 and made available to us in electronic form. (This inventory in part reflects an ongoing exchange of inventory information between the U.S. EPA and CBNS and, where possible, includes modifications reflecting changes from 1995 to 1996/97.) However, we have added several classes of sources that were lacking in the EPA inventory: iron sintering plants, electric arc furnaces, coal-burning power plants, and backyard trash burning. The methodology used to estimate dioxin emissions from backyard burning is similar to that summarized here. The inventory consists of eighteen source classes representing 22,439 sources, of which 3,735 are point sources and 18,704 are area sources.

While Canada and the United States have produced essentially complete inventories, in Mexico the Instituto Nacional de Ecología (INE) has only recently begun this process. As a result, we produced a preliminary inventory for Mexico, based on the numbers and operational characteristics of major sources provided to us by INE and emission factors derived from U.S. data.

Together, the three North American inventories comprise 44,091 sources, of which 5,343 are point sources and 38,748 are area sources. The total source inventory emits 4,713 grams toxicity equivalent quotient (TEQ): 364 grams TEQ (7.7 per cent) from Canadian sources, 2,937 grams TEQ (62.3 per cent) from U.S. sources, and 1,412 grams TEQ (30 per cent) from Mexican sources [1 gram = 0.035 ounce; 1 ounce = 28.35 grams]. Emissions from sources within Nunavut total 0.12 grams TEQ annually – only 0.003 per cent of total North American emissions. TEQ, or Toxicity Equivalent Quotient, is a measure of the overall toxicity of the dioxin and furan congeners, commonly grouped as "dioxin," based on their individual carcinogenic potency relative to that of the most potent congener, 2,3,7,8-TCDD. As shown in figure 5.3, emissions from the various source classes in each of the three national inventories vary over five orders of magnitude. Six of the twenty-three source classes referenced in figure 5.4 account for ninety-one per cent of the total emission: municipal solid waste incinerators, backyard trash burners, cement kilns burning hazardous waste, medical waste incinerators, secondary copper smelters, and iron sintering plants (figure 5.4). The combustion of domestic waste (in incinerators and backyard burning) accounts for more than one-half of the total emissions.

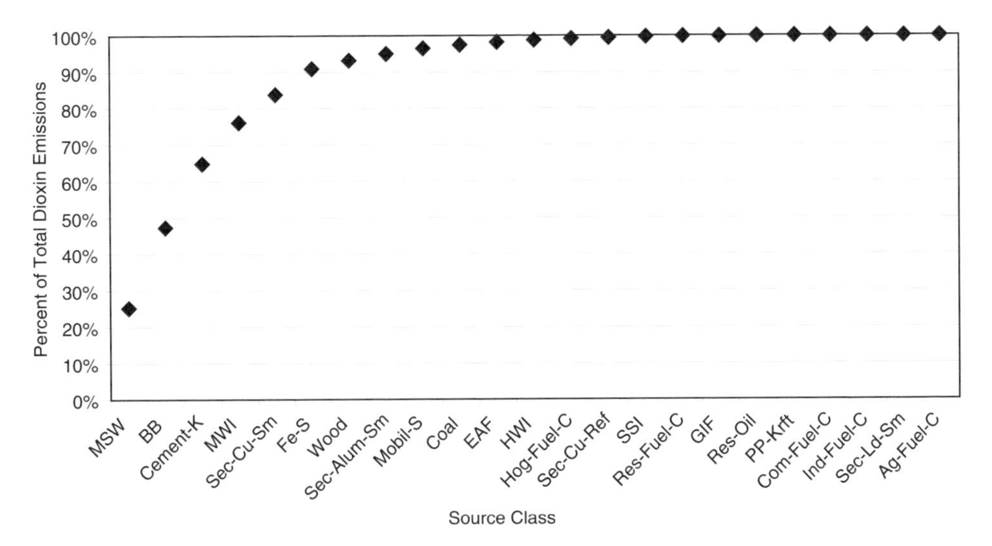

Figure 5.4 Total North American Dioxin Emission Inventory Cumulative Contributions of Source Classes
The emission values for all North American sources, in each source class, are ranked by their respective contributions to the total North American dioxin emission inventory.

RESULTS

The model's estimates of the total amounts of dioxin emitted from all North American sources that are deposited at each of the Nunavut receptors (figure 5.5) reveal considerable geographic variation. The deposition flux (picograms TEQ of dioxin per square metre) at the southernmost receptor, Sanikiluaq, is about ten times greater than at the most northern receptor, Arctic Bay, which is about 1,500 kilometres (930 miles) farther from the intense sources in the United States. There is a similar west-to-east gradient in dioxin deposition; thus, deposition at the easternmost receptor, Broughton Island, is about twice as great as deposition at the westernmost site, Ikaluktutiak. At each site, marine receptors uniformly receive more deposition per square metre than do adjacent terrestrial receptors because dioxin is more efficiently deposited onto water than land.

Overall, the greatest contribution to dioxin deposition in Nunavut is from U.S. sources, which account for seventy to eighty-two per cent, depending on the receptor. Canadian sources contribute eleven to twenty-five per cent, and Mexican sources five to ten per cent. Dioxin sources within Nunavut are responsible for only a very small fraction of the airborne dioxin deposited there. For example, based on the modelled estimates of deposition at a typical land receptor, Broughton Island, the total dioxin deposition flux from all North American sources is 8.90 picograms TEQ per square metre, of which Nunavut sources account for only 0.01 picograms TEQ per square metre, or 0.11 per cent.

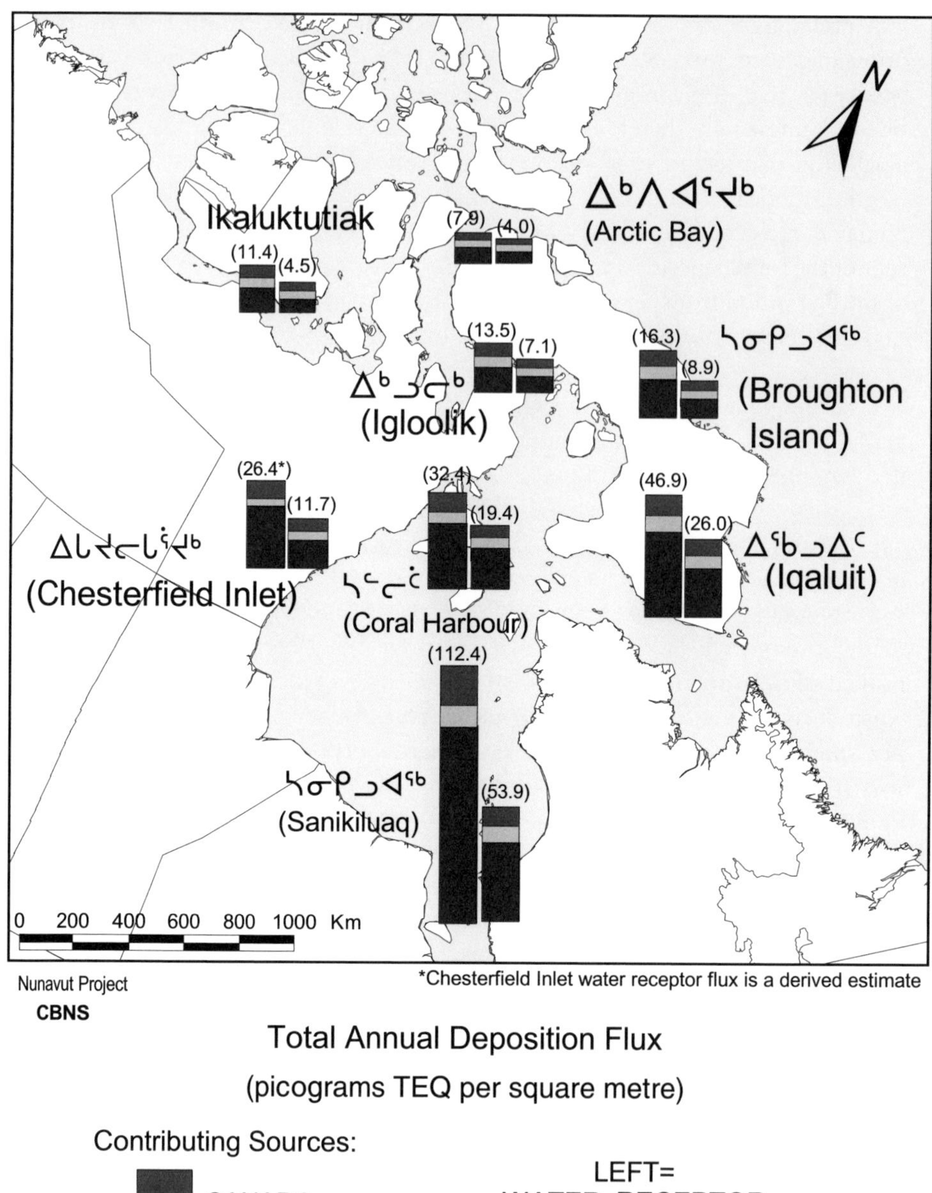

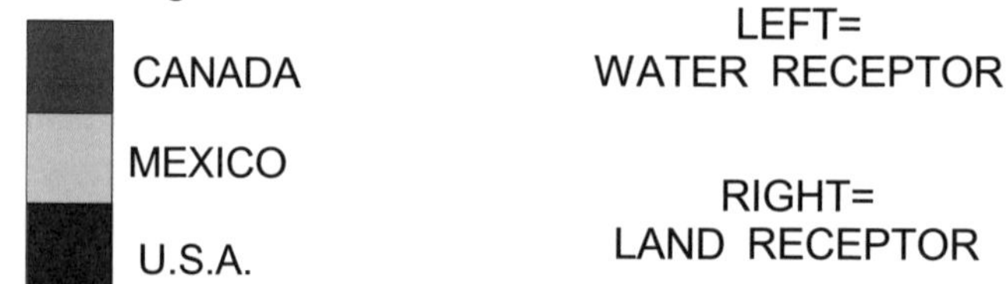

Figure 5.5 Dioxin Deposition at Nunavut Receptors
The heights of the bars represent the annual deposition flux at the adjacent water and land receptors at each of the sites in Nunavut. Deposition flux (picograms TEQ per square metre [1.2 sq. yd.]) values are in parentheses.

A preliminary estimate shows that the amount of the deposited dioxin that originates from sources outside North America is between two per cent and twenty per cent of the total deposition in Nunavut. North American sources outside of Nunavut are therefore responsible for the vast majority of the dioxin deposited on that territory. Moreover, only a very small proportion of the 44,091 North American sources accounts for most of the dioxin deposited at the Nunavut receptors. At Coral Harbour, a typical land receptor, thirty-five per cent of the total deposition is from just 0.04 per cent of the sources; fifty per cent of the deposition from 0.15 per cent; seventy-five per cent of the deposition from 1.54 per cent; and ninety per cent of the total deposition from 6.87 per cent of the sources.

The data generated by the air transport model can be used to rank the individual dioxin sources by the amount that *each* contributes to the dioxin deposited at *each* receptor. Consequently, the few individual sources that are responsible for most of the deposition can be identified by ranking the entire list of sources with respect to their contribution to the amount of dioxin deposited at the receptor. For example, the nineteen highest ranked individual sources – all in the United States and Canada – account for thirty-five per cent of the dioxin deposited at the Coral Harbour land receptor. The six highest ranked sources are in the eastern half of the United States: three are municipal waste incinerators, two are iron sintering plants, and one is a secondary copper smelter. The highest ranking Canadian source (seventh) is a municipal waste incinerator in Quebec (this ranking reflects its status in 1996/97; since then, modifications have significantly reduced the facility's emissions and consequently its deposition ranking). Three high-emitting Mexican sources are among the forty-one top-ranked facilities.

ANALYSIS OF SOURCE–RECEPTOR RELATIONSHIPS

The preceding sections quantify the relative contributions of different sources and classes of sources to the amounts of dioxin deposited at the Nunavut receptors. They show that the amounts vary among the eight receptor sites and between land and marine receptors at the same site. While these data suffice for ranking the sources by their contribution to the exposure of receptors to airborne dioxin, they do not explain the physical basis for the variations among sites and between receptors at the same site.

Several factors determine the amount of dioxin emitted from a specific source that is deposited at a particular receptor: the rate at which dioxin is emitted by the source; the geographic distance between the source and the receptor; the wind direction and velocity *en route*; the rate at which dioxin is destroyed and deposited *en route*; and the frequency and intensity of precipitation *en route* and at the receptor. The fate of airborne dioxin is also affected by meteorological conditions as it travels from source to receptor: temperature affects vapour/particle partitioning and, since these two states differ in their deposition rate, overall deposition; rain and snow will tend to increase deposition rates, especially of dioxin in the particle phase, both *en route* and at the receptor. Finally, all of the factors that influence the fate of airborne dioxin will be affected by the duration of transport from source to receptor, which depends on the distance travelled – a function of geographic source-to-receptor distance and wind direction and velocity. The distance travelled may be considerably greater than the geographic source–receptor difference, since wind directions are likely to vary considerably *en route*.

These complex processes combine to yield a *transport efficiency* – the Air Transport Coefficient (ATC) – which is expressed as the fraction of the dioxin emitted by a source that is deposited at the receptor. The amount of dioxin emitted by the source that is deposited at the receptor (the model output) can be expressed as:

Amount deposited at receptor = ATC x *amount emitted by source*

Figure 5.6 illustrates the variation, with source–receptor distance, in the relative amount of dioxin emitted by the sources and the relative amounts of dioxin deposited at several receptors. For this purpose, with each (land) receptor as the centrepoint, the total inventory of sources was segregated into a series of concentric zones 1,000 kilometres (620 miles) wide. It was then possible to estimate the total amount of dioxin emitted annually from the sources located within each concentric zone and their relative contribution to the total deposition at the receptor. Figure 5.6 shows that in each case there are virtually no emissions from sources less than 1,000 kilometres from the receptor. Except for Sanikiluaq, the southernmost receptor, most sources are 3,000 kilometres (1,860 miles) or more from the receptor. The ratio of emission to deposition confirms the sharp decrease in efficiency of transport with source–receptor distance.

To demonstrate more generally the interactions among emissions from *all* sources, air transport efficiency, and the resulting deposition of dioxin at all receptors, the relevant data on these factors have been collated on a common polar stereographic 100 x 100 kilometre (62 x 62 mile) grid. Each grid zone is treated as an individual source, with the total dioxin collective output of all the separate sources within the zone deemed to be emitted from its centroid.

In figure 5.7, the annual dioxin emissions from North America are plotted on the grid; each grid zone is colour-coded to indicate the total grams TEQ of dioxin emitted annually (1996/97) in that zone. Certain features of this map are noteworthy. First, it is evident that the area of Nunavut is entirely free of sources that emit more than 0.1 gram TEQ annually, and there appears to be only one such source within 500 kilometres (310 miles) of the Nunavut border. Second, the map clearly depicts the high concentration of dioxin sources in the eastern half of the United States, with particularly intense corridors

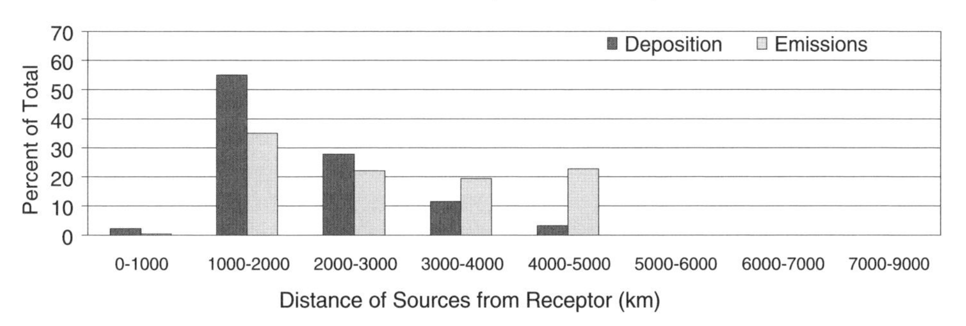

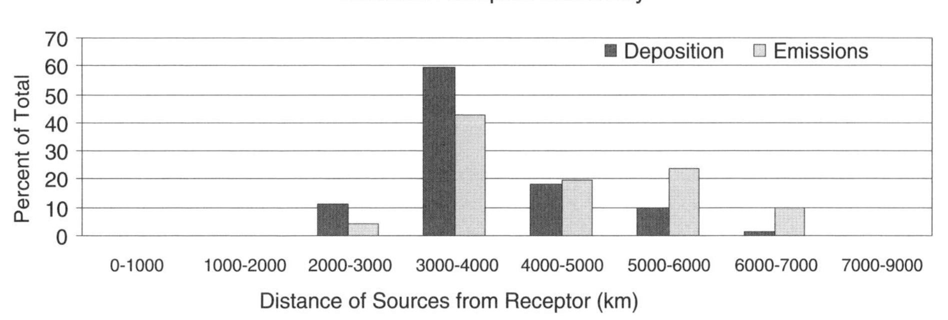

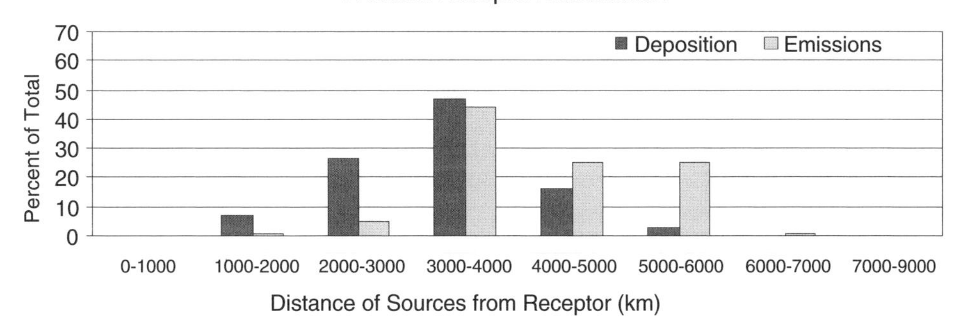

Figure 5.6 Dioxin Emissions and Deposition as a Function of the Distance of Sources from Selected Nunavut Land Receptors
These data were prepared by segregating the total inventory of sources into a series of concentric zones 1,000 kilometres (620 miles) wide with the indicated receptor as a centre point. The bars represent, as a per cent of their total values, the emissions of all sources within each concentric zone and their relative contribution to the total deposition flux at the receptor.

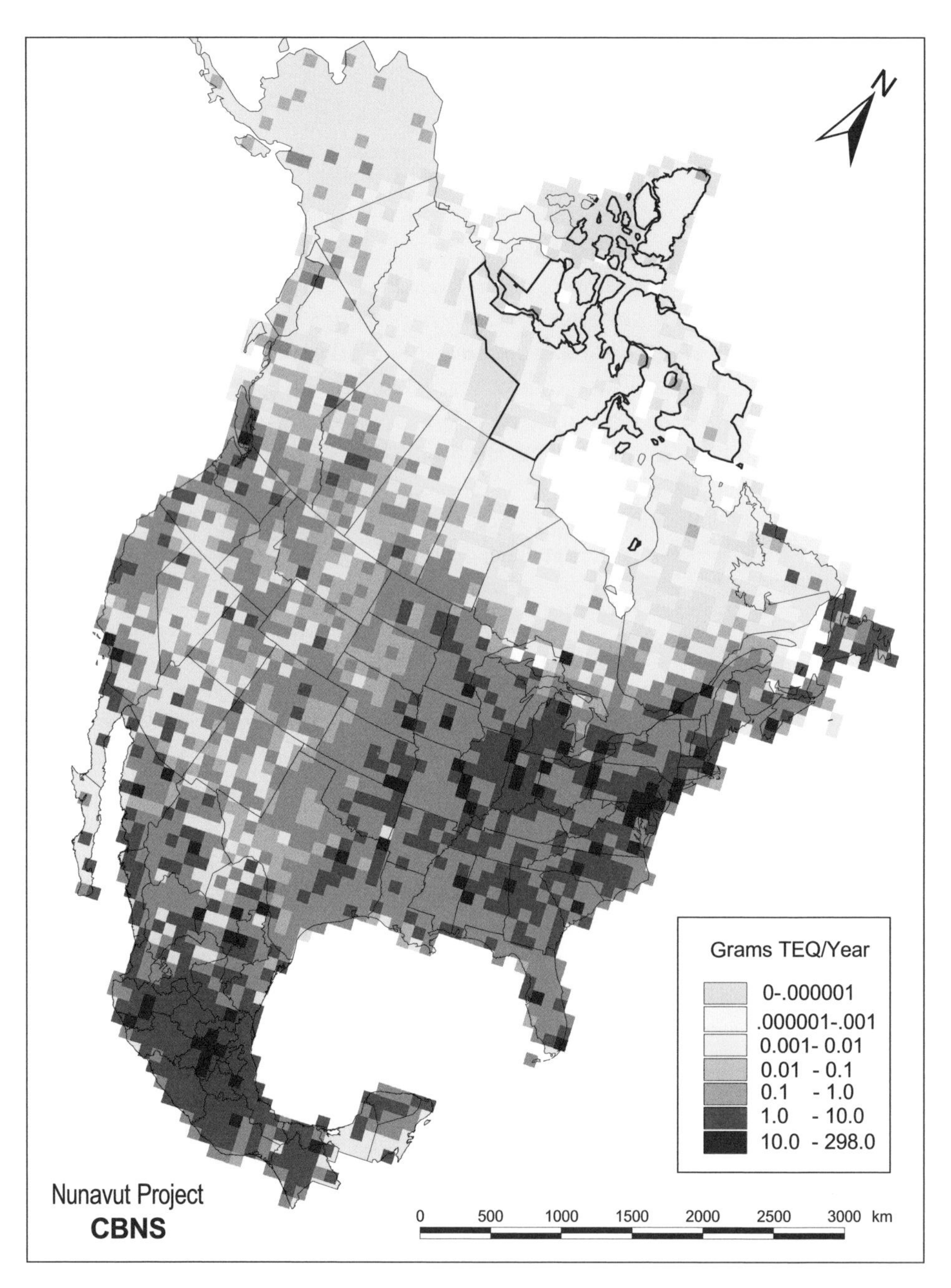

Figure 5.7 Geographic Distribution of Annual Dioxin Emission from North American Sources, 1996/97
This 100 x 100 kilometre (62 x 62 mile) grid of North America sums up the annual emissions from all sources within each grid zone. Each grid zone is colour-coded to indicate the total emissions.

along the entire eastern coast, in the Midwest, and between Minnesota and Texas. Finally, the map identifies intense levels of dioxin emission from central Mexican sources; these reflect population density and are largely a result of backyard trash burning, which accounts for about one-half of the country's total emissions.

The geographic distribution of the ATC can be mapped in a similar manner on the 100 x 100 kilometre polar stereographic grid. For this purpose, the model estimated the deposition, at a given receptor, of the fraction of dioxin unit amount emitted from each of 2,988 hypothetical sources uniformly distributed geographically on the same 100 x 100 kilometre grid that was employed to map the *actual* source emissions shown in figure 5.7.

Figure 5.8 compares the ATC maps of the most northern land receptor, Arctic Bay, and the most southern land receptor, Sanikiluaq. The maps define successive zones, centred around the receptor, of progressively smaller ATC values; so the most distant zones, e.g., in Mexico or Florida, represent the smallest fraction of the unit emission that is deposited at the receptor. Thus the map for deposition at Sanikiluaq shows the ATC value declining from 1×10^{-12} for sources in northern Quebec closest to the receptor (about 500 kilometres away), to 1×10^{-17} for the sources in southern Mexico (some 4,500 kilometres away). This exponential reduction in ATC with air transport distance is commonly observed and arises largely from the effect of diffusion and dispersion on the airborne dioxin concentration. The concentration decreases exponentially with increased transport time; hence, *en route* deposition decreases proportionally as well.

Figure 5.8 helps to account for the order of magnitude difference in the dioxin deposition at Sanikiluaq and Arctic Bay: fifty-three picograms TEQ per square metre at the Sanikiluaq land receptor and four picograms TEQ per square metre at the Arctic Bay land receptor. At each receptor, deposition is the product of the rate of dioxin emission from the source and the ATC value characteristic of the source's geographic location, summed for all sources. Thus, deposition at a receptor is maximized when the geographic locations of sources of high rates of emission and high values of ATC coincide. For the Sanikiluaq receptor, as can be seen in figure 5.8, the eastern half of the United States, where the bulk of the North American dioxin emissions originate, is characterized by high ATC values; most of that area is characterized by ATCs of 1×10^{-14} to 5×10^{-14}. In contrast, for air transport to Arctic Bay this same area is characterized by much lower ATCs, between 1×10^{-15} and 1×10^{-16}. Part of this difference in the ATC values of these two receptors is a result of the 1,500-kilometre (930-mile) difference in source-to-receptor distance.

The effect of the weather pattern alone on dioxin deposition at the receptor can be isolated by examining variations in the ATC map with time, since of the three factors that influence deposition – source emissions, source–receptor

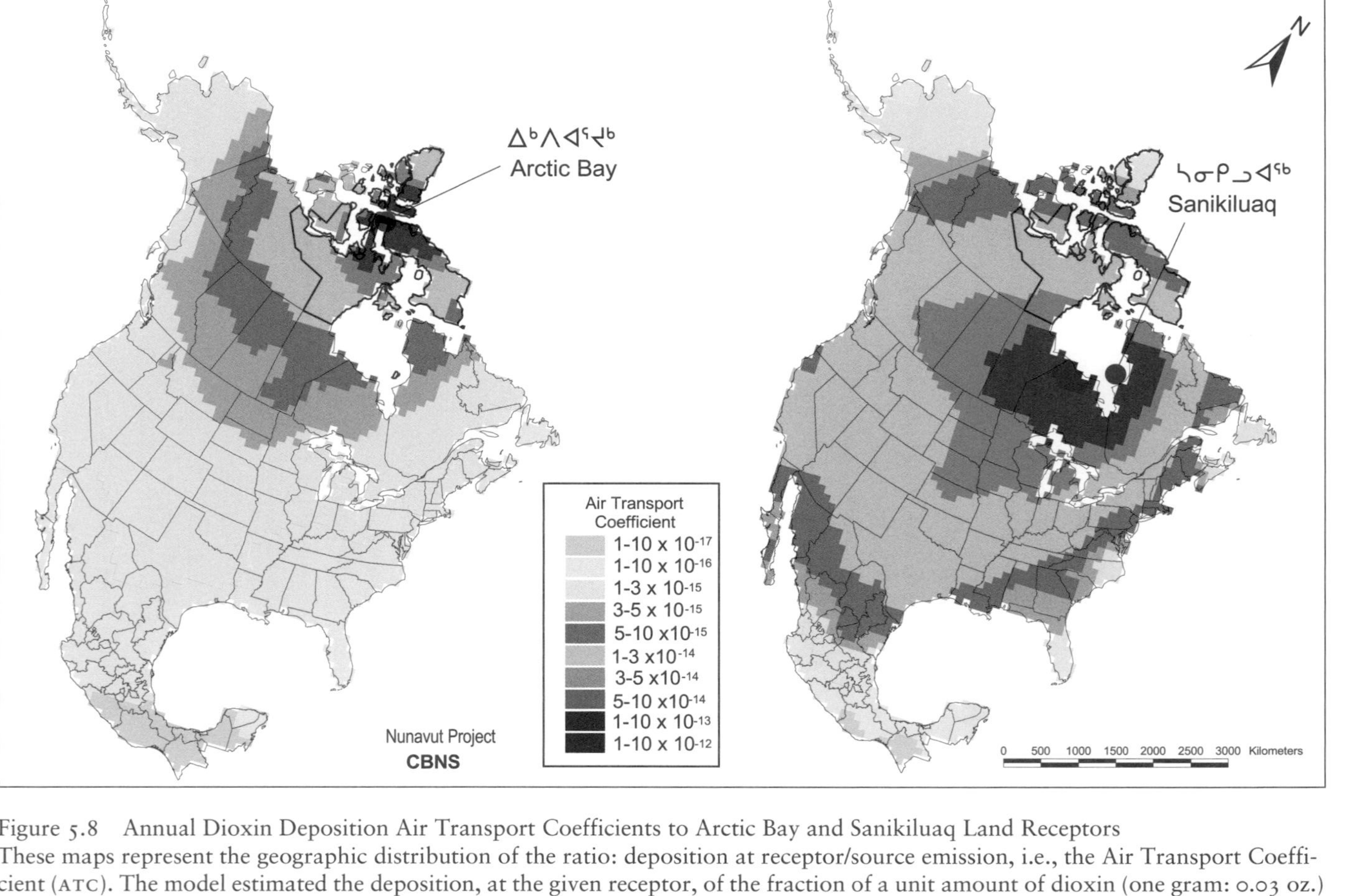

Figure 5.8 Annual Dioxin Deposition Air Transport Coefficients to Arctic Bay and Sanikiluaq Land Receptors
These maps represent the geographic distribution of the ratio: deposition at receptor/source emission, i.e., the Air Transport Coefficient (ATC). The model estimated the deposition, at the given receptor, of the fraction of a unit amount of dioxin (one gram: 0.03 oz.) emitted from each of the 2,988 "sources" created by the same 100 x 100 km grid used to map the actual source emissions shown in figure 5.7. Modelled estimates are for the one-year project period 1 July 1996 to 30 June 1997.

geographic distance, and weather pattern – only weather pattern varies over time. Figure 5.9 shows the successive monthly ATC maps for the Ikaluktutiak land receptor, together with the monthly model-estimated deposition values. Nearly one-half of the total annual dioxin deposition (4.5 picograms TEQ per square metre) occurs in September and October, which are characterized by weather patterns that efficiently, in comparison with the rest of the year, carry dioxin from the area of intensely emitting U.S. sources to Ikaluktutiak.

Without additional and more refined information, it is not possible to translate our modelled estimates of dioxin deposition at the receptors into levels of exposure in wildlife or people. Nevertheless, the significant geographic variation in dioxin deposition does provide a possible link to comparable variation in ecological exposure. For example, the *Canadian Arctic Contaminants Assessment Report* lists the results of a study of the dioxin content of caribou tissue in various Arctic areas.[10] The study included tissue samples taken from four herds in Nunavut (in 1991/92), of which three, in Bathurst, Southampton Island, and Lake Harbour, were in areas close to our receptors. Table 5.1 compares the model-estimated dioxin deposition flux at these land receptors with the tissue dioxin concentrations in the nearby herd locations. Although there are only three comparable locations, the geographic variation in the dioxin content of caribou tissue samples and our estimates of dioxin deposition exhibit a similar trend, with both sets of values increasing from west to east. This suggests that the differences in the dioxin content of the local biota reflect comparable differences in the levels of airborne dioxin deposited at the nearby receptors. This relationship lends credence to the view that the modelled deposition data are applicable to the basic goal of environmental policy.

In sum, the data generated by this project directly support the conclusion that the known occurrence of dioxin in Nunavut – in the Indigenous population, in the regional food chains, and in marine and terrestrial ecosystems – is a result of the deposition of airborne dioxin transported from distant sources, which are chiefly in the United States, to a lesser extent in Canada, and marginally in Mexico. These results show that the HYSPLIT air transport model is an effective means of estimating the relative rates of dioxin deposition among the Nunavut receptors and of ranking the contribution of the numerous sources to that deposition.

POLICY CONSIDERATIONS

Regulatory policy is based on the risk of dioxin exposure to human health, which has been extensively reviewed by the U.S. EPA. Over the last decade the EPA has produced a series of risk assessments of dioxin designed to estimate lifetime risk of cancer associated with dioxin in the general U.S. population. The most recent assessment concludes that the mean body burden of the

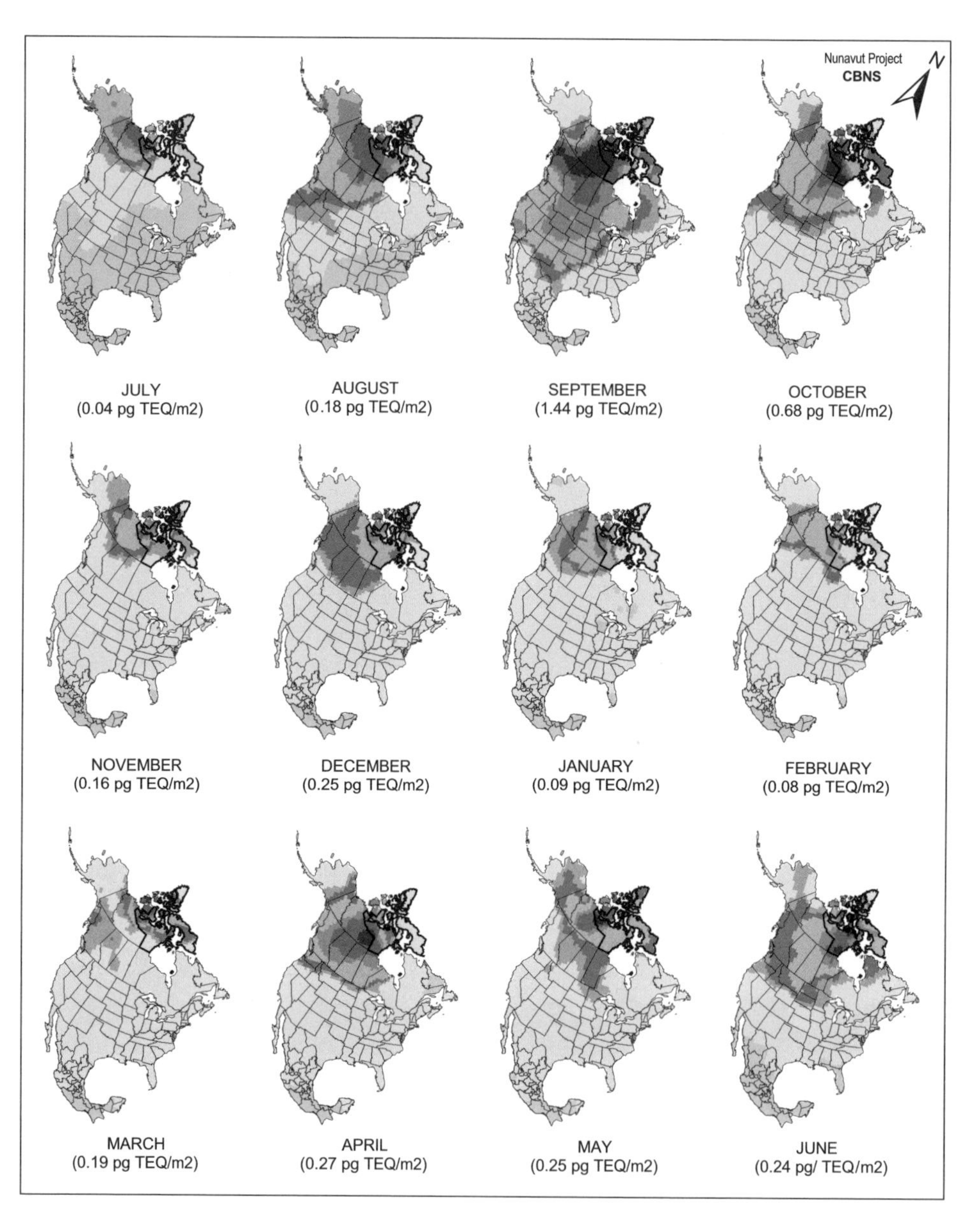

Figure 5.9 Monthly Dioxin Air Transport Coefficients and Deposition Flux (pg TEQ/m^2) to Ikaluktutiak Land Receptor (July 1996–June 1997)

These maps were prepared in the same manner as those in figure 5.8, except that the modelled values are for the separate successive months of the one-year project period. The July deposition value is low and does not represent a full month of activity because the project period began on 1 July 1996 and, especially for the most distant sources such as Ikaluktutiak, up to three weeks is required for dioxin to reach the receptors.

Table 5.1
Comparison of Dioxin Deposition at Nunavut Land Receptors with Tissue Dioxin Concentration in Nearby Herds of Caribou

Receptor	*Dioxin Deposition Flux (pg TEQ/m²/yr)*	*Herd Location*	*Dioxin Concentration Ng TEQ/kg tissue*
Ikaluktutiak	4	Bathurst	0.33
Coral Harbour	19	Southampton Island	0.85
*	–	Cape Dorset	1.23
Iqaluit	26	Lake Harbour	3.29

*There is no dioxin receptor near the location of the Cape Dorset caribou herd. Cape Dorset lies midway between the Coral Harbour and Ikaluktutiak receptors, and its caribou dioxin concentration is included for completeness in demonstrating the west-to-east trend in this measure of dioxin exposure.

general U.S. population is expected to lead to a lifetime risk of cancer of one in 1,000.[11] The "acceptable" risk is generally given as one in one million. Even considering that emissions have been reduced, it is fair to say that the current exposure of the U.S. population to dioxin would need to be reduced by at least ninety-nine per cent to produce "acceptable" levels of cancer risk. While the data on the dioxin body burden among the Inuit population of Nunavut are far too limited to support such a risk assessment, the *Canadian Arctic Contaminants Assessment Report* has compared these limited data with similar measurements made in southern Canada and the United States.[12] Samples of mothers' milk taken from Inuit women in the Nunavik region of Nunavut in 1989/90 averaged 19.0 picograms TEQ per gram lipid. In the same period, samples from women in southern Quebec averaged 9.0 picograms TEQ per gram lipid; and, as noted in the report, "similar levels" have been observed in the United States. It is reasonable to assume, therefore, that the remedial problem that confronts the Inuit is similar to that in the United States: current exposures to dioxin need to be reduced by ninety-nine per cent or more to bring the lifetime risk of cancer from dioxin exposure in the general population to the level of one in one million. Since there is no feasible way to protect food chains from the deposition of airborne dioxin, such a remedy must be directed at the sources.

Alternatively, human exposure could be reduced to a degree by moderating consumption of foods containing animal fat – a recourse that, certainly in Nunavut, would clash with the transcendent importance of the Indigenous diet in Inuit culture. Consequently, if remedial action is to be taken, the Inuit are faced with the daunting task of defining and implementing a policy that would effectively reduce the emissions collectively produced by 5,343 individual and

38,748 area sources, nearly all of them thousands of kilometres away, in other jurisdictions. The task is further complicated by the fact that there are considerable differences among the eight Nunavut receptor sites in the estimated level of dioxin deposition. Moreover, the data generated by the model are themselves numerous and complex, involving 44,091 sources grouped in twenty-three classes, in three different countries. Nevertheless, despite these difficulties, it is possible to assemble relatively simple sets of facts that illuminate the feasibility of alternative policy strategies.

As already indicated, most of the dioxin deposition at the receptors is due to a very small proportion of the total number of sources. Beyond that strategic generalization, the data can be organized to address two alternative policy strategies. One strategy is based on the regulatory approach common to most countries' environmental agencies: standards of allowable emissions are set for different source classes, such as municipal waste incinerators or cement kilns burning hazardous waste. In this case, the Inuit community at Coral Harbour, for example, could address to the United States its need for relief from exposure to the annual dioxin deposition (on land) of 19.24 picograms TEQ per square metre (1.2 sq. yd). Standards that virtually eliminated dioxin emissions from only five classes of U.S. sources (municipal waste incinerators, medical waste incinerators, cement kilns burning hazardous waste, iron sintering plants, and backyard trash burners) might reduce exposure at Coral Harbour by sixty-seven per cent. By adding improved regulation of municipal solid waste incinerators in Canada, and of informal trash burning in Mexico, the Coral Harbour community could address seventy-three per cent of its dioxin exposure, reducing it to 5.2 picograms TEQ per square metre. That would bring the level of exposure at the Coral Harbour land receptor to about the lowest level among all the Nunavut land receptors, which occurs at Arctic Bay and Ikaluktutiak. In this way, Inuit communities could target remedial policy initiatives towards those source class/country categories that might offer the best potential return in reduced exposure to dioxin.

An alternative approach to policy is directed towards specific individual sources rather than categories of sources subject to national regulations. This policy depends more on direct appeal for action to the operators of a particular facility and the people of the local community than to the national environmental agency. Such a direct appeal has the advantage of immediacy, avoiding the intricacies and delays inherent in international, and even national, administrative actions. It has the disadvantage of dealing independently with the sources and, with that, the inherent uncertainties associated with emission estimates from individual sources. Thus, if the community of Coral Harbour adopted this approach, total exposure to dioxin could be reduced by thirty-five per cent, if the nineteen individual sources – again, most in the United States – could be induced to virtually eliminate their dioxin emissions. To go beyond that target sharply increases the overall effort needed; 680 sources must be targeted to reach a seventy-five per cent reduction, and 3,031 sources to reach a ninety per cent reduction. However, experience in the United States has shown that public appeals for action can often succeed by stimulating the necessary administrative response.

In sum, the project's results provide Inuit communities with the basic information about exposure to airborne dioxin needed to support the development of alternative remedial policies. Clearly, these data need to be strengthened, particularly by systematically relating the levels of dioxin deposition to dioxin concentration in the local marine and terrestrial food chains.

Our experience in employing the national dioxin inventories in this project suggests a need for considerable improvement in this important data bank of transboundary airborne pollution. The results of this project clearly reinforce the precept that it would be helpful to establish a common set of priorities regarding source classes, notwithstanding the difficulties of compiling them and the limited resources available for the task (especially in developing countries). Clearly, quite workable inventories can be restricted to the six to eight source classes that are responsible for most of the airborne dioxin emissions. Our effort to help establish an initial inventory of dioxin sources in Mexico suggests that informal burning of domestic waste may be the most important single source class of airborne dioxin in developing countries.

Finally, the results of this project emphasize the importance of the Stockholm Convention on Persistent Organic Pollutants, of which dioxin is a major component. The results show that the atmospheric and ecological processes that carry airborne dioxin from its sources, through the food chain, to human beings is a problem of continental, if not global, dimensions. Remedial policy, directed at the virtual elimination of dioxin sources, must achieve a large scale as well.

NOTES

1 B. Commoner et al., "Long-range Air Transport of Dioxin from North American Sources to Ecologically Vulnerable Receptors in Nunavut, Arctic Canada," Final Report to Commission for Environmental Cooperation, October 2000.

2 J. Jensen, K. Adare, and R. Shearer, eds, *Canadian Arctic Contaminants Assessment Report* (Ottawa: Indian and Northern Affairs Canada 1997), p. 320.

3 U.S. Environmental Protection Agency, "Estimating Exposure to Dioxin-like Compounds," 1, 1994.

4 R.R. Draxler and C.D. Hess, "An Overview of the HYSPLIT-4 Modeling System for Trajectories, Dispersion and Deposition," *Australian Meteorological Magazine* 47 (1998): 295–308.

5 Commoner et al., "Dioxin Sources, Air Transport and Contamination in Dairy Feed Crops and Milk," Final Report to the Joyce Foundation, Jessie B. Cox

Charitable Trust, and the John Merck Fund, Center for the Biology of Natural Systems and New England Environmental Policy Center, September 1998.

6 M.A. Bergmann, Fisheries and Oceans Canada, telephone communication with B. Commoner, March 1999.

7 R. Riewe, *Nunavut Atlas* (Edmonton: Canadian Circumpolar Institute 1992).

8 Environment Canada and Federal/Provincial Task Force on Dioxins and Furans, "Dioxins and Furans and Hexachlorobenzene, Inventory of Releases," prepared for the Federal–Provincial Advisory Committee for the Canadian Environmental Protection Act, 1999.

9 P.M. Lemieux, J.A. Abbott, and K.M. Aldous, "Emissions of Polychlorinated Dibenzo-p-dioxins and Polychlorinated Dibenzofurans from the Open Burning of Household Waste in Barrels," *Environmental Science and Technology* 34 (2000): 377–84.

10 Jensen, *Canadian Arctic Contaminants*, p. 219.

11 U.S. Environmental Protection Agency, "Exposure and Health Reassessment of 2, 3, 7, 8 Tetrachlorodibenzo-p-dioxin (TCDD) and Related Compounds," part III, 2000.

12 Jensen, *Canadian Arctic Contaminants*, p. 320.

SECTION TWO
Regional and Global POPs Policy

6
Regional POPs Policy: The UNECE CLRTAP POPs Protocol

HENRIK SELIN

NEW CHEMICALS: FROM BLESSING TO CURSE

The United Nations Economic Commission for Europe (UNECE), through the Convention on Long-range Transboundary Air Pollution (CLRTAP), started investigating persistent organic pollutants (POPs) in the late 1980s as a result of changing scientific understanding and actions taken by Canada and Sweden.[1] Some of the chemicals now classified as POPs had been originally synthesized in the late nineteenth century, but large-scale manufacturing and use of most of them – to produce more and better food and cash crops, protect public health, and facilitate industrial development – began only after World War II. The first reports of the new substances were almost all positive, but by the early 1960s the scientific community was voicing concern for local wildlife.[2]

Early attempts to analyze environmental problems related to chemicals such as dichlorodiphenyl trichloroethane (DDT) and polychlorinated biphenyls (PCBs) were plagued with difficulties.[3] Many of the chemicals are not a single substance but consist of a multitude of congeners that exist in various formulations of commercial mixtures where environmental, physical, chemical, and biological processes differ for each congener. In addition, sensitivity to most chemicals has proven to be species- and gender-specific. Despite these difficulties, by the late 1960s and early 1970s risk assessments on a set of hazardous substances revealed effects thought severe enough to justify progressive introduction of domestic regulations mainly in northern industrialized countries.[4]

The first set of domestic regulations on hazardous chemicals was supplemented with only minor international regulations, made primarily in northern regional fora. Early global efforts were principally aimed at information sharing. In 1976, the United Nations Environment Programme (UNEP) established the International Register for Potentially Toxic Chemicals to compile and circulate information on chemical hazards. By the early 1980s, discussions were taking place within the United Nations Food and Agriculture Organization (FAO) and UNEP; these resulted in the 1985 International Code of Conduct on the Distribution and Use of Pesticides and the 1989 London Guidelines for the Exchange of Information on Chemicals in International Trade. Both included procedures to improve access to information about hazardous chemicals so that countries could better assess risks associated with chemical use.

With the first domestic and international actions in place, many northern industrialized countries began to believe they were gaining control over their problem with hazardous chemicals, even though little had been done in developing countries, where the necessary infrastructure generally was lacking. The optimism in the northern countries was supported by scientific studies in the late 1970s and early 1980s showing a decline in environmental concentration levels of several previously detected hazardous chemicals and signs of recovery in affected local wildlife. But by the late 1980s, new and unanticipated scientific information from the Arctic and *sub-Arctic* raised new concerns.

In the mid-1980s, Indian and Northern Affairs Canada (INAC), suspecting that abandoned military radar stations across northern Canada contained PCBs, conducted surveys to assess local PCB contamination around the radar stations. Comparative background control sites – remote areas thought to be pristine in Arctic Canada – were chosen. However, samples from these control sites contained surprisingly high levels of PCBs. This result was initially suspected to be a measurement error, but a second test confirmed the high levels. This discovery of hazardous chemicals was not the first in the Arctic environment; the existence of such substances has been known since the early 1970s, but these levels were alarmingly high.[5] In response, INAC officials undertook a multidisciplinary review of recent scientific findings. Facilitated by both qualitative and quantitative improvements in data samples, measurement, and analytical techniques since the early 1970s, three connected and formerly unidentified factors relating to a set of hazardous persistent organic substances with similar characteristics were revealed: systematic long-range atmospheric transport of emissions to the Arctic; high environmental contamination levels throughout the Arctic region; and actual and potential environmental and human health implications. Examples of substances that showed such characteristics included DDT, PCBs, toxaphene, chlordane, hexachlorocyclohexane (HCH), endrin, dieldrin, hexachlorobenzene (HCB), and mirex.

SEEKING AN INTERNATIONAL RESPONSE

The new Arctic scientific findings generated a wide range of additional scientific studies to examine emissions sources, emissions transport patterns, and environmental and human health risks, but they sparked only limited political reaction. INAC, the only major Canadian organization to seek political action, expressed particular concern for the Arctic Indigenous population. Arctic Native people are sustained by local food chains and are mainly exposed to hazardous chemicals through their diet: an Arctic meal has been described as consisting of "seal meat, corn chips and PCB's."[6] Studies also demonstrated alarmingly high levels of hazardous chemicals in breast milk and blood samples taken from the Arctic Indigenous population, some levels among the highest measured in the world.[7]

The indications of extensive long-range transnational transport suggested that international measures were needed. INAC officials contacted the Organisation for Economic Co-operation and Development (OECD), FAO, and UNEP, none of which showed any interest in pursuing the issue. Then, efforts were directed towards the UNECE and CLRTAP, which is a regional agreement that covers North America, Europe, and the former Soviet Union.[8] Until the late 1980s, the primary focus of the CLRTAP had been on acidification and eutrophication abatement. Although the CLRTAP was more regional in scope than OECD, FAO, or UNEP, its experience with long-range transboundary air pollution in much of the northern hemisphere made it an interesting option.

In August 1989, an INAC official's presentation of a report on the Arctic problem of toxic, persistent, and bioaccumulating organic chemicals persuaded the CLRTAP Working Group on Effects, the group responsible for assessment activities, to include the issue of hazardous organic substances in its working plan. Many important uncertainties remained, and to gain a better understanding of the problem, the Working Group on Effects requested that government-designated experts, under Canadian supervision, prepare a more extensive report on the effects of hazardous airborne organic compounds, including sources, transportation, biological uptake, ecosystem (including human) health, and economic factors, and investigate the magnitude, extent, and impact of the accumulation of such substances. The report was to be completed by spring 1991 and considered by the Working Group on Effects at its session in August 1991.[9]

Coincidentally, officials at the Swedish Environmental Protection Agency (SEPA) in the late 1980s began to show concern over the levels of hazardous organic substances in Sweden and the Baltic region.[10] Concentration levels of DDT and PCBs seemed to level off rather than decrease, and levels of other hazardous substances, such as chlordane, were rising in the Baltic environment. Supported by new studies showing long-range atmospheric transport to be an essential – and probably major – pathway by which several hazardous substances entered Baltic ecosystems, officials at SEPA began to believe this phenomenon was at least partly responsible for the new environmental trends.[11] SEPA then undertook an assessment of the situation for Sweden and the Baltic region similar to that done by INAC for the Canadian Arctic and reported similar results: evidence of inflow of emissions through long-range atmospheric transport; slowing decline rates of environmental DDT and PCB levels in combination with rising concentration levels of other hazardous organic substances; and concern regarding the environmental and human health aspects of this situation.

With these common conclusions, INAC and SEPA pooled resources and collectively voiced a concern over long-range transport of hazardous persistent organic substances in the northern environment. At the meeting of the Working Group on Effects in August 1990, Canada presented a draft of the paper requested a year earlier and together with Sweden argued that CLRTAP action on such substances was needed. To that end, Sweden proposed to form an Intergovernmental Task Force on Persistent Organic Pollutants to further investigate the POPs problem. The proposal was endorsed by the Working Group on Effects and approved by the CLRTAP Executive Body in November 1990.[12] Under the joint leadership of Canada and Sweden, the task force was to prepare a state-of-the-art report on the POPs situation in the CLRTAP region, including proposals for international actions.

THE TASK FORCE

The task force on POPs met four times, with from ten to fourteen states attending meetings and participating in its work.[13] Active states included the traditional "green" European CLRTAP countries such as Germany, the Netherlands, Norway, and Sweden, as well as Canada, Spain, the United Kingdom, and the United States. Intergovernmental organizations (IGOs) such as the OECD and the Helsinki Commission (HELCOM) attended some task force meetings, but no industry or environmental non-government organizations (NGOs) were present. The task force operated as a largely independent group, focusing on areas of uncertainty deemed crucial for assessing the need for CLRTAP to take regulatory action on POPs. First, it assessed the physical characteristics of the POPs problem, focusing on emission sources, long-range transport, and distribution among media. The assessments did not involve any laboratory testing or field studies, but efforts were directed at compiling and evaluating available scientific information. Second, it identified and evaluated possible abatement options and policy responses.

Assessing the POPs problem

POPs cannot be easily identified by a single chemical or even a small number of distinct substances, which makes it difficult to get a clear picture of all POPs and their sources. To define POPs for the purposes of CLRTAP POPs work, the task force agreed on certain physical, chemical, and biological characteristics that a substance should possess. This working definition characterized POPs as "organic compounds that, to a varying degree, resist photolytic, biological, and chemical degradation. Many POPs are halogenated and characterized by low water solubility and high lipid solubility, leading to their bioaccumulation in fatty tissues. They are also semi-volatile, enabling them to move long distances in the atmosphere before deposition occurs."[14] Suspected POPs were divided into three categories: pesticides, industrial chemicals, and unintentional by-products. The task force worked concurrently on establishing screening criteria, selecting substances for screening, collecting risk-assessment data, and conducting the screenings.

The task force found the use of several suspected POPs to be extensive, although obsolete and incomplete records of domestic production and use made exact reviews impossible. It also identified a number of important emission sources, including agricultural use, manufacturing and use of goods, spills and dumping of substances, waste incineration, and combustion. Environmental concentrations were found to be highest close to emission sources, but data also showed extensive long-range transport of emissions causing significant environmental accumulation in remote regions having no local emission sources. Although some hazardous industrial chemicals and pesticides had been regulated in a number of countries, their long-range transportation continued to be a problem because of incomplete regional control measures on use and disposal. Domestic controls of unintentional by-products were even less comprehensive.

Consistent with the earlier findings of INAC and SEPA, the task force identified the atmosphere as the primary transport medium for emissions, although non-atmospheric transport mechanisms were also discovered to be of importance. The physical and chemical properties of POPs in combination with predominant global atmospheric circulation patterns cause frequent, systematic migration of POPs to cooler latitudes, especially the North, exposing the sensitive Arctic and sub-Arctic environment and its human population to their effects. Several negative effects in wildlife were linked with POPs, including immune and metabolic dysfunction, neurological deficits, reproductive anomalies, behavioural abnormalities, and carcinogenic effects. Although the data are sparse and human effects more difficult to prove, human data were shown to be consistent with effects reported in both laboratory and wildlife studies. Based on its total assessment of the POPs problem, the task force concluded in its final report in 1994 that the weight of evidence "clearly indicates that action to address POPs is warranted now."[15]

Identifying policy options

To investigate the possible role of CLRTAP in international POPs abatement, the task force surveyed existing domestic and international regulations and other activities under the premise that effective controls on POPs could be achieved only by regulating the long-range transport of POPs from widely separated emission sources and receptors. The survey was carried out by considering the extent of domestic regulations and the objectives, functional scope, and geographical coverage of available international fora to deal with POPs. Based on the survey, the task force found that domestic controls were scattered and incomplete and also concluded that other international regimes and organizations were unlikely means to help achieve necessary POPs reductions because they had either too narrow a geographical domain and membership or a restricted mandate that could address only parts of the POPs problem.[16] In contrast, the task force found the CLRTAP, with its large geographical area and broad mandate, to be the most appropriate international mechanism for POPs controls and recommended the creation of a legally binding POPs agreement with mandatory regulations.

As a way to move forward with actions on POPs under CLRTAP, the task force suggested a simultaneous two-track approach because it would be too difficult to deal with all potential POPs in one series of negotiations and such an approach would also strengthen the long-term effectiveness of an agreement. The first track was to achieve a quick protocol on a small, manageable list of substances that had already been acted upon in many UNECE countries or for which available risk information was considered sufficient. Agreeing on measures under track one proved an arduous task, however, because opinions diverged on what was considered sufficient risk information and appropriate measures. Track two was to set up a mechanism for the future screening of additional substances for possible controls under the protocol once it entered into force.

The task force report was presented to the CLRTAP Executive Body in late 1994. Before the meeting, both the Working Group on Effects and the Working Group on Technology – the CLRTAP group active in identification and diffusion of pollution control technology – expressed support for the conclusions and recommendations of the task force, including the suggested two-track approach to future CLRTAP work on POPs.[17] The executive body was generally supportive of the task force work and agreed that work on POPs should continue, but opinions diverged on how to proceed. Some states favoured starting protocol negotiations based directly on the work of the task force; however, a larger group of countries argued that further assessments, particularly with regard to developing more detailed policy alternatives, were needed before protocol negotiations could start. For that purpose, the executive body set up an Ad Hoc Preparatory Working Group on POPs to work directly under the Working Group on Strategies, the permanent CLRTAP political negotiation committee.

THE PREPARATORY WORKING GROUP

The Preparatory Working Group (PWG) met four times between March 1995 and October 1996.[18] Attendance at PWG meetings was approximately twice that of the task force meetings; however, work continued to be dominated by roughly the same small group of states that had dominated in the task force – Canada, Denmark, Germany, Italy, the Netherlands, Norway, Spain, Sweden, the United Kingdom, and the United States. Delegations ranged in size from as many as ten delegates from Canada and the United States, to two to four delegates from active western European states such as Germany, the Netherlands, Sweden, and the United Kingdom, and one or two delegates from eastern and southern European countries.

Both IGOs and NGOs sent observers to PWG meetings. UNEP Chemicals, which by then had become active in the global POPs process, attended all four meetings, while the World Health Organization (WHO) and the World Meteorological Organization (WMO) attended a couple of meetings. For the first time, NGOs – dominated by industry representatives – also took an interest in the CLRTAP POPs work. In regular attendance were the International Chamber of Commerce (ICC), the Chemical Manufacturers Association (CMA), the Edison Electric Institute, the European Chemical Industry Council (Cefic), and Eurochlor. Industry was generally passive at meetings; industry lobbying primarily took place between meetings with the governments of Canada, Germany, France, the United Kingdom, and the United States, which also have the largest chemical companies. The only environmental NGOs present at PWG sessions were the International Union for Conservation of Nature and Natural Resources (the second meeting) and Greenpeace (the third meeting); after that, both organizations decided to direct their primary attention towards the global UNEP POPs work.

While the task force had concentrated on assessing the physical character of the POPs problem, the PWG focused on generating policy options and drafting a protocol text. PWG activities addressed three broad areas: establishing screening criteria and screening potential substances for initial inclusion in the protocol; identifying control options; and developing a mechanism for future evaluation of additional suspected POPs.

Screening of substances

PWG believed that continuing to establish screening criteria and subsequent screening of possible substances – work that began in the task force but which had not been concluded in time for the final task force report – was a key issue.

By spring 1995, the PWG chose the so-called Modified Task Force Methodology as a principal screening model for identifying possible protocol POPs: substances were screened individually using a common set of criteria to identify those that were subject to long-range transboundary atmospheric transportation within the CLRTAP region and which, at the time of deposition, were posing risks to wildlife and human health.[19] Screening information was collected from member states and from other international fora.[20] The Modified Task Force Methodology consisted of three stages.[21] In stage one, consistent with CLRTAP's focus on long-range air pollution, 107 candidate POPs were screened for their propensities for atmospheric transportation. Three different criteria were used: persistence (measured as atmospheric half-life and biodegradation in water and soil), vapour pressure, and monitoring evidence in remote regions. Twenty of the 107 substances were screened out in stage one. At stage two, substances were tested for harm potential by screening them for bioaccumulation potential and mammalian and aquatic toxicity. Seventeen substances for which data were insufficient were initially screened out, leaving only seventy to be assessed in stage two. A numerical scoring system was applied to the substances, and those with less than fifty per cent of the maximum score were eliminated. Thirty-two priority substances remained after stage two. In stage three, a risk assessment of the remaining thirty-two substances was conducted using several factors: production/use/emissions; volatilization/vapour pressure; chemical/biodegradation; bioaccumulation; aquatic/ecotoxicity; mammalian toxicity; measured environmental levels; direct evidence of long-range atmospheric transport; partitioning in environmental compartments; and risk from degradation products. In the end, fourteen substances were identified.

Although the Modified Task Force Methodology was the primary means to identify possible protocol POPs, states were not bound by its results, and opinions diverged on what were acceptable risk levels and on interpretation of the precautionary principle. Proposals for substances to be initially regulated therefore varied. Four substances that were screened out in the Modified Task Force Methodology – short-chain chlorinated paraffins (SCCPs), heptachlor, chlordecone, and lindane – were still supported by some states and together with the other fourteen substances became subject to negotiations in the Working Group on Strategies (table 6.1).[22]

Identifying control options

While the work of screening POPs for initial regulations was proceeding, discussions took place on possible formats for their controls. The PWG reached a political understanding that controls would, in general, be mandatory and regulated substances would be listed in annexes together with their controls. While some states favoured total bans on all regulated pesticides and industrial chemicals, others argued that they needed exemptions on some substances. With these opposing opinions, the issue of exemptions came to play a central role in the negotiations.

Several control options were considered for pesticides and industrial chemicals, with a focus on production, use, import, export, and management of

Table 6.1
The eighteen substances under consideration are grouped in three categories: pesticides, industrial chemicals, and unintentional by-products. HCB is listed as a pesticide, an industrial chemical, and an unintentional by-product.

Pesticides	*Industrial chemicals*	*Unintentional by-products*
Aldrin	Hexabromobiphenyl	Dioxins
Chlordane	PCBs	Furans
Chlordecone	PCP	HCB
DDT	SCCP	PAHs
Dieldrin	HCB	
Endrin		
HCB		
Heptachlor		
Lindane		
Mirex		
Toxaphene		

stockpiles and waste. Discussions on production and use concerned how to best regulate domestic activities to minimize emissions. Import and export deliberations addressed the potential conflicts with the GATT/WTO and the free-trade principle, and whether CLRTAP, as a regional environmental forum, should design trade restrictions, which had never been done before in a regional environmental agreement. Some European states believed effective POPs controls would depend on trade restrictions; however, the United States and Canada argued that it would be inappropriate for the region-specific CLRTAP to impose legally binding import and export restrictions. Not surprisingly, this opinion was also shared by the chemical industry. A similar discussion was held on stockpiles and wastes. Some states – principally from western Europe – argued in favour of designing guidelines for national waste destruction and mandatory controls on transnational transportation of stockpiles and wastes, while others such as Canada and the United States viewed these as inappropriate trade restrictions.

Two initiatives – identifying mobile and stationary emission sources and developing options for control of the major sources – were undertaken to address the unintentional by-products. Combinations of best available techniques (BAT) and emission level values (ELV) were elaborated in developing controls for the major emissions sources. ELV addresses individual emission sources and pollutants; some states preferred specific ELV controls on each source as the most effective way to reduce emissions. In contrast, the BAT system sets limits for overall emission levels within a given area, allowing flexibility in emission amounts from individual sources as long as the overall goal is achieved. Proponents of BAT stress its flexibility and cost-effectiveness.

The PWG work on control options generated a few prominent policy alternatives for all three categories of substances, but offered no final decisions on which activities should be covered by the protocol or on the design of the control options.

Designing the track-two mechanism

Under the two-track approach recommended by the task force, part of the PWG mandate was to design a mechanism for the future evaluation of additional suspected POPs. Two different proposals – partly reflections of variations in domestic regulatory traditions – for designing the mechanism were discussed. The proposals differed in their degree of formalization and concerned the question of predictability versus flexibility. Some states, including Canada and the United States, argued in favour of setting detailed specifications of the substance data, including numerical values for the scientific criteria, together with clear stipulations for how the substance assessments should be performed and by whom. This approach – seen to provide clarity and conformity in data requirements and assessment procedures – was also preferred by the chemical industry, for which the mechanism for future evaluation of additional substances was important. Most substances under consideration for the initial list either were completely phased-out or were in only limited production and use in the CLRTAP region, but substances with higher commercial value could come up for future evaluation.

Viewing this approach as too rigid, a group of mainly northern European states argued that having detailed numerical criteria specified in the protocol could be counter-productive. Scientific knowledge of POPs advances continuously and any criteria specified in detail could become outdated, binding the parties to obsolete requirements. By having less-detailed criteria, each future evaluation could be based on the latest available scientific information and criteria developments. Also, some states wanted to avoid specified numerical criteria because they believed that it would suggest that there is a clear line between hazardous and harmless substances, which they wanted to avoid. In the end, the PWG could not reach consensus, but options for assessing additional substances became more clearly identified.

NEGOTIATING AN AGREEMENT

By autumn 1996, sufficient detailed policy alternatives were available to allow for agreement on the need for protocol negotiations. This is not to say that all states equally desired an agreement. States differed in the importance they attached to reaching agreement, yet they generally shared a belief that an agreement was desirable.

Five POPs negotiating sessions were held between January 1997 and February 1998: four in the Working Group on Strategies and one Head of Delegations meeting in December 1997 before the final Working Group on Strategies session.[23] Participation patterns remained largely consistent with those of the PWG, with negotiations dominated by Canada, Italy, the Netherlands, Germany, Spain, Sweden, the United Kingdom, and the United States. Active states also included Denmark, Norway, Switzerland, Poland, the Russian Federation, and Ukraine. Towards the end of the negotiations, the European Commission gained a partial negotiating mandate for European Union (EU) member states and became more active, while the individual EU member states became less so. Delegation sizes remained largely consistent with those in PWG. IGO and NGO observers continued to follow the process during the negotiations. UNEP Chemicals, preparing for global intergovernmental negotiations on POPs for late June and early July 1998, was the predominant IGO. The NGOs were still overwhelmingly dominated by industry representatives from the same organizations that observed PWG sessions. Greenpeace was the only environmental NGO to follow the negotiations onsite, attending the October 1997 session. Representatives from the Inuit Circumpolar Conference attended a few meetings, lobbying successfully for the position that the Arctic and its population should be recognized as particularly vulnerable to POPs.

Starting negotiations

When the POPs negotiations started, the preparatory work had generated a shared understanding of the physical character of the POPs problem: the sources, pathways, and distribution mechanisms in long-range transportation of POPs and their negative environmental and human health effects were fundamentally perceived in a similar way. It had been agreed that the aim was a legally binding agreement and the two-track approach would guide the design of the regulatory system. There was also a general acceptance of a draft composite protocol text, setting the framework for a possible agreement. The protocol would contain a number of definitional articles, with annexes listing the initially regulated substances and their controls. Further, the preparatory work had clarified three main sets of issues in need of solutions: the initial list of substances, the format for their controls, and the mechanism for future evaluation of additional substances.

Initial list of substances. The substance-screening work in the PWG had identified eighteen substances to be considered for the initial list (table 6.1). Among the fifteen pesticides and industrial chemicals, seven substances were relatively unproblematic: aldrin, chlordane, dieldrin, endrin, mirex, hexabromobiphenyl, and toxaphene were no longer in production or use in the

UNECE region by the late 1990s and would therefore not require any new domestic limitations for any state. The other eight pesticides and industrial chemicals were still in production or use and would require negotiations to find consensus on joint regulatory actions. There was general acceptance that chlordecone, HCB, DDT, and PCBs should be included in the protocol; negotiations addressed whether they should be completely banned or whether exemptions should be given under the protocol. Inclusion of lindane, heptachlor, pentachlorophenol (PCP), and SCCP was questioned outright. There was agreement that the four unintentional by-products should be in the protocol, and negotiations concerned possible control options for stationary and mobile emission sources.

At the January 1997 meeting, negotiators reached political consensus on a total ban on chlordecone. New scientific data demonstrated that chlordecone had more severe bioaccumulation and toxicity potential than previously believed. Moreover, its use had declined steadily since the late 1970s; with few remaining uses in the region by the time of negotiations, there were no strong interests against a complete phase-out. Early progress was also made on lindane, which is developed out of technical hexachlorocyclohexane (HCH), an old, first-generation pesticide used for seed and wood treatment. During negotiations, Canada and the United States claimed that there was no evidence supporting the bioaccumulation of lindane; western European states argued that notwithstanding any lack of evidence on bioaccumulation, its chemical and biological characteristics warranted its inclusion on the basis of the precautionary principle.[24] A further complexity was that addressing only lindane could lead to increases in use of technical HCH. A compromise struck at the October 1997 meeting saw HCH included in the protocol in two categories: technical HCH and lindane.[25] No exemptions on technical HCH were given, and certain identified uses of lindane were to be reassessed no later than two years after the protocol entered into force. Thus, as most Europeans desired, lindane was included in the protocol, but permitted exemptions covered almost all known uses in the region (e.g., in Canada and the United States); so, in practice, few new lindane restrictions were introduced.

Controversies on HCB, DDT, PCBs, heptachlor, PCP, and SCCP took longer to resolve. HCB played a minor role in the negotiations to the end. HCB has been used as an industrial chemical to make fireworks, ammunition, and synthetic rubber, but its main application has been as a pesticide for seed treatment, and the Russian Federation, the Ukraine, and the United States were against a total HCB ban as a pesticide. A majority of countries were strongly in favour of a complete ban on DDT, but Italy, the Russian Federation, and the United States sought exemptions. Italy desired an exemption for the use of DDT as an intermediary in the manufacturing of the pesticide dicofol. The Russian Federation wanted to continue to produce and use DDT domestically against malaria. The

United States, referring to the WHO policy recommendation that DDT in some cases was the best available means to fight malaria, wanted an exemption for the export of DDT for malaria treatment.

While the DDT problem was present throughout the negotiations, PCBs were not addressed until the latter stages. Having maintained that both the manufacturing and the use of PCBs had ceased during the Soviet era, the Russian Federation suddenly claimed, at the October 1997 session, that it needed exemptions on the use of PCBs in transformers, suggesting that domestic PCB production may still exist. The Russian PCB announcement caused great surprise and sparked several activities. One was to try to formulate text on PCBs in the protocol that would be acceptable to both the Russian Federation and those who wanted a quick and all-encompassing ban on PCBs. Others included initiating several multilateral and bilateral processes to put pressure on the Russian government to agree to phase out PCBs and exploring ways to assist the Russian Federation to cease domestic production and use.

CLRTAP action on heptachlor was affected by the UNEP POPs work. In May 1995, the UNEP Governing Council identified heptachlor – the only UNEP POP that was not on the CLRTAP list at the time – as warranting global action. Under the Modified Task Force Methodology, heptachlor did not meet the set criteria; however, when it was identified by UNEP, many European voices within CLRTAP were quick to suggest that it should also be included in the CLRTAP list. Uncertainty remained about the propensity of heptachlor for long-range transport, and, in addition, the United States insisted that it needed heptachlor to combat fire ants in industrial electrical junction boxes.

PCP has been used primarily as a wood preservative. While some European states and the European Commission stated the precautionary principle to argue in favour of including PCP on the initial list, the United States and Canada argued that it did not qualify as a CLRTAP POP since available data suggested that its bioaccumulation potential was too low.[26] On SCCP, used in metalworking fluids, in sealants, as flame retardants, and in paint and coatings, states differed in whether they saw its propensity for long-range transport to be high enough to meet the criteria.[27] Again, different interpretations of the precautionary principle were invoked to support diverging positions; as with PCP, the largest opposition to including SCCP came from Canada and the United States, while the Netherlands and Sweden were among the strongest advocates for its inclusion.

Format for controls. At the onset of negotiations, parties agreed that a combination of total bans and specifically listed exemptions would shape the design of controls for the pesticides and industrial chemicals. In designing controls for the production and use of substances scheduled for elimination, states agreed to ban their use and only the United States refused to accept a ban on production, because it wished to retain the possibility of producing regulated substances for export to countries outside the CLRTAP region.

Trade regulations played an important role during the negotiations. As a production ban (with the U.S. reservation) was proposed, the trade issue related mostly to existing stockpiles of substances that were banned or for which exemptions on production were given. The United States and Canada argued that legally binding import or export restrictions imposed by a regional environmental forum would constitute a violation of the higher priority WTO free-trade principle. Accordingly, they argued, trade measures should be global and the then impending global POPs negotiations offered the most appropriate forum. Others, including most western European states, took the position that an export ban was especially desirable if only because its absence could lead to accusations of a double standard: the substances were considered too dangerous to be used by the parties but not for export to non-parties and developing countries. Also, because of frequent and extensive long-range atmospheric transportation, inflow of emissions of regulated substances used outside the region could have negative environmental and human health effects within the CLRTAP region.

With production and use of some POPs being phased out, waste management – involving identifying domestic POPs wastes, designing plans for their destruction, and examining the possibility of transnational waste transport for destruction – emerged as a central issue. European countries were generally in favour of introducing restrictions on the transboundary transport of waste, believing that states have a responsibility for their own waste and that transportation – especially to developing countries with often unsafe storage facilities and destruction capabilities – could increase the potential for leakage and emissions. The United States and Canada opposed transport restrictions, arguing that allowing transnational transport would ensure that waste could be disposed of in locations where it could be done most safely and cost-effectively.

Negotiations on unintentional by-products addressed two main issues identified by the PWG. The first was the type of control measures that should be used, i.e., BAT, ELV, or some combination of the two; the second was the combination of mandatory and voluntary measures. In parallel, work continued on the development of detailed technical annexes, covering both stationary and mobile emission sources.

Mechanism for future evaluation of additional substances. How to design a mechanism to allow substances to be evaluated and possibly included in the protocol in the future was a key issue of the negotiations. Canada and the United States, with support from the chemical industry, restated preference for a formal approach by which a substance would have to meet detailed numerical scientific criteria – listed in the protocol – for assessing harmfulness (e.g., persistence, toxicity, and bioaccumulation). The advantages of this approach were said to include transparency and consistency by having clearly defined criteria against which to assess all substances. A second group, mostly western European states, advocated a more flexible mechanism outside the protocol that could be more easily amended. Proponents reiterated that embedding detailed criteria could render the mechanism obsolete as knowledge and assessment methodologies advanced.

The two-track procedure proved not as efficient as originally thought. Design and content for the track-two mechanism was, in part, precedent-setting work with a potentially strong impact on the future direction of international chemicals-screening work. The first two negotiation meetings gave clarity to the two options (stringent criteria vs. flexibility) but made no real progress. At the October 1997 meeting, Canada presented a compromise proposal to specify assessment criteria in a separate CLRTAP Executive Body decision before the signing of the protocol. As well as offering the predictability sought by Canada, the United States, and the chemical industry, the proposal offered the flexibility of allowing for amendments to be made by the executive body without having to renegotiate the whole protocol. The October meeting did not achieve a final decision, but the Canadian proposal served as a basis for continued negotiations.

Concluding negotiations

Notwithstanding the slow progress of the first three negotiation meetings, there was optimism; signing of a protocol was tentatively set for the fourth Environment for Europe Ministerial Conference, in June 1998. Three main coalitions emerged after the October 1997 meeting: the European Union (with the European Commission having a partial negotiating mandate for all EU member states), supported by Norway and Switzerland; the United States, often together with Canada and closest to the industry position; and the Russian Federation, often backed by the Ukraine. At this stage of the negotiations, much of the actual negotiating took place among the three coalitions, with other states largely passive.

Each of the three coalitions had well-defined preferences on the outstanding issues. In general, the European Union sought the tightest regulations, but strove to find a broad agreement. Three sets of issues remained as obstacles to a POPs protocol. The first was determining the substances to be on the initial list and their control measures. Solving the HCH/lindane problem was an important step that made it more feasible to reach an overall compromise, yet possible exemptions on DDT, PCBs, HCB, and heptachlor remained open, as did the possible inclusion of PCP and SCCP. Italy, the Russian Federation, and the United States wanted DDT exemptions. The Russian Federation and Ukraine wanted exemptions for both PCBs and HCB. The United States resisted a total phase-out of heptachlor. The European Union supported regulations on both PCP and SCCP, but significant resistance to PCP inclusion came from the United States, which, together with Canada and the Russian Federation, also opposed inclusion of SCCP. On the by-products, the drafting of technical annexes continued, with some minor controversies remaining on the use of ELV or BAT and the combination of mandatory and voluntary control measures. The second issue was whether import and export should be addressed in the protocol. At the October meeting, the European Commission unexpectedly spoke in favour of both import and export controls. This was supported by the Russian Federation, but the United States and Canada continued to firmly oppose trade regulations. Thirdly, activities based on the Canadian compromise proposal for the mechanism for future evaluation of additional substances continued, but the previous adversarial positions were maintained.

A two-day Head of Delegation meeting was scheduled for December 1997 to solve the remaining differences and allow a protocol to be ready by early 1998. This official meeting was preceded by an informal meeting held in late November in London, where representatives from the United Kingdom, the European Commission, and the United States met with the chair of the Working Group on Strategies to try to outline a possible compromise. That aim was not fully achieved, but several important agreements moved the process forward.[28] An agreement on heptachlor was reached: the U.S. position that it needed an exemption to use heptachlor for the control of fire ants in closed industrial electrical junction boxes was accepted, with a compromise to re-evaluate such use no later than two years after the protocol enters into force. Agreement was also reached on which activities should be covered by the protocol as Canada and the United States managed to convince the Europeans that import and export controls should not be included in the protocol. In turn, the United States lifted its reservation on introducing production bans. An important step was also achieved on the procedure for adding substances: it was agreed that amendments should be regulated through a special executive body decision. The European Union accepted the numbers for the numerical criteria proposed by the United States, but disagreement remained on using the word "indicative" before the numerical criteria, i.e., whether the set criteria should be seen as absolute or merely guiding.

Despite progress at the December meeting, a few crucial issues remained unresolved. Trying to clarify the details of a final compromise, representatives from the United Kingdom, the European Commission, and the United States met with the chair of the Working Group on Strategies for a second time in London in early February 1998. Following that meeting, the chair opened the February 1998 session of the Working Group on Strategies by declaring that it should be the final negotiation meeting on POPs. The meeting would tackle several issues: agreement on PCBs; possible DDT and HCB exemptions; inclusion or exclusion of PCP and SCCP; the format for controls of the by-products; the wording of the executive body decision for future evaluations; and possible regulations on transboundary waste transport.

Based on a solution that had been put together at the second London meeting, the European Union won acceptance for some of its preferred regulations and options while taking into consideration U.S. and Canadian resistance to regulations on certain substances and transnational transport of waste, as well as agreeing to some of the exemptions the Russian Federation and Ukraine desired. A five-year extra extension on PCB phase-out was granted to countries with economies in transition (i.e., former communist states). Italy won an exemption for the use of DDT as an intermediate in the production of dicofol to be re-evaluated two years after the protocol enters into force. DDT exemptions were granted for public health protection as a component of an integrated pest management strategy against malaria and encephalitis in response to the Russian Federation and U.S. positions. Again in consideration of Russian preferences, HCB exemptions were granted on both production and use for specified purposes. Both PCP and SCCP were deleted from the initial list, but an EU-U.S. deal identified PCP in the article addressing research, development, and monitoring as a substance that warranted special attention. This signalled that it may be one of the first substances up for re-assessment under the track-two mechanism.

In the protocol, substances are grouped into three annexes (table 6.2). Annex I contains pesticides and industrial chemicals for which production and use are to be eliminated. Annex II contains pesticides and industrial chemicals scheduled for restrictions on use. Stockpiles of Annex I substances are to be destroyed or disposed of in an "environmentally sound manner." The parties committed themselves to "endeavour to ensure" that the disposal of Annex I substances is carried out domestically, but the protocol contains no ban against the transboundary movement of wastes. States managed to resolve differences over which by-product regulations should be mandatory and which should be voluntary; by-products are regulated in Annex III through a combination of ELV and BAT specified in detailed technical annexes to the protocol. Finally, meeting European wishes that the scientific criteria for the assessment of additional substances should be guiding rather than absolute, the word "indicative" was kept in the executive body decision.

Implementing the agreement

In March 1998, the CLRTAP Executive Body, by Decision 1998/2, confirmed information to be submitted and procedures for adding substances to the POPs protocol. Three months later, at the Fourth Environment for Europe Ministerial Conference in Århus, Denmark, thirty-three states and the European

Table 6.2
The sixteen substances initially included in the CLRTAP POPs Protocol. The substances are grouped according to category and the annex in which they are listed. DDT and PCBs are listed in both Annex I and Annex II; HCB is listed in both Annex I and Annex III.

Annex	*Pesticides*	*Industrial chemicals*	*Unintentional by-products*
I	Aldrin	Hexabromobiphenyl	
	Chlordane	PCBs	
	Chlordecone	HCB	
	DDT		
	Dieldrin		
	Endrin		
	Heptachlor		
	HCB		
	Mirex		
	Toxaphene		
II	HCH/Lindane	DDT	
		PCBs	
III			Dioxins
			Furans
			HCB
			PAHs

Community, represented by the European Commission, signed the POPs protocol. As of August 2002, there were thirty-six signatories and ten ratifications to the POPs protocol.[29] At the Århus meeting, the European Commission and the EU member states issued a joint ministerial declaration to show that, although the European Union had signed the POPs protocol, it would have preferred a more stringent agreement. This also was an attempt to ensure that some of the issues excluded from the protocol, including controls on SCCP and PCP and a ban on the export of Annex I substances, remained alive for future revisions.

The CLRTAP Executive Body is the primary decision-making body in putting the POPs agreement into effect. The CLRTAP Secretariat functions in its normal role as a coordination centre for the gathering of information on international and domestic implementation developments. The Co-operative Programme for Monitoring and Evaluation of the Long-range Transmission of Air Pollutants in Europe (EMEP) will guide and monitor domestic implementation. To that end, EMEP is developing standard operating procedures and quality control routines for sampling and chemical analysis and is creating an international measurement program on POPs.[30] In developing such a program, the two EMEP centres – the Chemical Coordinating Centre (CCC) and the Meteorological Synthesizing

Centre-East (MSC-E) – will specifically focus on a small group of selected POPs: lindane, PAHs, seven specific PCB congeners, and some dioxins and furans.

The continued political support of Canada, the Netherlands, Sweden, the United Kingdom, and the United States, all of which played crucial political roles during the preparatory work and the negotiations, will be important in helping the CLRTAP POPs agreement achieve its potential as a long-term effective instrument. Canada, in December 1998, was also the first state to ratify the POPs protocol, quickly followed by, among others, Denmark, Norway, Sweden, and Switzerland. However, the Russian Federation and a few other eastern European countries hold a key to reducing POPs emissions and improving environmental status in the CLRTAP region in the short term; it is primarily in these countries that the production and use of the initially regulated thirteen pesticides and industrial chemicals still occur. The weak domestic regulatory structures in many of these countries will make implementing the protocol difficult. As the POPs protocol contains no commitment to the transfer of technology or monetary funds to aid implementation, efforts to assist eastern European countries and the Russian Federation are organized chiefly outside the protocol with targeted bilateral and multilateral programs. In addition, effective control of by-products and waste management will be a significant challenge for all states. Controlling by-products involves identifying emission sources and implementing effective technical solutions. Managing the waste problem involves the arduous task of identifying POPs waste and waste contaminated by POPs and ensuring that such waste is disposed of in a safe manner.

The long-term effectiveness of the agreement is tied to the success of the track-two mechanism, which has the potential to make a large impact on future POPs abatement in the northern hemisphere. Some initial steps, designed to save time so that the mechanism can be quickly used once the protocol enters into force, have been taken. Executive Body Decision 1998/2 established an ad hoc Experts Group, which first met in November 2000.[31] The Experts Group will consider new information both for substanccs covered by the protocol and for new substances.[32] As such, the track-two mechanism can be used both to tighten exemptions for substances on the initial list and to assess additional substances that are not initially covered by the protocol, including PCP and SCCP. Brominated flame-retardants, used in computers and other technical equipment, have recently been demonstrated to exhibit PCB-like characteristics and may be considered in future.

The CLRTAP POPs work has been primarily reactive; i.e., controls have been applied to substances only when they have been found in the environment and been discovered to be hazardous for wildlife and human health. Management would be improved by actively seeking to identify additional POPs and their sources before they cause environmental and human health damage. This would require developing more effective methods of analyzing toxicity of existing and new chemicals when they are synthesized and developing a better understanding of the whole life cycle of pesticides and industrial chemical POPs through their production, trade, use, and disposal. It will be equally important to improve our understanding of when POPs are generated as by-products and to develop better and cheaper prevention techniques to apply to stationary and mobile sources. Finally, the CLRTAP POPs work is but one case of co-operation regarding international chemicals; a growing number of international initiatives raise the need for better international coordination to avoid overlapping or even counterproductive activities.[33]

NOTES

1 This chapter builds on H. Selin, *Towards International Chemical Safety: Taking Action on Persistent Organic Pollutants (POPs)*, Ph.D. diss., Linköping, 2000. *Linköping Studies in Arts and Sciences* 211, particularly chapters 4 to 6.

2 R. Carson, *Silent Spring* (Boston: Houghton Mifflin 1962).

3 N. Shifrin and A. Toole, "Historical Perspective on PCBs," *Environmental Engineering Science* 15, no. 3 (1998): 247–57, p. 248; and The Riverside Press and K. Mellanby, *The DDT Story* (Farnham: The British Crop Protection Council 1992).

4 E. Voldner and Y.-F. Li, "Global Usage of Selected Persistent Organochlorines," *Science of the Total Environment* 160/161 (1995): 201–10.

5 G. Bowes and C. Jonkel, "Presence and Distribution of Polychlorinated Biphenyls (PCB) in Arctic and Subarctic Marine Food Chains," *Journal of the Fisheries Research Board of Canada* 32, no. 11 (1975): 2111–23; Ross Norstrom et al., "Organochlorine Contaminants in Arctic Marine Food Chains: Identification, Geographical Distribution, and Temporal Trends in Polar Bears," *Environmental Science and Technology* 22, no. 9 (1988): 1063–71.

6 A. DePalma, "An Arctic Meal: Seal Meat, Corn Chips and PCB's," *New York Times*, 5 February 1999.

7 E. Dewailly et al., "High Levels of PCBs in Breast Milk of Inuit Women from Arctic Quebec," *Bulletin of Environmental Contamination and Toxicology* 43, no. 1 (1989): 641–46; Dewailly et al., "Inuit Exposure to Organochlorines through the Aquatic Food Chain in Arctic Quebec," *Environmental Health Perspectives* 101, no. 7 (1993): 618–20; and UNECE, EB.AIR/WG.6/R.20/Add.1, *Draft Executive Summary of the State of Knowledge Report of the Task Force on Persistent Organic Pollutants led by Canada and Sweden*, para. 40, 25 April 1994.

8 E. Chossudovsky, *"East-West" Diplomacy for Environment in the United Nations* (New York: UNITAR 1989).

9 UNECE, EB.AIR/WG.1/12, *Report of the Eighth Session*, para. 18, 18 September 1989.

10 Swedish Environmental Protection Agency, *Persistent Organic Compounds in the Marine Environment*, Report 3690, 1990.

11 P. Larsson and L. Okla, "Atmospheric Transport of Chlorinated Hydrocarbons to Sweden in 1985 compared to 1973," *Atmospheric Environment* 23, no. 8 (1989): 1699–1711.

12 UNECE, ECE/EB.AIR/24, *Report of the Eighth Session of the Executive Body*, para. 38(f), 11 December 1990.

13 The four meetings were held in Stockholm, 20–22 March 1991; Port Stanton, Canada, 17–20 May 1992; Berlin 17–19 May 1993; and The Hague, 21–25 February 1994.

14 UNECE, *Draft Executive Summary*, para. 3.

15 Ibid., para. 92.

16 Existing fora that were deemed to have too narrow a geographical domain and membership included the Oslo-Paris Commission, the North Sea Task Force, the Helsinki Commission, and the Great Lakes abatement and control agreements. In contrast, the OECD had a much broader geographical coverage and membership, but its mandate was believed to be too restricted.

17 UNECE, EB.AIR/WG.1/22, *Report on the Thirteenth Session*, 22 July 1994.

18 The four meetings were held in Geneva, 9 March 1995; Geneva, 3–5 July 1995; Geneva, 8–10 May 1996; and Aylmer, Canada, 21–23 October 1996.

19 United Kingdom, AEA/CS/RCEC 16419225, Issue 3, *Selection Criteria for Prioritising Persistent Organic Pollutants*, July 1995.

20 Information from other fora was collected primarily from the Oslo-Paris Commission for the Protection of the Marine Environment of the North-East Atlantic (OSPARCOM), the European Union (EU), OECD, WHO, and the Intergovernmental Forum on Chemical Safety (IFCS).

21 For more on the Modified Task Force Methodology, see H. Selin and O. Hjelm, "The Role of Environmental Science and Politics in Identifying Persistent Organic Pollutants for International Regulatory Actions," *Environmental Reviews* 7, no. 2 (1999): 61–8. See also, B. Rodan et al., "Screening for Persistent Organic Pollutants: Techniques to Provide a Scientific Basis for POPs Criteria in International Negotiations," *Environmental Science and Technology* 33, no. 20 (1999): 3482–8.

22 United Kingdom, "Review of Risk Characterisation Information on Selected Persistent Organic Pollutants," presented at the CLRTAP ad hoc Preparatory Working Group Persistent Organic Pollutants Meeting, 21–23 October 1996, Aylmer, Canada.

23 All five meetings were held in Geneva and took place 20–24 January 1997; 16–20 June 1997; 20–24 October 1997; 14–15 December 1997; and 9–13 February 1998.

24 Sweden, "Risk Assessment of Lindane," presented at the Working Group on Strategies Meeting, 20–24 January 1997, Geneva.

25 Netherlands, "A Compromise Proposal for Lindane," presented at the Working Group on Strategies Meeting, 20–24 January 1997, Geneva.

26 United States, "Environmental Fate and Transport of Pentachlorophenol," "Review of Screening Criteria Data for Persistent Organic Pollutants," and "What Does the Science Show Regarding Pentachlorophenol as a UNECE LRTAP POP?," presented at the Working Group on Strategies Meeting, 20–24 October 1997, Geneva.

27 Sweden, "Short Chain Chlorinated Paraffins," presented at the ad hoc Preparatory Working Group on Persistent Organic Pollutants Meeting, 21–23 October 1996, Aylmer, Canada, and "Short Chain Chlorinated Paraffins – Additional Risk Information," presented at the Working Group on Strategies Meeting, 20–24 January 1997, Geneva.

28 UNECE, EB.AIR/WG.5/R.94, *Changes to the Preliminary Draft POPs Protocol*, 22 December 1997.

29 Updated information on signatories and ratifications can be found at http: // www.unece.org/env/lrtap/.

30 UNECE, EB.AIR/1999/7/Add.1, *Draft 2000 Work-Plan for the Implementation of the Convention on Long-Range Transboundary Air Pollution*, section 2.5, 12 October 1999.

31 UNECE, EB.AIR/WG.5/66, *Report of the Thirty-Second Session*, paras 38–39, 12 September 2000.

32 UNECE, EB.AIR/WG.5/2000/1, *Further Assessment of Persistent Organic Pollutants*, 1 December 2000.

33 J. Krueger and H. Selin, "Governance for Sound Chemicals Management: The Need for a More Comprehensive Global Strategy," in *Global Governance* 8 (2002): 323–42.

7
Global POPs Policy: The 2001 Stockholm Convention on Persistent Organic Pollutants

DAVID LEONARD DOWNIE

The Stockholm Convention on Persistent Organic Pollutants (POPs) represents a significant achievement for international environmental policy. Adopted by consensus at a diplomatic conference in May 2001, the convention is the first global treaty that seeks to eliminate substances specifically toxic to the environment and human health. It is, in a very real sense, the centrepiece of global POPs policy.

The Stockholm Convention bans or limits the production, use, release, and trade of twelve particularly toxic POPs, often referred to as the "dirty dozen." These are the pesticides aldrin, chlordane, DDT, dieldrin, endrin, heptachlor, mirex, and toxaphene; the industrial chemicals PCBs and hexachlorobenzene (which is also a pesticide); and the two unintentionally produced by-products of industrial processes and combustion: dioxins and furans. The convention promotes the use of best available techniques and practices for replacing existing POPs and managing and disposing of POPs wastes; establishes scientifically based criteria and a specific procedure for establishing controls on additional POPs; and seeks to prevent the development and commercial introduction of new POPs. Although most of the dirty dozen are subject to an immediate ban, certain country-specific and time-limited exemptions are permitted. Most importantly, a broad health-related exemption is granted for DDT, which is still used in many countries to control malarial mosquitoes.

The Stockholm Convention establishes a financial mechanism to help developing countries and countries with economies in transition meet their treaty obligations. Developed country parties will provide new and additional financial resources as part of this process. Technical assistance will also be provided through global, regional, and bilateral mechanisms. A Conference of the Parties (COP), made up of all governments party to the convention, will be the principal decision-making authority for the convention, and a professional secretariat will act as the day-to-day administrator. The treaty also includes important provisions for, *inter alia*, information exchange, monitoring, national implementation plans, public education, reporting, and research.

This chapter provides an overview of the development and content of the Stockholm Convention.

DEVELOPING GLOBAL POPs POLICY

The Stockholm Convention is the most recent product of more than two decades of effort by the international community to address toxic chemicals. This work has been informed by a growing body of scientific information concerning the threats posed by toxic chemicals and slowly expanding political consensus that global co-operation is necessary to address them. The development of this policy can be usefully divided into three interrelated stages of action: coordinating national policies; regulating international trade in chemicals; and mandating the minimization, if not the elimination, of the production, use, and emission of specific toxic chemicals. Although unplanned as an overall strategy, each stage has involved several intergovernmental organizations and treaty initiatives; together they have produced a path that leads logically to the goals and contents of the Stockholm Convention.[1]

Coordinating national chemical controls

Four decades ago, Rachel Carson's *Silent Spring* introduced the general public to the risks of DDT. The resulting media attention led several countries to take legal action to ban or severely restrict the use of DDT in the early 1970s. Since then, more and more governments have enacted national legislation to regulate the production and use of other hazardous chemicals as growing scientific evidence demonstrated their toxicity to humans and animals. However, because some chemicals were known to circulate across national borders, concern increased that no country acting alone could fully protect its citizens or environment.

Before 1992, international co-operation regarding chemicals primarily involved developing tools for risk assessment and conducting assessments of selected chemicals. This work was done largely by the Chemicals Programme of the Organisation for Economic Co-operation and Development (OECD); by the International Programme on Chemical Safety (IPCS) – a joint program of the United Nations Environment Programme (UNEP), the World Health Organization (WHO), and the International Labour Organization (ILO); and by the Food and Agriculture Organization of the United Nations (FAO).

In 1992, governments at the United Nations Conference on Environment and Development (UNCED, or the Rio Earth Summit) adopted *Agenda 21*, which includes Article 19 on the "Environmentally Sound Management of Toxic Chemicals Including Prevention of Illegal International Traffic in Toxic and Dangerous Products." This article called for the creation of an Intergovernmental Forum on Chemical Safety (IFCS) to promote the coordination of international work on chemicals.[2]

Since its establishment in 1994, the IFCS has provided policy guidance and strategies for harmonizing risk assessment methods and chemical classification. It has also attempted to help strengthen information exchange, risk reduction, chemical management capacity-building, and related activities. Meanwhile, the Inter-Organization Programme for the Sound Management of Chemicals (IOMC) was established to promote coordination among international organizations involved in implementing Article 19. The IOMC's current membership includes UNEP, ILO, FAO, WHO, the UN Industrial Development Organization (UNIDO), the UN Institute for Training and Research (UNITAR), and the OECD.[3]

Regulating international trade

Steps to regulate international trade in hazardous chemicals started with the FAO's International Code of Conduct on the Distribution and Use of Pesticides (as amended in 1989) and the London Guidelines for the Exchange of Information on Chemicals in International Trade (amended 1989). Together, these instruments led to the voluntary Prior Informed Consent (PIC) procedure, jointly administered by FAO and UNEP.[4]

The goal of the PIC procedure is to promote a shared responsibility between exporting and importing countries for protecting human health and the environment from the harmful effects of toxic chemicals. The procedure formally obtains and disseminates decisions of importing countries regarding their desire to receive future shipments of a certain chemical and seeks to ensure compliance with these decisions by exporting countries. The voluntary PIC system has worked well; it addresses twenty-two pesticides and five industrial chemicals (including seven POPs) and enjoys the participation of 154 countries. By the mid-1990s, however, many governments saw the need for a legally binding treaty to govern trade in these hazardous chemicals. Treaty negotiations began in 1996, and the Rotterdam Convention on the Prior Informed Consent Procedure for Certain Hazardous Chemicals and Pesticides in International Trade was adopted and opened for signature in Rotterdam in September 1998. As of 27 August 2002, seventy-three countries have signed and twenty-seven have ratified the Rotterdam Convention, which will enter into force when ratified by fifty countries.[5]

A mandate for global controls

In May 1995, UNEP's Governing Council called for an international assessment of twelve POPs – the dirty dozen (Decision 18/32). In response to the decision, the IOMC summarized the scientific literature on POPs; consolidated the available information on their chemistry, toxicity, environmental dispersion, and other relevant properties; and established a UNEP/IFCS ad hoc Working Group on Persistent Organic Pollutants.[6]

In November 1995, the Intergovernmental Conference to Adopt a Global Programme of Action for Protection of the Marine Environment from Land-based Activities (GPA) considered POPs as part of its agenda. Following extensive deliberation, the "Washington Declaration" issued from the conference called, in part, for talks on a legally binding treaty to reduce or eliminate the discharge, manufacture, and use of the twelve POPs.[7]

In June 1996, the IFCS concluded that sufficient evidence existed to warrant international action – including a global legally binding instrument – to reduce the risks posed by POPs to human health and the environment. These recommendations were forwarded to the UNEP Governing Council and the WHO World Health Assembly.[8]

In February 1997, at the UNEP Governing Council, dozens of governments formally agreed that "international action, including a global legally binding instrument, is required to reduce the risks to human health and the environment arising from the release of the twelve specified persistent organic pollutants." The governing council endorsed the "conclusions and recommendations" of the IFCS report and specifically requested that UNEP, together with WHO and other relevant international organizations, "prepare for and convene an intergovernmental negotiating committee, with a mandate to prepare an international legally binding instrument for implementing international action initially beginning with the twelve specified persistent organic pollutants" (paragraph 8).[9]

In May 1997, governments attending the fiftieth World Health Assembly of the WHO echoed the UNEP Governing Council. Resolution WHA50.13 endorsed the recommendations of the IFCS, called for additional international action on POPs, and specifically requested that WHO participate actively in the POPs treaty negotiations.[10] With the explicit endorsements by government members of the IFCS, UNEP Governing Council, and the WHO World Health Assembly, the call for international negotiations was now perfectly clear.

NEGOTIATING THE STOCKHOLM CONVENTION

Based on this clear mandate, UNEP prepared to convene the Intergovernmental Negotiating Committee for an International Legally Binding Instrument for

Implementing International Action on Certain Persistent Organic Pollutants (POPs INC). As part of this process, UNEP and the IFCS convened eight regional/sub-regional awareness-raising workshops on POPs, in which 138 countries participated, between July 1997 and June 1998. The workshops increased awareness of POPs issues among people in developing countries and in countries with economies in transition in preparation for the upcoming global negotiations.[11]

Almost three years of global POPs negotiations began with INC-1 in Montreal in June 1998. Although often complex and at times quite difficult, the negotiations were also characterized by commendable inclusiveness, transparency, diligence, and, ultimately, success. The negotiations included five official week-long meetings of the INC; two meetings of the Criteria Expert Group, which focused on criteria and procedures for identifying and potentially adding new chemicals; a formal inter-sessional consultation on financial assistance issues; numerous formal contact groups; countless informal consultations; and a final Conference of Plenipotentiaries. Nearly 1,400 delegates from 151 countries participated in these meetings. More than 600 experts from intergovernmental organizations and industrial, environmental, and academic non-government organizations attended as observers. Delegates completed work on a draft convention in Johannesburg in December 2000 after an extraordinary all-night plenary session that closed INC-5.[12]

On 22 May 2001, the Conference of Plenipotentiaries formally adopted, by unanimous approval, the final text of the convention and seven important resolutions concerning the treaty and its interim implementation. A "Global POPs Policy" had been achieved.

IMPLEMENTING THE STOCKHOLM CONVENTION

The Stockholm Convention enjoys strong global support. The UNEP Governing Council, the World Health Assembly of the WHO, the IFCS, the Global Environment Facility (GEF), and other international bodies have all endorsed a legally binding instrument to eliminate or control POPs.[13] By September 2002, 151 countries had signed the convention, 21 countries had completed their ratification procedures and become parties to the convention, and many others were at various stages of the ratification process.[14]

Despite this support, however, the goals of the Stockholm Convention will not be achieved unless a vast majority of countries both ratify and fully implement it – an accomplishment that has eluded several multilateral environmental agreements (MEAs) in recent years. Indeed, the convention requires ratification by fifty countries to enter into force. The impact of a significant delay would be magnified by the particular qualities of POPs; the persistence of these chemicals in water and soil and their ability to accumulate in living tissue mean that each year that passes without a global solution means decades of additional exposure.

Rather than wait for the fiftieth ratification, the Conference of Plenipotentiaries appealed to states, UNEP (as the interim secretariat), and the GEF (as the interim financial mechanism) to voluntarily implement the provisions of the convention and to initiate efforts to achieve its rapid and effective entry into force. As part of this process, the conference specifically invited UNEP to convene additional meetings of the INC so that governments could begin detailed examinations of the issues upon which the COP will need to make decisions at its first or subsequent meetings (Resolution 1.3). As a result of these instructions, UNEP, the GEF, other international organizations, and some national governments have begun to implement the Stockholm Convention.

THE STOCKHOLM CONVENTION

The contents of the Stockholm Convention can be usefully divided into nine interrelated components that can help us to summarize and understand the most important aspects of the treaty:

- Control measures on twelve specific POPs
- Controlling additional POPs
- Financial and technical assistance
- Implementation plans
- Monitoring, reporting, and information exchange
- Effectiveness evaluations
- Amending the convention
- Treaty administration
- Unfinished business

Control measures

The stated objective of the Stockholm Convention is "to protect human health and the environment from persistent organic pollutants" (Article 1). To do so, the convention follows four related strategies: (1) eliminate the production and use of specific POPs; (2) eliminate trade of specific POPs; (3) minimize emissions of POPs that cannot be eliminated; and (4) avoid production and use of new POPs. The specific requirements of the control measures (which give specificity to general strategies outlined previously) are contained in articles 3, 4, 5, and 6 and annexes A, B, and C.

Eliminating the production and use of specific POPs. The Stockholm Convention requires that all parties prohibit the production and use of the intentionally produced POPs listed in Annex A (aldrin, chlordane, dieldrin, endrin, heptachlor, hexachlorobenzene, mirex, PCBs, and toxaphene; more may be added in future). Parties must also severely restrict the production and use of the POPs listed in Annex B; at present, DDT is the only POP listed.

The clarity and conciseness of these control measures are seen as strengths of the convention (Article 3.1), yet it provides for a variety of exemptions that allow the continued production and use of specific POPs, by specific parties, for narrowly defined applications. These exemptions are not eternal; provisions exist for their gradual elimination.

The most important exemption is the broad, health-related "acceptable use" granted for DDT, which is still needed in many countries to control malarial mosquitoes. The DDT exemption in Part II of Annex B will permit governments to protect their citizens from malaria – a major killer in many tropical regions – until they are able to replace DDT with chemical and non-chemical alternatives that are cost-effective and environmentally friendly.[15] Another important exemption granted is for PCBs. While new equipment no longer uses PCBs, hundreds of thousands of tons (1 ton = 1.02 metric tonnes) of it are still in use in electrical transformers and other equipment. The provisions in Part II of Annex A permit governments to maintain existing equipment containing PCBs until 2025 in a way that prevents leaks, giving them time to arrange for PCB-free replacements.

More broadly, all parties can exercise several *general exemptions* for some or all of the POPs listed in annexes A and B. These exemptions address POPs in objects already in use; e.g., utility poles treated with particular pesticides (Annex A, note ii); POPs that exist as unintentional trace contaminants of other processes (Annex A, note i); and uses that involve very small amounts of POPs, such as those needed as a reference standard for laboratory-scale research (Article 3(5) and Annex A, note iii).

Finally, *country-specific exemptions* allow parties to use particular POPs for certain uses for a limited time. Any state, on becoming a party, may register for any of the specific exemptions listed in Annex A or Annex B by notifying the secretariat in writing (Article 4(3)). All such registrations expire after five years unless a party indicates an earlier date in the *Register of Specific Exemptions* or the CoP grants an extension (Article 4(4)). When there are no longer any parties registered for a particular type of specific exemption, no new registrations may be made with respect to it (Article 4(9)).

Eliminating trade in specific POPs. Parties must enact measures to eliminate all trade in the ten intentionally produced POPs, except for the environmentally sound disposal of the substances; use by a party with specific exemptions or acceptable purposes; and use by a non-party that would qualify as a specific exemption or an acceptable purpose. In the latter case, there exist significant conditions, accountability, and requirements to report to the exporting country that is a party (Article 3(2)).[16]

Minimizing emissions of POPs that cannot be eliminated. With surprisingly strong language, the convention seeks to minimize the release of by-product POPs or POPs that have exemptions or are not yet listed in annexes A and B. Article 5 and Annex C specifically address dioxins and furans, the dangerous, unintentionally produced POPs. Each party is required to take measures to reduce its total releases, from anthropogenic sources, of dioxins and furans "with the goal of their continuing minimization and, where feasible, ultimate elimination" (Article 5). Parties must promote the use of "best available techniques" and "best environmental practices" to reduce emissions, and each party must develop an action plan for reducing emissions within two years of entry into force of the convention and then implement that plan as part of its overall implementation plan (Article 5 and Annex C).

The convention contains a number of provisions, both general and specific, that compel or guide parties to limit releases of intentionally produced POPs having a general or special-use exemption (Article 3(6)). Parties must also attempt to assess other pesticides and industrial chemicals currently in use so that the proper controls can be put into effect to limit releases of any that would be classified as POPs in accordance with Annex D (Article 3(4)). And, all parties must take a number of steps to minimize releases of POPs from stockpiles and wastes (Article 6(1)(a-e)).

Avoiding production and use of new POPs. Looking forward, the convention requires parties to take measures to prevent unnecessary or inadvertent development and commercial introduction of chemicals that would be classified as POPs. Although the efficacy of these provisions is as yet unproven, their goal is important. Such measures would benefit human health and the environment and would help the chemical industry avoid wasting valuable resources on products and processes that would be subject to elimination under the convention (Article 3(4), Article 3(6), and Preamble).

Controlling additional POPs. One of the most important factors contributing to the effectiveness of an MEA is its process for adjusting the scope and strength of its environmental protection measures in response to new information. This is certainly true for the Stockholm Convention, which addresses a broad class of very toxic chemicals on which scientific knowledge is expanding rapidly. The UNEP Governing Council recognized this need before negotiations began: its mandate for the global POPs negotiations specifically instructed the INC to develop criteria and procedures for identifying and controlling additional POPs.[17]

In an effort to ensure that the treaty remains a dynamic and living instrument, the Stockholm Convention establishes a specific procedure and criteria for identifying, evaluating, and adding POPs to the control regime. Article 8 sets out a fifteen-step process that parties will follow when proposing, evaluating, and making decisions on candidate POPs. This innovative, if elaborate, procedure integrates the use of specific scientific criteria for identifying candidate POPs (Annex D[18]) and specific information requirements for developing a risk profile for candidate POPs (Annex E[19]) with consideration of socio-economic impacts of controlling a POP (Annex F) and a strong perspective of precaution. Precaution informs the process in such a way that the absence of strict scientific certainty shall not prevent parties from controlling a potentially hazardous substance (articles 8(7)(a) and 8(9)).

Through this process (table 7.1) parties will be able to consider and thoroughly evaluate chemicals to determine if they possess POP properties and behaviours to a degree sufficient to warrant inclusion in the treaty. Any party may submit a proposal to the secretariat for adding a chemical to the treaty. A POPs Review Committee (PRC) will then work on behalf (and under the oversight) of the parties to apply the screening criteria, develop and evaluate risk profiles, and develop and evaluate risk assessments. There are safeguards to ensure that all parties have opportunities to get a full hearing on any nominated candidate POP, and the parties themselves will hold final decision-making authority.

Financial and technical assistance

The global mandate for the POPs negotiations stated that any future treaty should take into account "the special needs of developing countries and countries with economies in transition." This statement by the UNEP Governing Council recognized that many of these states do not always have the financial resources and technical capacity to adopt alternative products and processes. They cannot always afford or gain access to new and cleaner industrial technologies. They often do not have the infrastructure or resources to distribute alternatives to POPs, to disseminate information about such alternatives, or to develop, monitor, and enforce effective restrictions on the use of POPs.

Thus, one of the most important aspects of the Stockholm Convention is its explicit recognition that these countries will need financial assistance to help them meet their obligations as parties to the convention (Article 13(4)). To this end, and although the convention does not specify target levels, developed countries have agreed to provide new and additional financial resources to help meet these needs (Article 13(2) and 13(6)). This provision of financial assistance should also take into account the need for adequacy, predictability, and the timely flow of funds and the importance of burden-sharing among the contributing parties (Article 13(2)).

The Stockholm Convention establishes a "financial mechanism" to facilitate the distribution of financial assistance and encourages bilateral and other

Table 7.1 The Process for Adding New Chemicals

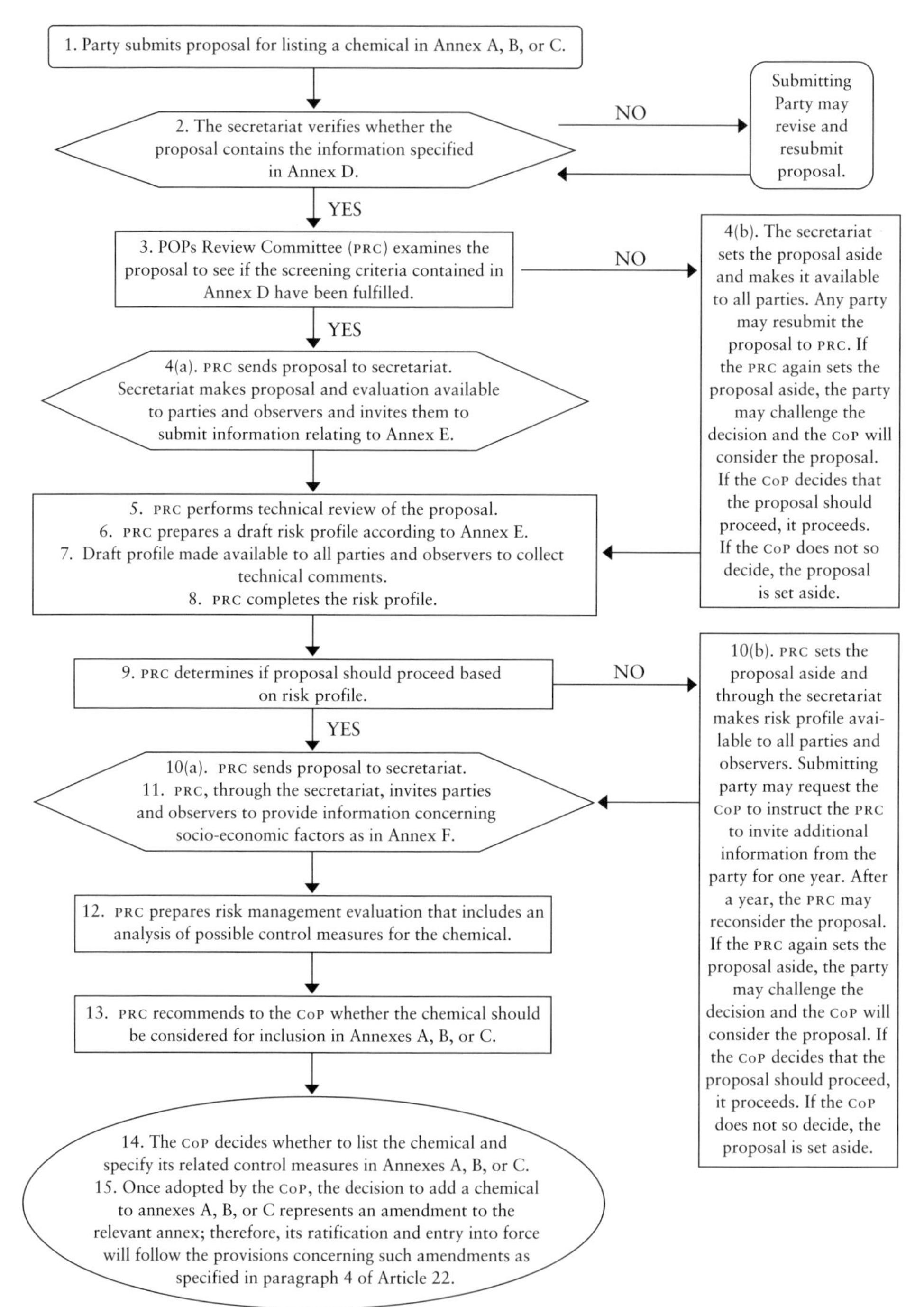

multilateral assistance through other mechanisms (Article 13). The convention requires the CoP to provide guidance to the financial mechanism and to review its operation. The first review will take place no later than the second meeting of the CoP and regularly thereafter (Article 13(8)).

The Global Environment Facility (GEF) will serve as the principal entity entrusted with the operations of the financial mechanism (at least until the CoP decides upon the final institutional structure; see Article 14) and is expected to organize itself to fulfil this function through measures related specifically to POPs (Article 14). The Conference of Plenipotentiaries requested that the GEF Council establish an operational program specifically for POPs (Resolution 2.2) and that the GEF Assembly consider the broader step of establishing a new focal area to support implementation of the Stockholm Convention and the GEF's role as the financial mechanism (Resolution 2.1). The plenipotentiaries also requested that donors contribute adequate additional financial resources to enable the GEF to perform its mandate effectively (Resolution 2.4) and that the GEF report to the first session of the CoP on the measures it has taken to ensure its approval process is transparent and that the access procedures are simple, flexible, and expeditious (Resolution 2.3).

The Stockholm Convention also recognizes that providing technical assistance to developing country parties and parties with economies in transition is essential to successfully implement the convention (Article 12). Developed country parties, and other parties in accordance with their capabilities, have a responsibility to provide developing country parties and parties with economies in transition with appropriate technical assistance for capacity-building relating to implementation of the convention. Resolutions approved by the Conference of Plenipotentiaries strongly urge states, regional economic integration organizations, and international organizations to initiate the provision of assistance through bilateral, regional, and global mechanisms to help developing countries build capacity and reduce the production, use, and release of POPs (resolutions 1.2, 1.11, and 2). In response, the GEF, UNEP, and several states began to make financial resources and technical assistance available in 2001, particularly to support enabling activity, capacity-building, and the development of implementation plans.

Implementation plans

Each party must develop an implementation plan (IP) setting out its strategy and plans to implement its obligations under the convention (Article 7(1)(a)). Within two years of the convention's entry into force for a party, the party must communicate its IP to the CoP (Article 7(1)(b)). Parties must subsequently review and update their IPs as appropriate (Article 7(1)(c)) and report on their implementation of the convention (Article 15).

Although national, regional, and sub-regional circumstances do vary, the general outlines of the development process and contents of the IPs will be similar for nearly all parties. This process will consist of four broad steps: (1) understanding a country's obligations under the Stockholm Convention; (2) identifying the national situation; (3) determining general solutions; and (4) delineating specific plans, programs, and strategies (table 7.2). Countries have been encouraged to begin consulting stakeholders and developing IPs even before they have completed their ratification process or the convention enters into force (Resolution I, paragraph 11). To facilitate this, the GEF, UNEP, and a number of countries in Europe and North America began to provide financial and technical assistance for the creation of IPs by developing countries in 2001.

The requirement to develop and report on IPs may at first appear simplistic, but it is an important one and it follows the practice set in other MEAs, particularly during the last decade. IP-type requirements are included in the Montreal Protocol, Climate Change Convention (FCCC), Desertification Convention (CCD), Biodiversity Convention (CBD), Basel Convention, and CLRTAP POPs Protocol.[20] Its inclusion in the Stockholm Convention is not simply paying homage to precedent; experience has shown that IP mechanisms can act as both incentives and procedural avenues to assist parties to

- gain a better understanding of the environmental problem, the relevant issues and inventories in their countries, the views of national stakeholders, their national obligations under the convention, the necessity for a comprehensive and integrated political and regulatory response, and potential frameworks for such a response;
- begin developing and implementing concrete analytical and policy responses;
- achieve greater transparency through equal access to relevant and standardized information concerning implementation issues in other countries;
- learn from model diffusion (e.g., from the successful actions of other parties); and
- improve the information they use to review the implementation and strategic objectives of the MEA.

Monitoring, reporting, and information exchange

Some of the most important but overlooked components of an effective international environmental treaty are monitoring, reporting, and providing information. Indeed, many participants in the POPs negotiations saw effective provisions in these areas as essential to the treaty's long-term success.[21]

Understand national obligations under the convention

- Examine Stockholm Convention
- Delineate obligations that apply to one's country

+

Identify national situation

- Make inventories of POPs sources and estimate releases
- Assess regulatory and other mechanisms
- Assess needs and develop plans
- Assess capacity to manage or dispose of stockpiles or wastes

+

Determine solutions

- Designate a national focal point
- Set national priorities
- Identify needs for enhancing regulatory and other mechanisms
- Develop strategies, laws, and policy for POPs
- Promote information, awareness, and education measures
- Encourage research and development

+

Delineate specific national plans and programs

- Delineate plans that fully implement the convention
- Reflect national circumstances

=

What does a developing country need to do it?

Technical and financial assistance to develop the National Implementation Plan

Technical and financial assistance to build capacity

Technical and financial assistance to implement solutions

Table 7.2 National Implementation Plans

Article 15 of the Stockholm Convention contains most of the requirements for reporting, although other specific provisions are scattered throughout the treaty. Indeed, the presence of reporting requirements in so many sections of the convention, while by no means unprecedented, creates a rather confusing array of obligations. Although the volume of reporting may at first appear significant, many of the specific items are linked, reducing the overall burden and making it possible for the CoP or the secretariat to devise integrated formats and procedures.

The Stockholm Convention requires that all parties report on their production, use, import, and export of POPs; their use of any exemptions; and their overall implementation of the convention. Such reporting can be critical to the successful implementation of an MEA. For example, it is important that parties report their baseline and then, regularly, their subsequent production, use, and trade of such substances so that all parties and the CoP can understand the progress, or lack of it, in meeting the objectives of the MEA. It is also important that parties report on their measures to implement the treaty and their success so that all parties and the CoP can understand patterns of compliance, areas requiring additional attention, and areas that could require additional financial and technical assistance. Table 7.3 summarizes the key reporting obligations in the Stockholm Convention.

Article 11 contains specific provisions related to research, development, and monitoring. Under the convention, parties must monitor the presence and impact of POPs and candidate POPs; conduct research on POPs, candidate POPs, and alternatives to POPs; and attempt to develop and promote the use of environmentally safe alternatives to POPs.

The Stockholm Convention also contains specific provisions outlining what governments should do to facilitate the exchange of information among parties (Article 9) and to provide for or enhance public information, awareness, and education (Article 10). Developing and exchanging information among parties and with the public on the threats posed by POPs and on POPs management, elimination, and replacement can increase support for the goals, augment capacity to implement the convention, and further enhance an international atmosphere of co-operation and shared responsibility for global POPs policy. Through the convention, parties should exchange or facilitate the exchange and awareness of information relevant to the threat posed by POPs – including specific chemicals or products, the reduction or elimination of production, use, and release of POPs, and alternatives to POPs. Parties are encouraged to exchange such information directly (with one another, national stakeholders, and the public) via nationally designated focal points or through the secretariat, which will serve as a central clearing-house on POPs, including information provided by parties, intergovernmental organizations, and non-government organizations.

Table 7.3 Key Reporting Obligations in the Stockholm Convention

Production and Use of POPs
All parties must provide statistical data on total quantities of production of each of the chemicals listed in annexes A and B, or a reasonable estimate of such data, to the secretariat.
All parties producing a chemical listed in annexes A or B as a closed-system site-limited intermediate must provide data on the total production and use of the chemical, or a reasonable estimate thereof, to the secretariat. The secretariat shall make such notifications available to the CoP and to the public. Such production and use shall not be considered a production or use-specific exemption and shall cease after a ten-year period.
All parties requesting extension of production or use as a closed-system site-limited intermediate must submit a new notification regarding the nature of the closed-system site-limited process including the amount of any non-transformed and unintentional trace contamination of the POPs starting material in the final product to the secretariat. The period will be extended for an additional ten years unless the CoP, after a review of the production and use, decides otherwise.
All parties seeking exemption for chemicals occurring as constituents of articles manufactured or already in use before date of entry into force of relevant obligation must provide notification that a particular type of article remains in use within that party to the secretariat. The secretariat shall make such notifications publicly available.

Import and Export of POPs
All parties must provide statistical data on total quantities of import and export of each of the chemicals listed in annexes A and B, or a reasonable estimate of such data, to the secretariat.
All parties must provide to the secretariat a list of states from which each substance was imported and to which each was exported.
All parties must provide annual certification for the countries to which it has exported chemicals within sixty days of receipt to the secretariat.

Chemical Use Exemptions
All states requiring exemptions (except DDT) must, upon becoming a party, register for one or more types of specific exemptions listed in annexes A or B by means of notification in writing to the secretariat. Registrations of specific exemptions shall expire five years after the date of entry into force of the convention with respect to a particular chemical, unless an earlier date is indicated in the register by a party.
The CoP may, upon request from the party concerned, decide to extend the expiry date of a specific exemption for a period of up to five years. The review of registration shall be carried out on the basis of all available information and a report submitted by the requesting party justifying the continuing need for registration of exemption.
All parties requiring DDT exemptions must provide information on the amount used, conditions of use, and relevance to disease management strategy to the secretariat and the World Health Organization every three years in a format to be decided by the CoP and WHO.

Implementation of the Convention
All parties shall develop and transmit a plan for the implementation of its obligations under the convention to the CoP within two years of the date on which the convention enters into force. This implementation plan is to be reviewed and updated regularly in periods to be determined by the CoP.
All parties shall provide to the CoP a report every five years on progress in eliminating PCBs.

Effectiveness evaluations

The Stockholm Convention requires that the CoP formally review the effectiveness of the convention. Without such evaluations, even stringent provisions on monitoring and reporting can be ineffectual. The first review will take place four years after the Stockholm Convention enters into force, and the CoP will determine intervals for subsequent evaluations (Article 16(1)). Reviews will be conducted using available scientific, environmental, technical, and economic information, including reports and other monitoring information concerning the presence, transport, and impact of POPs (Article 16(2) and Article 16(3)(a)); national reports submitted pursuant to Article 15 on reporting (Article 16(3)(b)); reports on national implementation plans; and information provided pursuant to the non-compliance procedures to be established by the CoP under Article 17 (Article 16(3)(c)).

The CoP will also specifically review the effectiveness of the financial mechanism with a view to improving its operation and impact (Article 13(8)). The first of these evaluations will occur no later than the second meeting of the CoP. The reviews will examine the extant ability of the financial mechanism to help eligible parties implement the treaty, the level of funding, the operation of the GEF or other constituent institution(s), and the overall effectiveness of the financial mechanism. Requiring such reviews was an important component of the final compromise on the measures to provide financial assistance to help parties implement the convention.

Amending the convention

For an environmental treaty to be effective over time, parties must be able to amend it and to do so quickly enough to achieve the intended impact on protecting the environment and human health. And, as shown through the expansion of the Montreal Protocol and its consequential success in protecting the ozone layer, the methods chosen to expand the strength and scope of an environmental treaty can be critical to its success or failure.

It can be difficult, however, to craft effective amendment procedures because governments often have competing concerns in this area – e.g., concerns for rapid flexibility versus those for strict sovereign control of the amendment procedure; concerns for a strong precautionary approach versus those for a strong risk-analysis approach; or concerns for continual expansion of environmental protection versus those for stable expectations for governments and corporations.

The Stockholm Convention addresses this issue with different amendment procedures for different parts of the convention. The most traditional and slowest amendment procedures are reserved for amendments to the convention articles and the list of chemicals subject to the control measures – texts that require the most careful consideration before they are changed. Where speed of implementation could be important to environmental protection and where change would not have an impact on sovereign authority, the amendment procedure is more flexible. Six different procedures exist for amending or otherwise altering the contents of the Stockholm Convention. They are found primarily in articles 21 and 22, although related elements are contained in Article 25(4) and Article 4(9).

Amending the convention: The main body

Any party may propose an amendment to the preamble or the articles (including adding articles) in the convention. The parties are to make every effort to reach agreement by consensus; however, the amendment can, as a last resort, be adopted by a three-fourths majority of the parties present and voting (Article 21(3)). Once an amendment is approved by the COP, a party must formally ratify it to become a party to it, and the amendment will enter into force only after it has been ratified by at least three-fourths of the parties to the convention (Article 21(5)).

Thus, a party is bound by an amendment to the main body of the convention only if it specifically and intentionally ratifies or otherwise approves the amendment. This type of opt-in procedure preserves the maximum possible sovereign authority over the amendment process and is the most common and most traditional amendment procedure found in international treaties.

Amending the convention: Adopting additional annexes

Any party may propose the addition of an annex that is limited to procedural, scientific, technical, or administrative matters (Article 22(2)). The parties are to make every effort to reach agreement by consensus, but an annex can be adopted by a three-fourths majority vote of the parties present and voting. The additional annex will enter into force one year after its approval by the COP for all parties except those that have submitted a notification of non-acceptance (Article 22(3)(c)). That is, any party unable to accept an additional annex approved by the COP can notify the United Nations in writing within one year of the adoption of the additional annex (Article 22 (3)(b)).

Thus, a party is bound by an additional annex unless that party specifically, intentionally, and formally notifies the United Nations that it will not be so bound. This opt-out procedure preserves sovereign authority over the amendment process – states become parties to the amendment only if they choose to – while also ensuring that additional annexes concerned with purely procedural, scientific, technical, or administrative matters can enter into force in a timely manner.

Amending the convention: Adding new chemicals or otherwise amending Annex A, B, or C

Any party may propose an amendment to Annex A, B, or C, applying the fifteen-step procedure for developing the amendment (Article 8). The parties are to make every effort to reach agreement by consensus, but the amendment can be adopted by a three-fourths majority vote of the parties present and voting. Once approved by the COP, the amendment (including the addition of a new chemical) will enter into force one year after it arrives at the treaty depositary. At that time, it will be binding on all parties except those that have made either of two choices.

A party's choice depends on whether that party had previously made a specific declaration in its instrument of ratification, acceptance, approval, or accession of the convention. Paragraph 4 of Article 25 provides that, when ratifying the convention, a party may declare that it will not be bound by any amendment to Annex A, B, or C unless it formally and specifically ratifies or otherwise approves the amendment (Article 22(5) and Article 25(4)).

Choice 1: An amendment to Annex A, B, or C will not enter into force for a party that has made such written declaration until ninety days after the party deposited its instrument of ratification, acceptance, approval, or accession for the specific amendment.

Choice 2: For a party that made no declaration, an amendment will enter into force one year from the date of notification by the depositary of the adoption of the amendment and will be binding unless a party has submitted a notification of non-acceptance concerning that amendment (Article 22(3)(c)). A party may withdraw a previous notification of non-acceptance at any time and the relevant annex would then enter into force for that party (Article 22(3)(b)).

Thus, the procedure for amendments to Annex A, B, or C is either opt-in or opt-out, depending on a choice each party made when ratifying the convention. This process allows amendments to Annex A, B, or C to take effect in the shorter "opt-out" time frame that allows for more effective protection of the environment and human health and also preserves the "opt-in" option for states with internal constitutions or other concerns that demand such procedures.

Amending the convention: Amending Annex D, E, or F

Any party may propose an amendment to Annex D, E, or F – the technical annexes concerning the criteria for evaluating additional POPs, developing risk profiles, and considering socio-economic factors (Article 22(5)(a) and Article 21(1)). A consensus decision by the COP to amend these annexes will be communicated formally to the parties and will enter into force for all parties on the date specified in the decision (Article 22(5)(c)).

Thus, all parties are bound by an amendment to Annex D, E, or F immediately and automatically on the date specified in the decision. This type of amendment preserves sovereign authority over the amendment process because the decision to amend the annexes must be reached by consensus. At the same time, having an amendment enter into force for all parties on the date specified ensures that the changes can enter into force as soon as possible.

Amending the convention: Adopting an additional annex or an amendment to an annex that is itself related to an amendment to the convention

Should the COP approve an additional annex or an amendment to an annex that is itself related to an amendment to a non-annex portion of the convention, the additional annex or the amendment cannot enter into force until the related amendment to the convention enters into force (Article 22(6)). The adoption procedures are those normally applied for the annex or amendment and the specific entry-into-force requirements would still have to be met. However, the final entry into force would be delayed until the related amendment to the convention entered into force – that is, until the ninetieth day after the date of deposit of instruments of ratification, acceptance, or approval by at least three-fourths of the parties (Article 22(6) and Article 21(5)). This linking procedure is a relatively simple method to ensure internal consistency within the convention.

Amending the convention: Eliminating unused specific exemptions

Any state, on becoming a party, may register for any of the specific exemptions listed in Annex A or Annex B (Article 4(3)). All such registrations expire after five years unless a party indicates an earlier date in the *Register of Specific Exemptions* or the COP grants an extension (Article 4(4)). When there are no longer any parties registered for a particular type of specific exemption, no new registrations will be allowed for that exemption (Article 4(9)). Although not technically an amendment procedure (as the actual text of the convention would not change), this represents a sixth procedure for changing the effective terms of the convention. This change is automatic and immediate and allows the convention to prevent the intentional or inadvertent use of a truly obsolete POP.

Treaty administration

The Stockholm Convention requires fifty parties to enter into force (Article 26). Once it does, the "Conference of the Parties" (COP) will be the main decision-making authority of the convention (Article 19). The COP will act to administer, oversee, review, and evaluate the implementation of the convention and consider its amendment or expansion. Article 19 contains an expansive list of specific functions and powers. The COP will be made up of states party to the convention. Other states, international organizations, and non-government organizations may attend the CoP as observers but will not have decision-making authority. The CoP will meet at regular intervals and, as much as possible, will make decisions by consensus. If consensus is not possible, the CoP could hold a vote according to rules of procedures that it will establish at its first meeting. Each state party to the convention will have one vote (Article 23(1)).[22]

UNEP will serve as the secretariat for the Stockholm Convention (Article 20(3)). Article 20 delineates a number of specific activities to be undertaken by the secretariat: it will act as the administrator of the convention, serve as a conduit for authoritative information, make necessary arrangements for the functioning of the CoP, and facilitate implementation of the convention. The secretariat will act on behalf of the parties and be overseen by the CoP. UNEP has assigned the task of acting as interim secretariat for the Stockholm Convention to UNEP Chemicals in Geneva.

This pattern of treaty administration ensures that the parties themselves will control the decision-making process, implementation, and future development of the convention. As such, it follows the precedent set in many other MEAs, including the 1987 Montreal Protocol on Substances that Deplete the Ozone Layer, the 1992 Convention on Biological Diversity (CBD), and the 1998 Rotterdam Convention on the Prior Informed Consent (PIC) Procedure for Certain Hazardous Chemicals and Pesticides in International Trade.[23]

In addition to the provisions on the CoP (Article 19) and the secretariat (Article 20), the Stockholm Convention contains a number of other articles that address treaty administration. These provisions, found in nearly identical form in most MEAs, include non-compliance (Article 17); settlement of disputes (Article 18); right to vote (Article 23); signature (Article 24); ratification (Article 25); entry into force (Article 26); reservations (Article 27); withdrawal (Article 28); depositary (Article 29); and authentic texts (Article 30).

UNFINISHED BUSINESS

No global environmental agreement emerges from its initial negotiation and ratification as a finished regime. Indeed, most MEAs contain specific instructions to their CoPs to finish important components of the treaty.[24] Reasons vary, but having such "unfinished business" usually reflects a decision by the negotiators that particular elements of the treaty would be best left to the parties themselves, although inadequate time or an inability to resolve differences over contentious issues can also be a factor.

The following table lists elements of the Stockholm Convention that the convention text specifically instructs the CoP to complete, most at its first or second meeting.[25] The Stockholm Convention exists, but many important elements of its final design rest with the parties themselves.

Table 7.4
Unfinished Aspects of the Stockholm Convention

Treaty Component[26]	*Required Action*	*Required Date*
Control measures	COP will decide upon its review process for entries in the Register of specific exemptions (Article 4, paragraph 5)	First COP meeting
Control measures	COP will adopt guidelines on products and processes to prevent the formation and release of the chemicals listed in Annex C (Article 5, paragraph 1(c))	Not specified
Control measures	COP should adopt guidelines on best available techniques and best environmental practices for prevention and release reduction measures (Article 5, paragraph 1(d))	Not specified
Control measures	COP will co-operate with the appropriate bodies of the Basel Convention to establish levels of destruction and irreversible transformation necessary to ensure that the characteristics of POPs as specified in paragraph 1 of Annex D are not exhibited (Article 6, paragraph 2)	Not specified
Control measures	COP will co-operate with the appropriate bodies of the Basel Convention to determine the concentration levels of the chemicals listed in annexes A, B, and C to define low POPs content, then develop the methods that constitute environmentally sound disposal (Article 6, paragraph 2)	Not specified
Control measures	COP will evaluate the continued need for the procedure contained in paragraph 2(b) of Article 3, including consideration of its effectiveness (Article 19, paragraph 7)	Third COP meeting
Control measures	COP should decide a format for each party using DDT to report information on the amount used, the conditions of such use, and its relevance to that party's disease management strategy (Annex B, part II, paragraph 4)	Not specified
Control measures	COP may develop guidance with regard to best environmental practices (Annex C, part V, section c)	Not specified
Financial and technical assistance	COP will provide guidance to developed country parties relating to technical assistance for capacity-building (Article 12, paragraph 3)	Not specified
Financial and technical assistance	Parties will establish arrangements for the purpose of providing technical assistance and promoting the transfer of technology to developing country parties and parties with economies in transition relating to the implementation of this convention (Article 12, paragraph 4)	Not specified
Financial and technical assistance	COP will provide guidance to parties for the purpose of providing technical assistance and promoting the transfer of technology to developing country parties and parties with economies in transition (Article 12, paragraph 4)	Not specified
Financial and technical assistance	COP will decide upon the entities, including existing international entities, to be entrusted with the mechanism for the provision of adequate and sustainable financial resources to developing country parties and parties with economies in transition (Article 13, paragraph 6)	Not specified

Table 7.4 (*Continued*)

Treaty Component	*Required Action*	*Required Date*
Financial and technical assistance	CoP will adopt appropriate guidance to be provided to the financial mechanism for the provision of adequate and sustainable financial resources to developing country parties and parties with economies in transition (Article 13, paragraph 7)	First CoP meeting
Financial and technical assistance	CoP will agree with the entity or entities participating in the financial mechanism upon arrangements to give effect to the financial mechanism (Article 13, paragraph 7)	First CoP meeting
Implementation plans	CoP should decide the manner in which parties will review and update their implementation plans (Article 7, paragraph 1(c))	Not specified
Implementation plans	CoP will decide the intervals and format of reporting by parties on implementation measures and the effectiveness of these measures (Article 15, paragraph 3)	First CoP meeting
Monitoring, reporting, and information exchange	CoP will initiate establishment of arrangements to provide itself with comparable monitoring data on the presence of the chemicals listed in annexes A, B, and C as well as their regional and global environmental transport (Article 16, paragraph 2)	First CoP meeting
Monitoring, reporting, and information exchange	CoP should specify intervals for receiving reports from parties on the results of monitoring activities on a regional and global basis (Article 16, paragraph 2(c))	Not specified
Effectiveness evaluation	CoP will decide the intervals at which to evaluate the effectiveness of the convention (Article 16, paragraph 1)	Not specified
Treaty administration	CoP will develop and approve procedures and institutional mechanisms for determining non-compliance with the provisions of this convention and for the treatment of parties found to be in non-compliance (Article 17, paragraph 1)	As soon as practicable
Treaty administration	CoP should adopt procedures for arbitration in an annex (Article 18, paragraph 2)	As soon as practicable
Treaty administration	CoP will adopt additional procedures relating to the conciliation commission to be included in an annex (Article 18, paragraph 6)	No later than second meeting
Treaty administration	CoP should decide regular intervals at which to hold ordinary meetings of the Conference of the Parties (Article 19, paragraph 2)	Not specified
Treaty administration	CoP will by consensus agree upon and adopt rules of procedure and financial rules for itself and any subsidiary bodies, as well as financial provisions governing the functioning of the secretariat (Article 19, paragraph 4)	First CoP meeting

Table 7.4 (*Continued*)

Treaty Component	*Required Action*	*Required Date*
Treaty administration	CoP will establish a subsidiary body to be called the Persistent Organic Pollutants Review Committee for the purposes of performing the functions assigned to that committee by this convention (Article 19, paragraph 6)	First CoP meeting
Treaty administration	CoP will decide on the terms of reference, organization, and operation of the POPs Review Committee (Article 19, paragraph 6(b))	First CoP meeting
Treaty administration	CoP will adopt rules of procedure for admission and participation of observers at meetings of the Conference of the Parties (Article 19, paragraph 8)	Not specified

NOTES

1 I drew on a number of sources while developing this section. These include (and for further information, please consult): UNEP Chemicals, POPs Information Kit (from which I borrowed key organizational elements) http: //irptc.unep.ch/pops/POPs_Inc/press_releases/infokite.html; *Earth Negotiations Bulletin* and *Linkages*, especially http: //www.iisd.ca/chemical/chemicalsintro.html; Rune Lönngren, *International Approaches to Chemicals Control – A Historical Overview* (Sweden: National Chemicals Inspectorate 1992), and David Downie, "War on POPs: The 2001 Stockholm Convention on Persistent Organic Pollutants," unpublished manuscript under review.

2 For the text of *Agenda 21*, see http: //www.unep.org/Documents/Default.asp?DocumentID=52. For Article 19 of *Agenda 21*, see http: //www.unep.org/Documents/Default.asp?DocumentID=52&ArticleID=67 For the text of the resolution establishing the IFCS, see http: //www.who.int/ifcs/fs_res1.htm. For more information on the IFCS, see http: //www.who.int/ifcs/.

3 For more information on the IOMC, see http: //www.who.int/iomc/.

4 For information on PIC, including the Rotterdam Convention introduced below, see http: //www.pic.int/. For the text of the London Guidelines, see http: //irptc.unep.ch/pic/longuien.htm. For the full text of the FAO Code, see http: //www.fao.org/waicent/FaoInfo/Agricult/AGP/AGPP/Pesticid/Code/PM_Code.htm.

5 For additional and more recent information on the convention, see http: //www.pic.int/.

6 For the text of UNEP GC Decision 18/32, see http: //irptc.unep.ch/pops/indxhtms/gc1832en.html. For more information on the IOMC, see http: //www.who.int/iomc/. For more information on the IPCS, see

http: //www.who.int/pcs/. For meeting reports and other official documents on the IFCS ad hoc Working Group on Persistent Organic Pollutants, see http: //www.who.int/ifcs/pop_home.htm.

7 For the Washington Declaration, see http: //www.gpa.unep.org/documents/gpa/wadeclaration/default.htm. For the official meeting report, see http: //irptc.unep.ch/pops/indxhtms/manwash.html. For additional documents from the GPA and Washington Conference, see http: //www.gpa.unep.org/documents/default.htm#gpa and http: //irptc.unep.ch/pops/WashConf.html.

8 Two meetings took place in June 1996: the Intergovernmental Experts Meeting on POPs and the IFCS Ad Hoc Working Group. For the official report of the Experts Meeting, see http: //www.who.int/ifcs/pop_home.htm#Experts. For additional documents, see http: //irptc.unep.ch/pops/ifcsvarious.html#EM. For the official report of the Ad Hoc Working Group, see http: //www.who.int/ifcs/pop_home.htm#WG_POPs or http: //irptc.unep.ch/pops/indxhtms/manwgrp.html. For additional working group documents, see http: //irptc.unep.ch/pops/ifcsvarious.html#AHWG.

9 For the full text of UNEP GC Decision 19/13C, see http: //irptc.unep.ch/pops/gcpops_e.html. The Governing Council Decision 19/13C called for work on developing and sharing information; evaluating and monitoring response strategies; determining alternatives to POPs; identifying and developing an inventory of PCBs; quantifying the available capacity for incinerating or destroying unwanted stocks; and identifying sources of dioxin and furan emissions. The decision also recommended that the INC give due consideration to the recent development by the UN Economic Commission for Europe (UNECE) of a POPs protocol under the Convention on Long-range Transboundary Air Pollution.

10 The text of resolution WHA50.13 of the 1997 World Health Assembly, 5–14 May 1997, Geneva, is reprinted in UNEP/POPs/INC.1/INF/6 at http: //irptc.unep.ch/pops/POPs_Inc/INC_1/inf6.htm.

11 These workshops included St. Petersburg, Russian Federation, 1–4 July 1997; Bangkok, Thailand, 25–28 November 1997; Bamako, Mali, 15–18 December 1997; Cartagena, Colombia, 27–30 January 1998; Abu Dhabi, United Arab Emirates, 7–9 June 1998; Kranjska Gora, Slovenia, 11–14 May 1998; Lusaka, Zambia, 17–20 March 1998; and Puerto Iguazú, Argentina, 1–3 April 1998. For official proceedings of the workshops, including others held after the start of the POPs INCs, see http: //irptc.unep.ch/pops/newlayout/prodocas.htm. For background on the workshops, see http: //irptc.unep.ch/pops/newlayout/wkshpintro.htm.

12 The negotiations included INC-1 in Montreal, 29 June to 3 July 1998; CEG-1 in Bangkok, 26–30 October 1998; INC-2 in Nairobi, 25–29 January 1999; CEG-2 in Vienna, 14–18 June 1999; INC-3 in Geneva, 6–11 September 1999; INC-4 in Bonn, 20–25 March 2000; a formal Inter-sessional Meeting on Financial Resources and Mechanisms in Vevey, Switzerland, 9–21 June 2000; INC-5 in Johannesburg,

4–9 December 2000; and the Conference of Plenipotentiaries in Stockholm, 23–24 May 2001. The Stockholm meeting also included a one-day negotiating session that finalized the important resolutions adopted at the Stockholm Conference. For official reports and other documents from the negotiations, see http: //irptc.unep.ch/pops/newlayout/negotiations.htm. For journalistic summaries see http: //www.iisd.ca/linkages/chemical/index.html or http: //irptc.unep.ch/pops/newlayout/press_items.htm. For specific numbers of delegates and countries, see David Downie et al., "Could Where We Stand Depend on Who Sits? A Preliminary Analysis of Participation Patterns in Global Environmental Negotiations" (Unpublished manuscript, 2002). For lists of the countries, intergovernmental organizations, and non-government organizations that participated in each of the five INCs that produced the Stockholm Convention, see the official reports of the INC meetings (UN Document UNEP/POPs/INC.5/7, UNEP/POPs/INC.4/5, UNEP/POPs/INC.3/4, UNEP/POPs/INC.2/6, and UNEP/POPs/INC.1/7) as well as the lists of participants from each INC at http: //irptc.unep.ch/pops/newlayout/negotiations.htm).

13 For references, examples, and further information, see UNEP Governing Council Decision 19/13c in 1997 and Decision 21/4 in 2001 (http: //irptc.unep.ch/pops/gcpops_e.html). For WHO, see Resolution WHA50.13 of the 1997 World Health Assembly, 5–14 May 1997 reprinted in UNEP/POPs/INC.1/INF/6 at http: //irptc.unep.ch/pops/POPs_Inc/INC_1/inf6.htm. For IFCS, see paragraph 6(a) in document IFCS/EXP.POPs/Report.1, reprinted in UNEP/POPs/INC.1/INF/11 (http: //irptc.unep.ch/pops/POPs_Inc/INC_1/inf11.htm). For the GEF, see paragraph 5 of UNEP/POPs/INC.5/7, *Report of the Intergovernmental Negotiating Committee for an International Legally Binding Instrument for Implementing International Action on Certain Persistent Organic Pollutants on the Work of Its Fifth Session,* Johannesburg, 4–9 December 2000 (http: //irptc.unep.ch/pops/POPs_Inc/INC_5/meetdocen.htm). For examples of concerns expressed by other international bodies, see *Global Programme of Action for the Protection of the Marine Environment from Land-based Activities*, in UNEP/POPs/INC.1/INF/9, and *Assessment Report on Certain Persistent Organic Pollutants Prepared by the International Programme on Chemical Safety*, in UNEP/POPs/INC.1/INF/10 (http: //irptc.unep.ch/pops/POPs_Inc/INC_1/Inc1_docs.html).

14 The ratification procedures in the United States and some countries of Europe were slowed by the attacks on New York and Washington on 11 September 2001. The numbers of signatories and parties to the convention are accurate as of 3 September 2002. For a current list, see http: //www.chem.unep.ch/sc/documents/signature/signstatus.htm.

15 To limit potential abuse of this exemption, parties using DDT must develop and implement, as part of their IPs, an action plan that includes (i) development of regulatory and other mechanisms to ensure that DDT use is restricted to disease vector control; (ii) implementation of suitable alternative products, methods and strategies,

including resistance management strategies to ensure the continuing effectiveness of these alternatives; (iii) measures to strengthen health care and to reduce the incidence of the disease (Annex B, Part II, paragraph 5(a)).

16 The principal trade provisions are contained in paragraph 2 of Article 3, but interrelated provisions are contained elsewhere in the convention.

17 UNEP Governing Council Decision 19/13c. See http: //irptc.unep.ch/pops/gcpops_e.html.

18 Persistence in different media with numerical cut-off values; bioaccumulation in organisms with numerical cut-off values; potential for long-range environmental transport; and evidence of toxicity or other adverse effects.

19 Sources, hazard assessment, environmental fate, monitoring data, local and long-range exposures, existing risk evaluations and related assessments, and the chemical's status under international conventions.

20 IP requirements go by different terms and have somewhat different requirements in existing MEAs. For example, the Montreal Protocol calls such requirements "Country Programmes." In the UNFCCC they are "National Communications"; in UNCCD, "Action Programmes"; in CBD, "National Biodiversity Strategy and Action Plans" and the related reporting requirements are called "national reports"; in the CLRTAP POPs Protocol and Basel convention there are no specific referents but IP-like provisions exist. For detailed analyses of the IP requirements in other MEAs, as well as the indicative points listed in this section, see UNEP/POPs/INC.3/INF/3, "Scope, content and development process of national action plans under the auspices of existing multilateral environmental agreements," http: //irptc.unep.ch/pops/POPs_Inc/INC_3/inc-english/Inc3_docs.html.

21 Personal observation and discussion with delegates during the POPs negotiations.

22 Regional economic integration organizations (REIOs), such as the European Union, may also become parties to the convention. An REIO will have a number of votes equal to the number of its member states but will vote only if its member states do not exercise their right to vote (Article 23(2)).

23 For articles dealing with the Conference of the Parties in other conventions, see Article 11 of the 1987 Montreal Protocol (http: //www.unep.ch/ozone/Montreal-Protocol/Montreal-Protocol2000.shtml#_Toc483027807); Article 15 of the 1989 Basel Convention (http: //www.unep.ch/basel/text/con-e.htm); Article 7 of the 1992 UNFCCC (http: //www.unfccc.de/resource/conv/conv_019.html); Article 23 of the 1992 Convention on Biological Diversity (http: //www.biodiv.org/convention/articles.asp?lg=0&a=cbd-23); Article 22 of the 1994 United Nations Convention to Combat Desertification (http: //www.unccd.int/convention/text/pdf/conv-eng.pdf; Article 18 of the Rotterdam PIC Convention (http: //www.pic.int/en/ViewPage.asp?id=104#18Article).

24 For example (and please note these lists are indicative rather than exhaustive), the "unfinished business" in the PIC Convention includes, but is not limited to, Article 17 on non-compliance procedure and Article 18(6)(b) on the terms of reference,

organization, and operation of the Chemical Review Committee (http: //www.pic.int/en/ViewPage.asp?id=104#18Article). In the Montreal Protocol it includes Article 8 on non-compliance and Article 10(3) on technical assistance (http: //www.unep.ch/ozone/Montreal-Protocol/Montreal-Protocol2000.shtml). In the United Nations Framework Convention on Climate Change (UNFCCC) it includes Article 7(3) on rules of procedure; Article 11 on the financial mechanism; and Article 12(7) on the mechanism for providing developing country parties with technical and financial assistance (http: //www.unfccc.de/resource/conv/index.html). In the United Nations Convention on Biological Diversity (UNCBD) it includes Article 18(3) on the clearing-house mechanism; Article 20(2) on the guidelines for financial support; Article 21(1) on the institutional structure for the financial mechanism; and sections of Article 23 on the CoP (http: //www.biodiv.org/convention/articles.asp).

25 The table highlights only those parts of the convention that the CoP must finish, not aspects of the convention that the CoP or the parties are required to implement, such as the effectiveness evaluation or developing and submitting IPs.

26 The table divides the most important unfinished business by "Treaty Component" (Treaty components are the eight categories used to outline the convention in this section of the Users Guide), the action required by the CoP, and the date by which the CoP is required to act.

8
The Stockholm Convention in the Context of International Environmental Law

NIGEL BANKES

INTRODUCTION

This chapter considers the Stockholm Convention on Persistent Organic Pollutants[1] (POPs) within the broader context of other multilateral environmental agreements (MEAs)[2] negotiated over the last couple of decades and focuses on three themes.

The first theme is recognition of principles. What are the principles underlying the Stockholm Convention? How are those principles made operational? One of the last issues to be settled in the early morning hours of the final negotiating session in Johannesburg was the way in which the text was to recognize and incorporate the precautionary principle.[3]

The second theme is that of relationships. What is the relationship between the global Stockholm Convention and the regional POPs protocol developed under the auspices of the UN Economic Commission for Europe?[4] What is the relationship between the Stockholm Convention and the other global MEAs, especially those with a focus on chemicals, the Rotterdam Convention and the Basel Convention?

The third theme is effectiveness and compliance. What processes does the convention put in place for monitoring its effectiveness and securing compliance?

This chapter reaches four main conclusions. First, the Stockholm Convention is an example of a successful set of MEA negotiations conducted relatively expeditiously. Second, one of the strengths of the text is the extent to which it not only articulates guiding principles but also seeks to operationalize them. This is true not only of precaution but also of another critical principle: common but differentiated responsibilities (CDR). Although the parties did not devote as much negotiating time to an explicit parsing of the CDR principle as they did to the precautionary principle, it is clear that a successful outcome to the negotiations depended on the adequacy of the provisions for financial and technical assistance in the text. The third conclusion is that the negotiators took considerable care to address the implementation of the convention. Although much remains to be done by the Conference of the Parties (COP) once the agreement enters into force, the necessary framework is now in place. Finally, the trade–environment issues were much easier to deal with in this convention than they have proven to be in other recent MEA negotiations, largely because the focus of this agreement was not trade but rather the banning of production and use of listed chemicals.

Working through the relationship between the POPs convention and the Basel Convention proved more difficult than it should have been, but compromises were found. One of the puzzles of the negotiations was the extent to which (apart from the financial and technical assistance provisions) some of the most difficult issues arose among developed country parties rather than along the anticipated north–south lines. This was primarily the case for the negotiations on both precaution and the Basel/POPs interface.

Following are accounts of how the Stockholm Convention treats particular issues and comparisons with how the issues were treated in other recent MEAs.

PRINCIPLES AND ORGANIZING IDEAS UNDERPINNING THE CONVENTION

What are principles?

Legal systems include both rules and principles. Rules operate in an all-or-nothing manner and prescribe certain conduct for members of the relevant community. Principles, while still normative, operate at a higher level of generality, stating "a reason that argues in one direction but does not necessitate a particular decision."[5] Principles help us organize rules thematically. They also guide the exercise of discretion and help resolve hard cases where it may be difficult to decide whether to apply rule A or rule B. There exist two especially authoritative sources of principles relevant to international environmental law, the Stockholm Principles of 1972[6] and the Rio Declaration of 1992.[7]

Even the most cursory review of the convention and the record of the negotiations shows the extent to which principles informed the negotiations.[8] This may be particularly evident in the preamble to the convention. Preambles traditionally record the shared understandings on which an agreement is based. While the preamble was not especially hard to negotiate, having been compiled by the chair based on country submissions, its text refers to a number of ideas or concepts that are often called "principles." That this term was not always used in the convention illustrates, perversely enough, the significance that negotiating parties attached to the label. Unlike the terms "concept," "idea," or "approach," the term "principle" is freighted with normative significance.

Thus paragraph 8 refers to "precaution," paragraph 10 to the *sic utere tuo* idea ["no state should use its territory in such a way as to cause significant harm to another state"], paragraph 16 to the related idea of the avoidance of harm, paragraph 13 to the concept of "common but differentiated responsibilities," and paragraph 18 to the "approach that the polluter should, in principle, bear the cost of pollution."

But while negotiating blocs like to be able to base their negotiating positions (in these and other negotiations) on agreed-upon principles, there are limits to the efficacy of principles. Because principles, unlike rules, do not require particular decisions or behaviour, an effective convention needs to make principles operational through the adoption of more prescriptive and determinate rules.

The following subsections deal successively with precaution, the CDR principle, and the duty not to harm and the related principle of liability.

Precaution

The idea of precaution suggests that scientific certainty should not be required before governments act to protect their citizens or the global community from serious and environmental harm and that before persons introduce new, potentially hazardous substances or engage in new, potentially hazardous activities, the onus should be on them to demonstrate that the new activity or substance will not be harmful.[9] Beyond these basic ideas, the terms of the debate on precaution are contentious but well known and often generate more heat than light. Is the concept a principle or just an approach? What is the correct articulation of precaution – is it the Rio 15 formulation[10] or some other formulation?[11] What does an application of precaution require in any defined scenario? What precisely must *X* prove before introducing new substance *Y*? Does the inclusion of the precautionary principle in a convention allow parties to modify or ignore their obligations under other instruments, e.g., trade instruments? These are some of the issues that have bedevilled recent MEA negotiations, in particular the Cartagena Protocol on Biosafety and the POPs convention.

In principle, the precautionary negotiations for the POPs agreement should have been easier than those for biosafety. The POPs agreement is primarily concerned with multilateral precautionary measures and only peripherally with trade.[12] By contrast, trade was front and centre for biosafety; the precautionary issues in that agreement primarily concerned the right of an importing party to take unilateral steps when it perceived a risk to its biological diversity. Yet in the end, the precautionary discussions were very difficult for the POPs agreement, perhaps primarily because of a split between the pragmatists who could agree to operationalize the concept of precaution and those parties (especially from the European Union (EU)) for whom specific "precautionary principle" formulations held continuing political, if not legal, importance.

How did the parties deal with "precaution" in the POPs convention? The concept of precaution is referred to no fewer than four times in the convention and its annexes and may also have informed, to some degree, the provisions on new chemicals (paragraphs 3 and 4 of Article 3) and monitoring (Article 16).[13] The first reference occurs in paragraph 8 of the preamble, where the foundations of the convention are outlined:

Acknowledging that precaution underlies the concerns of all the Parties and is embedded within this Convention,

The second mention follows immediately thereafter in the Objective article, Article 1:

Mindful of the precautionary approach as set forth in Principle 15 of the Rio Declaration on Environment and Development, the objective of this Convention is to protect human health and the environment from persistent organic pollutants.

The first phrase, while part of an operative clause of the convention, actually serves a preambular function for the protection objective that forms the real subject matter of the article. Consequently, not until Article 8, which deals with the listing of new chemicals, is it possible to speak of operationalizing precaution in any concrete manner.

While the initial list of banned and restricted chemicals addresses the present, Article 8 and the procedure for adding new chemicals is oriented to the future. A precautionary approach could be reflected in the drafting of Article 8 in a number of different ways; one might elect simply to set the bar so low that it would be relatively easy – through the statement of the technical criteria or through the procedure – to add chemicals of concern. For example, the agreed procedure might allow for a less technical and more politicized process. Such a balance would support the view that the determination of how much risk or level of protection an individual or a society demands (i.e., what to do in the event of scientific uncertainty) is ultimately a political question and not a scientific question.[14] Another approach might be to make explicit reference to the concept of precaution either throughout or at different places in the article. In the end, the article seems to have adopted most, if not all, of these techniques.

Depending upon how one views Article 8, there are five stages to the scheme for adding new chemicals, and the central actors are the Persistent Organic Pollutants Review Committee (POPRC, or "the committee") and the Conference of the Parties (CoP); subsidiary roles are accorded to the parties, the secretariat, and observers. The first stage requires a party to nominate a chemical for listing and to provide the information and evaluative criteria called for by Annex D.[15] The secretariat will then forward the proposal to the POPRC for the second stage. The committee reviews the proposal in light of the Annex D criteria and determines either that the screening criteria have been fulfilled or that the proposal should be set aside. If the screening criteria are deemed fulfilled, the secretariat invites all parties and observers to submit information that will permit the preparation of a risk profile.

The third stage requires the POPRC to draft a risk profile based upon the foregoing submissions. The draft is to be made available to parties and observers for technical comments that will help guide the committee in deciding whether or not to proceed with the proposed listing. The POPRC must assess whether or not "the chemical is likely, as a result of its long-range environmental transport, to lead to significant adverse human health and/or environmental effects such that global action is warranted."

The fourth stage involves the committee preparing a risk management evaluation in accordance with Annex F and the submissions of parties and observers. The evaluation should include an analysis of possible control measures. Following that, the committee recommends whether the chemical should be considered by the CoP for listing.

The final stage is the decision to list the chemical and to specify related control measures in annexes A, B, and/or C. This engages the procedure to propose, adopt, and have enter into force proposals to amend the annexes to the convention stipulated in Article 21 and especially Article 22. The CoP plays the central role in this procedure (described in greater detail below), but Article 8(9) purports to instruct the CoP, "taking due account of the recommendations of the Committee, including any scientific uncertainty," to decide "in a precautionary manner" whether to list the chemical.

Although the foregoing represents the basic scheme, the final round of negotiations resulted in the addition of several twists that permitted a chemical to move from one stage to the next even though it might not have passed the technical thresholds established by the annexes and applied by the committee. Thus, paragraph 5 provides that where the committee determines that a chemical has not met the screening criteria, any party may re-submit a proposal and, if the proposal is again rejected by the committee, the party may challenge the committee's decision at the next meeting of the CoP. The CoP may then decide to allow the proposal to proceed to the next stage.[16] Paragraph 8 contemplates a similar re-submission and challenge procedure where the committee, having developed and evaluated the risk profile, has decided that the proposal should not proceed.

Thus, while Article 8 contains only one explicit reference to precaution (Article 8(9))[17]and the related idea of scientific uncertainty, it seems fair to conclude that the notion of precaution suffuses the drafting of the entire article. We can see all of the different techniques being brought to bear in an effort to ensure that the case against a chemical does not have to be open-and-shut before the chemical can be listed. Although this fact is perhaps most obvious in relation to the procedures for re-submission and challenge, we can also see it in the criteria developed in the annexes and the instructions to the committee for its application of the annex. For example, much of Annex D (which stipulates the evaluative criteria) is framed in the alternative, thereby allowing for considerable flexibility in application.[18] Furthermore, Article 8(3) explicitly instructs the committee to act in "a flexible and transparent way" in carrying out its evaluation. This phrase originated with the report of the Criteria Experts Group (the CEG Report) established to advise the Intergovernmental Negotiating Committee (INC) on appropriate measures for adding new chemicals. While on its face the phrase may be far from clear, the negotiating record suggests it was intended to support an approach that would permit a proposal to proceed through the listing process even with some data gaps.[19]

There is one further and final reference to precaution in the convention. Annex C deals with listed chemicals (e.g., dioxins and furans) that are unintentionally released during common and widespread industrial processes, including waste incineration and thermal metallurgical processes. Article 5 of the convention contemplates that parties shall take measures to reduce the total releases of anthropogenic sources of these chemicals "with the goal of their continuing minimization and, where feasible, ultimate elimination." One such measure is the promotion of best available techniques (BAT) and best environmental practices, taking into account general guidance on prevention and release reduction measures listed in Annex C. Thus, Annex C contains a general discussion of BAT and lists a set of factors to be taken into account in determining BAT. The lead-in to the list suggests that, in determining BAT, consideration should be given to the list "bearing in mind the likely costs and benefits of a measure and consideration of precaution and prevention."[20]

Where was the convention silent on precaution? In addition to considering where the parties actually used precautionary language, we should also ask where the concept of precaution might have provided useful direction to the parties but was not used. The two obvious examples in the POPs convention are the requirements for the domestic treatment of proposed new chemicals and for domestic reviews of existing chemicals.

The convention has very little to say about the appropriate procedures for assessing new chemicals or about procedures for assessing or re-assessing existing chemicals. One of the unstated premises of the convention is that these procedures are essentially domestic and not international and will continue to be so. Article 3, paragraphs 3 and 4, deals with these matters:

3. Each Party that has one or more regulatory and assessment schemes for new pesticides or new industrial chemicals shall take measures to regulate with the aim of preventing the production and use of new pesticides or new industrial chemicals which, taking into consideration the criteria of Annex D [persistence, bio-accumulation, potential for long-range environmental transport and adverse effects], exhibit the characteristics of persistent organic pollutants.

4. Each Party that has one or more regulatory and assessment schemes for pesticides or industrial chemicals shall, where appropriate, take into consideration within these schemes the criteria in paragraph 1 of Annex D when conducting assessments of pesticides or industrial chemicals currently in use.

Although the drafting of these two paragraphs adopts the imperative "shall," in fact, the drafting is conditional in that the paragraphs do not require that a party develop an assessment scheme for new chemicals or a re-assessment scheme for existing chemicals. Instead, a party assumes obligations under these paragraphs only if it has in place such a regulatory assessment scheme. Different obligations are adopted for the two different types of chemicals. For new chemicals, the duty is to "regulate with the aim of preventing" the production of new chemicals with POPs characteristics, while for existing chemicals the duty is merely to take account of POPs characteristics when conducting re-assessments. There is no overall obligation of result (e.g., a duty not to introduce).[21]

These paragraphs make no reference to the application of precaution and yet if one were thinking about applying a reverse onus, which is often said to be a key component of a precautionary approach, then this would have been the place to do so in the convention text.[22] The negotiating record suggests that some parties did propose precautionary text for the new chemicals article but that the proposal was withdrawn at the insistence of one party.[23] While simply having provisions dealing with the domestic treatment of new and existing chemicals is itself precautionary, greater efforts could have been made to incorporate these provisions.

The amendment procedure and precaution. We have so far examined how the convention operationalized the concept of precaution and where it failed to do so. But we still need to know how the process for adding new chemicals to the convention concludes. How do we move from the science-based process of Article 8 to the political process of adopting an amendment to one of the annexes? We have already noted how the convention allows for a political challenge to the essentially scientific process of recommending new chemicals for listing in one or more of the annexes. But how does that process conclude? To what extent is the consent of parties a general consent to the procedures, institutions, and outcomes of the convention, and to what extent have the parties reserved the right to withhold or grant specific consent to each proposal to add a new chemical? Clearly, the parties have plumped for the latter option.

While some of the participants in the negotiations had, from the beginning, referred to the need to adopt a flexible amending procedure to ensure the convention would be "dynamic in nature,"[24] and while there was considerable attention paid to the scientific procedure for considering new chemicals (witness, for example, the CEG and lengthy consideration of CEG reports in plenary and in contact groups), rather less attention was paid to the related question of the adoption of committee proposals by the CoP and their subsequent entry into force.[25]

Once the committee has developed a risk profile and risk evaluation in accordance with Article 8, the committee recommends whether the CoP should list the chemical in an annex. At this point, articles 21 and 22 take over. In common with all amending procedures for international conventions, the POPs procedure for amending an annex is that of (1) proposal, (2) adoption, and (3) entry into force.

Proposals for amendment must be made by a party (Article 21) and communicated to other parties by the secretariat for consideration by the CoP.[26] When it comes to adoption, and following in part the precedent established by the United Nations Economic Commission for Europe (UNECE) POPs Protocol,[27] the convention distinguishes between amendments to the control annexes A, B, and C and those to procedural annexes D, E, and F. Proposals to amend the former are to be considered at the following CoP, which must attempt to achieve consensus on the proposal. Failing consensus, the amendment may be adopted by a three-fourths majority (Article 21(3) and Article 22(4)). Proposals to amend the procedural annexes can be adopted only by consensus (Article 22(5)). The distinction made here is a logical one, as it is imperative that all parties operate by the same procedural rules for adding new chemicals. Annexes D, E, and F form part of those rules and must apply completely, hence the ideas of adoption by consensus only and of entry into force without the need for ratification.

The differentiation is maintained with respect to entry into force. Once adopted, an amendment to a substantive annex (e.g., a proposal to add a new chemical) shall, with one important exception, enter into force for all parties except any party that has notified the secretariat within one year of adoption that it cannot accept the amendment. In short, this is an "opt-out" procedure. Additional chemicals can therefore be added without any need to seek domestic ratification, and there is no minimum level of support that must be attained before entry into force, which is simply fixed as one year after the date of adoption.

As mentioned, there was one important exception. Some negotiating parties (notably the United States) were unable to accept this procedure and sought the creation of an exception that may be characterized as an optional opt-in procedure. Under this scheme any party may, on deposit of its instrument of ratification or equivalent, indicate that amendments to the control annexes properly proposed and adopted shall enter into force for that party only upon deposit of its instrument of ratification or equivalent.[28]

An amendment to a procedural annex enters into force automatically and for all parties on the date specified in the CoP decision. There is no opportunity for opting out and no opportunity to insist upon domestic ratification. This is the only example in the convention of a general consent to future outcomes, and it is obviously quite limited in scope.

Assessment. In trying to assess the manner in which the Stockholm Convention uses the idea of precaution we need to determine where the idea is most relevant. Where can the concept or the principle actually add value to the articulation of rules? The answer is clear: Precaution has little to say about the basic prohibitions in the convention as applied to the chemicals listed from the outset in annexes A or B. We have done the science on these chemicals. We know that they are persistent, have high toxicity, and have a tendency to bioaccumulate. The real question then becomes how precaution is reflected in domestic procedures for approving new chemicals for use and for re-assessing those already in use and in international procedures for adding new chemicals to the annexes and for dealing with those situations in which listed chemicals continue to be used, either because a party has a permitted use or because these chemicals are produced as unintended by-products of common and essential industrial processes.

While references to an agreed articulation of the precautionary principle or approach in the preamble or the objectives clause of a treaty may have some interpretive significance, it is surely more important to have an understanding of the implications of precaution in a concrete case. In my view, one of the achievements of the POPs convention lies in its efforts to operationalize precaution in concrete instances,[29] especially in relation to the international listing process. Yet the convention fails to demand precautionary action in domestic approval systems for new and existing chemicals. This discrepancy is understandable in preserving state autonomy where there is no agreement to take collective action, but the root cause of international POPs problems lies in domestic approval schemes that have allowed chemicals with POPs characteristics to enter the market.

MEAs exhibit a range of amending procedures for their technical annexes.[30] Some MEAs contemplate that amendments may be made by the CoP or its equivalent and will enter into force upon a prescribed date without the need for ratification. The most notable examples are the Montreal Ozone Protocol (the power to make adjustments) and the Convention on International Trade in Endangered Species (CITES) (amendments to the appendices listing endangered species).[31] Other MEAs more conventionally require specific consent to each amendment. It is difficult to discern any particular trend among states to provide their consent to contemplated amendments in advance, thereby according the COP significant law-making powers. Instead, the contracting parties seem to set amending procedures on a case-by-case basis and the choice seems to depend on a variety of factors, including clear scientific consensus on the problem, efficacy, and the cross-sectoral nature of interests that may be affected.

In the end we can see that the POPs convention falls somewhere in the middle of the spectrum and illustrates how amending procedures are crafted to meet different concerns. On matters that are primarily technical (appendices D, E, and F) the parties were able to delegate to the COP significant law-making powers protected by the requirement of consensus (no resort to voting). On matters that might engage significant economic interests across different sectors of the economy (e.g., the addition of new chemicals to the annexes), specific consent was more jealously guarded. While some states were prepared to accept an opt-out formula, others insisted on an opt-in formula. Both preserve the concept of state consent, but the opt-in formula will demand a more formal domestic process that will likely offer greater procedural protection to affected industries and interested government departments.[32] Inertia alone will not be sufficient to carry an amendment forward if opt-in is the prevailing norm. It remains to be seen how many parties will avail themselves of the opt-in option. It also remains to be seen whether this relatively uncommon amending technique becomes more widely adopted in future MEAs. It would, in my view, be unfortunate if it did, for it effectively creates two different and quite unequal ways in which parties assume treaty obligations. It must also increase the risk of a patchwork of obligations.

Common but differentiated responsibilities

Elements of the CDR principle. As articulated in Principle 7 of the Rio Declaration on the Environment and Development,[33] the principle of common but differentiated responsibilities has three elements.[34] First, the duty of all states is to co-operate at a global level to conserve, protect, and restore the Earth's ecosystem. Second, the duties of states, while common, are also differentiated. Relative contributions to global environmental degradation provide the basis for differentiating those responsibilities. Third, developed states acknowledge a special responsibility for the global pursuit of sustainable development for two distinct reasons: (1) the pressures that developed countries place on the global environment, and (2) their capacity to attain sustainability through their technological and financial resources.

Operationalizing the CDR principle. Much as with precaution (although perhaps less explicitly so), the issue for the POPs negotiators was one of how the convention should recognize these elements of the CDR principle.[35] From the first meeting of the INC-1 in Montreal in 1998 until the final substantive negotiating session in Johannesburg in December 2000, the negotiating record is replete with calls for recognition of the principle.[36] Developed countries gave a mixed response. Certainly, there was strong acknowledgment that successful implementation by developing countries would require additional financial resources, but at least one developed country (the United States) also wanted to deliver the message that POPs constituted a local problem as well as a global problem and, to the extent that regulation and prohibition of POPs would deliver benefits locally, the entire financial burden should not fall on developed countries. There was also considerable disagreement as to the selection of the most appropriate financial mechanism, with developed countries generally preferring the Global Environment Facility (GEF) in preference to the creation of a new mechanism.

In the end, the only explicit reference to the CDR principle occurs in preambular paragraph 13, which notes the respective capabilities of developed and developing countries and notes the common but differentiated responsibilities of states as set forth in Rio Principle 7. However, as with the precautionary principle, the real issue is how the CDR principle should be operationalized in this convention. On that point the negotiating lines were drawn on the basis of conditional language and commitments to provide financing and technical assistance.

Early in the negotiations, the developing countries took care to argue that the fundamental control obligations of articles 3 to 5 should be conditional.[37] Although this language was bracketed, it provided continuing negotiating leverage for developing countries regarding their objective to secure strong technical assistance and financial resources and mechanisms – the provisions we now find in articles 12 and 13.[38] Although the explicit conditional language was ultimately dropped from the text at INC-5,[39] one can see several places in which the CDR principle (and particularly the third branch of the principle) continues to influence the text.

First, in addition to the core articles 12 and 13, a number of other articles contain less dramatic conditional language acknowledging that party obligations are limited by capabilities. This is the case, for example,[40] for the obligations of Article 10 (public information, awareness, and education) and Article 11 (research, development, and monitoring), where each contains a formulation along the lines of "Each Party shall, within its capabilities. ..." Second, one paragraph in each of articles 12 and 13, while not explicitly conditional in nature, recognizes in declaratory terms the reality that developing countries acting alone will not be able to meet their commitments under the convention. In Article 12 the relevant paragraph, the first in the article, serves as a preamble or an interpretive lead-in to the operative clauses:

> The Parties recognize that rendering of timely and appropriate technical assistance in response to requests from developing country Parties and Parties with economies in transition is essential to the successful implementation of this Convention.

The article then proceeds to call on parties to co-operate in providing technical assistance and, where mutually agreed, technical assistance for capacity-building. Recognizing that the actual technology itself will ordinarily be privately owned, paragraph 4 limits the duty of the parties to "promote" the transfer of technology.

Article 13's formulation of the "conditional" language is somewhat more complex:

4. The extent to which the developing country Parties will effectively implement their commitments under this Convention will depend on the effective implementation by developed country Parties of their commitments under this Convention relating to financial resources, technical assistance and technology transfer. The fact that sustainable economic and social development and eradication of poverty are the first and overriding priorities of the developing country Parties will be taken fully into account, giving due consideration to the need for the protection of human health and the environment.[41]

The commitments of the developed country parties will include not only the technical assistance provisions of Article 12 but also those contained in the previous paragraphs of Article 13. These paragraphs include the common duty of all parties to support national activities taken to achieve the objective of the convention (the first branch of the CDR principle) as well as the differentiated responsibility of developed country parties to provide "new and additional financial resources to meet the agreed full incremental costs" of taking measures to fulfil convention obligations. Although the article admits some flexibility in this respect to accommodate competing views, Article 14 contemplates that the GEF should be the principal entity for the operation of the financial mechanism, at least on an interim basis until the first meeting of the CoP. The decision to accord this responsibility to the GEF was made easier by its announcement at the opening of INC-5 that the GEF council meeting had decided that should the GEF be designated, new and additional financial resources would be made available through a third replenishment of the GEF.[42]

Assessment. An agreement on technical assistance and financial resources was crucial to a successful outcome of the negotiations and will prove equally crucial to the successful implementation of the convention. Whether or not the specific provisions of articles 12 and 13 are regarded as measures to operationalize the CDR principle is probably of far less significance to the developing countries concerned than is the actual text of these provisions and, in particular, the recognition of such fundamental ideas as "new and additional financial resources" and "to meet the agreed full incremental costs."

These POPs provisions fare well when compared with the similar provisions of other recent MEAs, especially the two other chemicals conventions, which, perhaps because of their focus on trade, have comparatively little to say about either financial assistance or the transfer of technology.[43] Neither refers explicitly to the CDR principle. More in line with the POPs convention, and using the same language of "new and additional financial resources" and "agreed full incremental costs" as well as the compromise quasi-conditional language of Article 13(4), is the Convention on Biological Diversity (CBD).[44] The POPs convention does not go quite so far as the CBD (Article 16) in incorporating the ideas of "fair and most favourable terms" and "concessional and preferential terms where mutually agreed" with respect to technical assistance. Similarly it seems fair to observe that the POPs convention does not go so far as the Framework Convention for Climate Change (FCCC) in recognizing the different elements of the CDR principle.[45] Nevertheless, the convention text does record the concrete obligations of the developed country parties to provide tangible financial and technical assistance to developing country parties and to parties with economies in transition.

The duty to prevent harm and the principle of liability for harm

Background. Rio Principle 2 (reproduced in paragraph 10 of the POPs Preamble), in common with Stockholm Principle 21, recognizes that states have a responsibility to ensure that activities within their jurisdiction or control do not cause damage to the environment of other states or of areas beyond the limits of national jurisdiction. The duty not to cause harm is not an absolute duty because the two articulations of the principle are qualified by an acknowledgment of the sovereign right of states to exploit their own resources. Both the Stockholm and Rio principles also spoke of the need to further develop the international law of liability for environmental damage. Rio Principle 13 puts the point this way:

> States shall also [i.e., in addition to developing national laws] cooperate in an expeditious and more determined manner to develop further international law regarding liability and compensation for adverse effects of environmental damage caused by activities within their jurisdiction or control to areas beyond their jurisdiction.

But, while one of the most basic ideas underlying the convention, it needed little reflection in the convention text. Indeed, much as with precaution, the most obvious places to see the idea reflected would be in the provisions dealing with domestic approval of new chemicals (and review of existing chemicals) and in the provisions dealing with POPs as unintended by-products. We have already canvassed those provisions in the context of precaution. As we saw, these provisions fall well short of an absolute duty to prevent the harm that might flow from the introduction and use of POPs.

Treatment of liability by the Stockholm Convention. While the issue of liability did not loom large in the POPs negotiations, it was persistently pursued by one country, Colombia.[46] Colombia was not rewarded with convention text, but the issue was identified as a concern in the diplomatic conference. Thus Resolution 4 of the conference acknowledges the international nature of the issue and recognizes that the time is ripe "for further discussions on the need for elaboration of international rules in the field of liability and redress resulting from the production, use, and intentional release into the environment" of POPs.[47] To that end, the conference invited the UNEP Secretariat to organize a workshop on liability and redress. The report of the workshop will be considered by the first meeting of the CoP "with a view to deciding what further action will be taken."

Assessment. The failure of the Stockholm Convention to agree on substantive provisions dealing with liability is not surprising and hardly a matter for regret. While, in principle, a comprehensive regime should include provisions dealing with liability for harm, the difficulty of negotiating an appropriate regime is amply illustrated by experience under other MEAs,[48] including Basel[49] and the Environment Protocol[50] to the Antarctic Treaty. If anything, one would expect an appropriate regime to be even more difficult to negotiate for POPs than for, say, the Cartagena Protocol on Biosafety[51] or Basel. Key difficulties include causal connection (given problems of long-range transport and diffuse sources); the nature of liability (strict or negligence-based); the need for a regime to differentiate among liability for POPs as by-products, prohibited POPs, and perhaps DDT;[52] and the channelling of liability to manufacturers and shippers.

RELATIONSHIPS

The proliferation of MEAs requires that we pay greater attention to the question of relationships among different instruments. There are different dimensions to such relationships.[53] One dimension is temporal. How does this agreement relate to earlier agreements? Another dimension is geographical. How does the global interact with the regional? And a third, and probably the most important dimension for present purposes, is that of subject matter. How does an agreement that focuses on the reduction or elimination of listed chemicals relate to other agreements addressing chemicals as hazardous wastes or to international trade in chemicals? Each of these dimensions offers room for debate and the possibility of confusion as to the interaction of different normative regimes.

In some situations we may also need to think about the relationship between treaty norms and norms of customary international law.[54] This is a familiar issue in the context of the law of the sea,[55] but it is not unusual to see MEAs grappling with the problem by means of savings clauses designed to recognize existing freedoms.[56] In the case of POPs, however, the existing customary norms can be articulated only at such a high level of generality that it is difficult to envisage practical problems arising from the interaction between customary law and the regime articulated in the convention text.[57] Issues may arise in future but, for the present, it is hard to identify any particular difficulties.

Temporal relationships

The default rule for determining the relationship between later and earlier agreements is prescribed in Article 30 of the Vienna Convention on the Law of Treaties[58]: in short, the later agreement will prevail over the earlier in the absence of any provision to the contrary. This is an oversimplification[59] but this pithy rendering has caused many difficulties in negotiating some recent MEAs, especially those that have the potential (because of their subject matter or because of measures that the MEA may require parties to take vis-à-vis non-parties or vis-à-vis non-compliant parties) to conflict with existing trade agreements. MEAs often address the issue of temporal relationships either through an operative article titled something like "Relationship with other agreements"[60] or through preambular language recording the intent of the contracting parties.[61]

The Stockholm Convention does not contain such a clause principally because the parties were convinced that it was unnecessary. The trade issues to which the convention might give rise were not considered to be that significant and the detailed articulation of import and export measures in Article 3(2) and elsewhere in the agreement put to rest any concerns about operational conflicts with general trade rules. Thus, at INC-5 the parties agreed to delete draft bracketed text that had been inserted by the Australian delegation as a marker at the close of INC-2.[62] This clause had been designed to ensure that earlier trade law rules would prevail over the later POPs agreement and had garnered significant opposition from both environmental groups and some major negotiating blocks like the European Union (EU).

Geographical relationships

The UNECE Convention on Long-range Transboundary Air Pollution (CLRTAP) Protocol on POPs informed the negotiation of the Stockholm Convention by providing both a conceptual approach to the problem and a possible text for the main control provisions. However, delegations were sensitive to referring to protocol text directly since developing countries, in particular, had no input into its negotiation. In the end, there is no direct reference to the regional agreement in the convention, even as a preambular recital. Furthermore, the convention contains no references to the possible interaction of global and regional regimes other than to contemplate regional implementation through economic integration organizations such as the EU. This is hardly surprising as it is difficult to contemplate effective regional solutions to the global problem of POPs, in much the same way as it is impossible to contemplate regional solutions to problems like climate change or ozone depletion. Thus, the models offered by regional seas agreements, for example, are simply not appropriate here.[63] However, to the extent that parties to the protocol have assumed more stringent obligations under the regional agreement, they will of course be required to fulfil those obligations as well as the more general obligations under the convention.[64]

Subject-matter relationships

At least two other global MEAs, the Rotterdam Convention and the Basel Convention, address some elements of POPs. The Rotterdam Convention on Prior Informed Consent (PIC) deals with trade in hazardous chemicals and pesticides, some of which are POPs or could have POPs characteristics. The Basel Convention deals with hazardous wastes. POPs that are waste are therefore subject to the Basel regime, at least for those states that are parties to Basel.[65] The interrelationship between PIC and POPs does not seem to have been problematic. Thus the Stockholm Convention refers to the Rotterdam Convention in the preamble (and also to the Basel agreement), and Article 3(2)(b) requires that a party shall take measures to ensure that a POP listed in either Annex A or Annex B can only be exported subject to a number of constraints and "taking into account any relevant provisions in existing international prior informed consent instruments," a general reference that certainly includes both PIC and Basel.

The relationship between the POPs and Basel regimes was much more contentious and was one of the last issues to be settled at INC-5. During the course of negotiations, the secretariat prepared a number of papers on the Basel agreement,[66] and the main negotiating group sought advice from the legal drafting group on Basel issues.[67] Negotiating parties disagreed on the scope of the Basel Convention[68] and on the legal status of the Basel technical guidelines.[69] Those who took a minimalist view on both of these questions tended to argue that there were gaps in the legal regime that the POPs convention needed to occupy. They also questioned the effectiveness of Basel and in some cases wanted to give the POPs CoP the power to make legally binding decisions that would trump Basel technical guidelines. Those who took a more expansive view were concerned that much of the field was already occupied by the Basel Convention and therefore any attempt by the POPs convention to occupy the field would lead to duplication, confusion, and potentially inconsistent obligations, especially with respect to the disposal options available for POPs and other wastes with POPs contamination.

Ultimately, the parties were able to agree on a compromise. While we cannot explore the details of that compromise here, we can note some of the main elements. First, stockpiles for which there could be no lawful use under the POPs convention are deemed to be waste (Article 6(1)(c)), with the result that the material becomes subject to the Basel regime for those countries party to Basel. Second, obligations to dispose of waste are to take account of international rules, standards, guidelines, and relevant global and regional regimes (Article 6(1)(d)(ii)), which references must clearly include both Basel and the Bamako Convention.[70] Third, the POPs COP is directed to co-operate closely with the relevant Basel bodies on issues of environmentally sound disposal (Article 6(2)). This request for co-operation was further amplified in Resolution 5 of the diplomatic conference, which called on the bodies of the Basel Convention to continue their work on the development of appropriate technical guidelines for the environmentally sound management of POPs wastes. Fourth, the meeting record expresses the understanding of the parties that the text did not establish a conflict with the Basel Convention.[71]

Assessment

The proliferation of MEAs suggests complex questions of interrelationships, yet the POPs convention did not pose significant difficulties. There were no serious issues surrounding the relationship between the proposed convention and customary law. The convention largely tracks the approach taken by the regional UNECE CLRTAP protocol, and thus the negotiators found it unnecessary even to refer to that agreement. Trade issues were not considered to be major, and it was therefore possible to work through a list of import and export restrictions to which all parties could agree. Because of this approach, the overall question of the relationship between this agreement and earlier trade agreements proved, in the end, to be non-contentious. Finally, possible conflicts with the Basel regime were resolved by recognizing the need for co-operation between the two bodies and the need to respect Basel's existing work in the area of the environmentally sound management of waste POPs. There is no guarantee that future conflicts will be avoided, but, since most countries are party to Basel, we can assume that the respective COPs and other subsidiary bodies will work hard to avoid any such conflicts.

IMPLEMENTATION, EFFECTIVENESS, COMPLIANCE, AND MONITORING

As international environmental law matures, and as the global community puts in place framework conventions and ever more prescriptive protocols to deal with a range of issues (global warming and ozone depletion, biological diversity, trade in endangered species, and the regulation of different aspects of trade in chemicals and waste), attention has turned to matters of implementation, effectiveness, compliance, and monitoring. For existing instruments this has meant that institutions established by the convention, including the CoP, have had to think through, for example, how to put in place a compliance procedure when one had not been anticipated at the outset. In the case of a new instrument it is possible to anticipate these needs, and then the question becomes one of how much to put in place as part of the convention text and how much can safely be postponed until entry into force for elaboration by successive meetings of the CoP.

Implementation

The premise of the law of treaties and of the law of state responsibility is that no state should become a party to an international agreement unless it is already in a position to comply with the terms of the agreement.[72] While unassailable as a theoretical proposition, it is not without its practical difficulties, especially for a developing country. Such a country may require technical and financial assistance to implement treaty obligations and yet ratification will often be a condition precedent to securing that assistance. It was for just these reasons that Canada contributed $20 million to assist developing countries in preparing for ratification. Treaty implementation provisions of MEAs must therefore tread a fine line: one that respects the legal rule yet which recognizes the practical difficulties faced by many potential parties.

The POPs convention addresses this set of issues through several articles.[73] Article 5 requires each party to develop action plans for Annex C chemicals, and Article 7 requires each party to develop and "endeavour to implement" plans for the implementation of its convention and to file an implementation plan with the secretariat within two years of entry into force. Still more practical are the two articles (articles 12 and 13) dealing with technical and financial assistance, which we have discussed in the context of the CDR principle.

In addition to implementation plans, each party shall also report periodically to the CoP on the measures it has taken to implement the provisions of the convention and the effectiveness of those measures in meeting the objectives of the convention (Article 15).

Effectiveness

While full implementation is a necessary condition of an effective MEA, it is not itself a sufficient condition, or at least cannot be so unless we accept the premise that the measures adopted by the MEA are themselves sufficient (and the best possible measures) to deal with the mischief or problem that the convention was set up to address. Even leaving to one side questions as to whether or not the measures adopted are the most cost effective,[74] other MEAs (and perhaps the best example is the Vienna Ozone Convention and Montreal Protocol) suggest that we also need a mechanism for assessing and re-assessing whether the convention is successful in attaining its objective and, depending upon the outcomes of that assessment, a procedure for the adoption of new control measures where necessary to better meet that objective.

The first pre-condition to such a mechanism is, of course, a clear statement of an objective. Article 1 of the convention fulfils this role and, although clear, suffers from being too broad. Other possible formulations of an objective, while narrower and more measurable (e.g., eventual elimination) were ultimately rejected as being impractical or simply unattainable. Article 16 of the convention adds an innovative second step by requiring the parties, by the second COP (and periodically thereafter at intervals to be decided) to "evaluate the effectiveness of this Convention." That same article also recognizes that any such procedure requires the collection and analysis of data and therefore requires the first COP to make arrangements to provide itself with comparable monitoring data on the presence of listed chemicals as well as on their regional and global environmental transport. The article goes on to note that any evaluation of effectiveness should also take account of national reports to the secretariat (Article 15) as well as information on non-compliance.

While Article 16 demands a comprehensive assessment of the effectiveness of the convention, other articles demonstrate similar concerns.[75] This is particularly true of Article 5, which requires that parties review their success in meeting the obligations of the article and also review the efficacy of domestic laws and policies relating to the management of the release of Annex C chemicals.

Monitoring

Monitoring refers to the systematic collection of data. The data, which will be most useful if collected and reported in accordance with common standards, may be used for a variety of purposes, including assessment of effectiveness and assessment of compliance. The convention contains two references to monitoring. The first is in Article 11, which is a general article dealing collectively with research and development in addition to monitoring. The article encourages all three activities (i.e., research, development, and monitoring) in relation to a broad range of matters including sources and releases, environmental transport, fate, and transformation, but it fails to establish any coordinating or report-receiving role for the COP or any other subsidiary body to be designated by the COP (except as part of a general duty to report to the COP (Article 15(1)) on all measures taken to implement the convention).[76]

The second reference to monitoring occurs in Article 16, which deals with effectiveness evaluation. It is clear that the monitoring contemplated here is more focused and more clearly integrated into the institutions created by the convention.

Compliance

There has long been recognition that the standard MEA dispute-resolution mechanisms lack teeth.[77] In recent years, therefore, attention has focused on measures that may be taken to foster compliance without resorting to formal dispute resolution.[78] The convention text effectively postpones the issue of compliance by delegating to subsequent CoPs the need to develop a compliance mechanism:

Article 17: Non-compliance
The Conference of the Parties shall, as soon as practicable, develop and approve procedures and institutional mechanisms for determining non-compliance with the provisions of this Convention and for the treatment of Parties found to be in non-compliance.

Notwithstanding a busy schedule for the first meeting of the CoP,[79] the negotiating parties agreed (Resolution 1) at the diplomatic conference to call upon the INC to focus, *inter alia*, upon "modalities and procedures relating to non-compliance." This represents an important step, for some states would have preferred to have seen this issue indefinitely postponed and the resolution does therefore add an urgency to what was otherwise fairly standard MEA drafting.

Assessment

The POPs convention breaks some new ground on issues of implementation, compliance, and monitoring. It has certainly taken advantage of the evolution in the drafting of MEAs over time to include certain provisions on which earlier agreements were completely silent, most notably with respect to compliance,[80] but it also adds requirements for national implementation plans and, where appropriate, consultations on them. It also obliges parties to create a monitoring system. The key question here is whether or not the convention has gone far enough. On the question of a compliance regime, the convention is not as detailed as the Kyoto Protocol (Article 18), and neither does it require public involvement as does, for example, one important regional agreement.[81] Attention must now turn to an elaboration of an appropriate and robust compliance mechanism. There are certainly some helpful examples on which to draw, notably the compliance scheme of the Montreal Ozone Protocol,[82] and the parties will be helped in this regard by the explicit mandate accorded by Article 17. Without such a clear mandate and clear political commitment, experience with the Basel agreement shows how difficult it could be to put such a mechanism in place.[83]

The convention's key achievement in these fields is the incorporation of a procedure for examining the effectiveness of the convention and the recognition that such a procedure requires that monitoring and reporting schemes be designed, at least in part, with this objective in mind. While other agreements have provisions to assess their effectiveness and, as a result, the parties to those agreements have taken further measures to achieve the objectives of those agreements, this is one of the first MEAs to devote a separate article explicitly requiring an assessment of effectiveness.[84]

CONCLUSIONS

The POPs negotiations stand as one of the most successful MEA negotiations of recent years. They were completed in the space of just five substantive sessions spread over some thirty months. Unlike the Biosafety Protocol negotiations, they never broke down in acrimony, and unlike the climate change negotiations and the Kyoto Protocol, there is no suggestion that a major state will fail to sign and ratify the agreement.

In this paper I have selected a few themes to allow us to draw some comparisons with other recent MEAs. In common with other MEAs, the POPs convention shows a continuing concern with principles and the implementation of principle. The POPs convention follows the lead of the Biosafety Protocol in operationalizing the idea of precaution and the example of the Convention on Biological Diversity and the Framework Convention for Climate Change in operationalizing the CDR principle. In common with other MEA negotiations, the parties decided to postpone further negotiations on liability rather than hold the convention hostage to a successful resolution of that issue, and, in this case unlike Biosafety and Basel, the convention text is completely silent on the question of liability.

Like other recent agreements, the text also shows the negotiators' preoccupation with the successful implementation of the agreement. This healthy preoccupation is reflected not only in the technical and financial assistance provisions but also in the suite of provisions dealing with monitoring, effectiveness and compliance, and national implementation plans. Much remains to be done to build a compliance regime for the POPs agreement, but the basic building blocks are there. One can only hope that the parties to the convention will accord this greater priority than they accord to the creation of a liability regime.

As this MEA takes its place alongside other regional and global MEAs and trade regimes there will be questions about how these various instruments relate to one another. There seems to have been general recognition that the POPs convention was filling a defined niche between the existing PIC and Basel regimes. Unlike the Biosafety Protocol, the POPs agreement was never subject to the charge that it was simply occupying a field already covered by existing trade arrangements, albeit looking at that field from a different perspective. Consequently, the relationship between this new convention and existing agreements – with the exception of Basel – was never particularly contentious or especially difficult to work out.

Indeed, while difficult in relation to the issues of financing (a north–south issue) and precaution (an issue among developed countries), within the MEA field this set of negotiations was not as daunting as some. Much of the intellectual work needed to conceptualize the problem as well as its solutions had already been done over the previous decade within the UN Economic Commission for Europe and through the subsequent adoption of the CLRTAP POPs Protocol. Thus there was no real argument about the science or the reality of the problem, which distinguishes this convention from early climate-change discussions. The most significant difficulty facing the negotiations was undoubtedly the issue of technical and financial assistance. Once a solution to that had been found, the parties were able to reach agreement on some of the more semantic differences relating to issues like the treatment of precaution and the shared goal of ultimate elimination. It is a tribute to the negotiators that, in the end, they preferred pragmatic steps to operationalize principles in favour of endless arguments over different formulations of those principles.

ACKNOWLEDGMENTS

Thanks to Angela Ovens, LLB candidate, the University of Calgary, for her research assistance and special thanks to Yves Lebouthillier (DFAIT) and Anne Daniel (Senior Counsel, Department of Justice and Legal Services, Environment Canada) for their very valuable comments on a first draft of this chapter.

NOTES

1 Stockholm Convention, 22 May 2001, reproduced in *International Legal Materials* 40 (2001): 1 and available on the UNEP Chemicals website at http: //www.chem.unep.ch. Negotiations followed a now-standard format for MEAs. The mandate was provided by UNEP/GC.19/13C, which asked the executive director to convene an intergovernmental negotiating committee (INC) to prepare a legally binding international instrument beginning with twelve specified POPs. There were five INC negotiating sessions: INC-1, Montreal, June 1998, UNEP/POPs/INC.1/7; INC-2, Nairobi, January 1999, UNEP/POPs/INC.2/6; INC-3, Geneva, September 1999, UNEP/POPs/INC.3/4; INC-4, Bonn, March 2000, UNEP/POPs/

INC.4/5; and INC-5, Johannesburg, December 2000, UNEP/POPs/INC.5/7 (hereafter referred to as INC-1, etc.). The INC remains in place to deal with issues that require attention before the convention formally enters into force. I disclose that I attended INC-3 and INC-4 as a member of the Canadian delegation. For an excellent review of the process aspects of the POPs negotiations that emphasizes the participation of NGOs and others see Peter Lallas, "The Role of Process and Participation in the Development of Effective Environmental Agreements: A Study of the Global Treaty on Persistent Organic Pollutants (POPs)," *University of California at Los Angeles Journal of Environmental Law and Policy* 19 (2001): 83.

2 The main candidates for comparison purposes are two chemical conventions: the Rotterdam Convention on the Prior Informed Consent Procedure for Certain Hazardous Chemicals and Pesticides in International Trade, Rotterdam, 10 September 1998 (hereafter the PIC Convention or the Rotterdam Convention), reproduced in *International Legal Materials* (*ILM*) 38 (1999): 1; the Basel Convention on the Control of the Transboundary Movements of Hazardous Wastes and their Disposal, Basel, 22 March 1989, reproduced in *ILM* 28 (1989): 657. See also the Vienna Convention for the Protection of the Ozone Layer, Vienna, 22 March 1985, reproduced in *ILM* 26 (1987): 1516; the Framework Convention on Climate Change, New York, 9 May 1992 (hereafter the FCCC), reproduced in *ILM* 31 (1992): 849; the Convention to Combat Desertification, Paris, 17 June 1994 (hereafter the CCD), reproduced in *ILM* 33 (1994): 1328; the Convention on Biological Diversity, Rio de Janeiro, 5 June 1992 (hereafter the CBD), reproduced in *ILM* 31 (1992): 819; the Cartagena Protocol on Biosafety, Montreal, 29 January 2000, reproduced in *ILM* 39 (2000): 1027; and the Convention on International Trade in Endangered Species, Washington, 3 March 1973 (hereafter CITES), reproduced in *ILM* 12 (1973): 1085.

3 See the summary of the final round published in *Earth Negotiations Bulletin* 15 (2000): 54, http: //www.iisd.ca/chemical/pops5/, hereafter *ENB*, for INC-1 and similar citations for the successive INCs.

4 Protocol to the Convention on Long-range Transboundary Air Pollution on Persistent Organic Pollutants, Århus, 24 June 1998, *International Legal Materials* 37 (1998): 505.

5 Ronald Dworkin, "Is Law a System of Rules," in Dworkin, ed., *The Philosophy of Law* (Oxford: Oxford University Press 1977), 47.

6 Declaration of the United Nations Conference on the Human Environment, A/CONF.48/14/Rev.1. Commentators disagree on the extent to which these principles codify or reflect customary international law, but that is a debate we need not engage in here.

7 The Rio Declaration on Environment and Development, A/CONF.151/5/Rev.1, reproduced in *International Legal Materials* 31 (1992): 876.

8 The negotiating record also shows an early and continuing emphasis on the importance of principles. For example, at INC-1, the African Group of Countries tabled a list of relevant principles, Annex VII, para. 7, while at INC-2, the Group of 77 and China suggested that the convention needed a section on principles "which would

provide the guiding principles for the convention and its future evolution." The principles would include common but differentiated responsibilities; the right to development; the primacy of the protection of human health and the environment; the special situation and needs of developing countries; the polluter pays principle; and the principle that measures to implement the convention should not constitute a means of arbitrary or unjustifiable discrimination in international trade.

9 The literature on precaution is massive. On the underlying ideas see in particular Andre Nolkaemper, "'What You Risk Reveals What You Value' and Other Dilemmas Encountered in the Legal Assault on Risks," in David Freestone and Ellen Hey, *The Precautionary Principle and International Law: The Challenge of Implementation* (Kluwer: The Hague 1996), 73–97. On the question of the status of the principle as a rule of customary international law, see Peter Cameron and Julie Abouchar, "The Status of the Precautionary Principle in International Law," in Freestone and Hey, *The Precautionary Principle*, 52; and David Vanderzwaag, "The Precautionary Principle in Environmental Law and Policy: Elusive Rhetoric and First Embraces," *Journal of Environmental Law and Policy* 8 (1998): 355. In general, the academic community seems more inclined to view the principle as having attained the status of a rule of customary international law than do international judicial and arbitral tribunals, which generally support the view that the principle has yet to attain that status.

10 Rio Declaration, principle 15. See www.unep.org/unep.rio.htm.
"In order to protect the environment, the precautionary approach shall be widely applied by States according to their capabilities. Where there are threats of serious or irreversible damage, lack of full scientific certainty shall not be used as a reason for postponing cost-effective measures to prevent environmental degradation."

11 One commonly cited formulation, in addition to Rio 15, is the 1990 Bergen Ministerial Declaration on Sustainable Development in the ECE Region, A/CONF.151/PC/10, *Yearbook of International Environmental Law* 429 (1990): 431. The version of the principle incorporated in the Canadian Environmental Protection Act, 1999, SC 1999, c.33, s.2(1)(a) draws upon both the Bergen formulation and the cost-effectiveness element of the Rio formulation.

12 Negotiators on all sides of the POPs negotiations seemed to be satisfied that the trade provisions of the convention (articles 3(1)(b), 3(2), and 6(1)(c) and (d)) did not create conflicts with general trade rules.

13 We should also note that some agreements are considered to have incorporated precautionary ideas without using the term. The best-known example is Article 5(7) of the Sanitary and Phyto-Sanitary Measures Agreement (the SPS Agreement), but see also articles 10(6) and 11(8) of the Biosafety Protocol.

14 See in particular Nolkaemper, "What You Risk Reveals What You Value."

15 This will include information on the basic POPs qualities of persistence, bioaccumulation, and potential for long-range transport and evidence of or potential for adverse effects on human health or the environment based on toxicity or ecotoxicity data.

16 Articles 8(5) and (8) tell us that CoP is to make its decision based on the committee's views, the relevant annex, and any additional information provided, but it is difficult to read these as serious constraints on the CoP's discretion. Evidently, the challenge provisions were included precisely because it must have been the prevailing view that a primarily political body (the CoP) should, under some circumstances, be able to trump a primarily scientific body (the POPRC). Presumably, the CoP will be moved to trump the POPRC for political reasons and not for scientific reasons.

17 Note that the negotiating text for INC-5 also included a bracketed reference to the precautionary principle in what was then Article F.3, pertaining only to the application of the screening criteria.

18 For example, for "persistence," Annex D requires information on the half-life of the chemical in water, air, or sediment or "evidence that the chemical is otherwise sufficiently persistent to justify its consideration within the scope of this Convention."

19 See INC-4, para. 58, making the point that the term "flexible" had been defined in the CEG Report "as meaning that a proposal could be considered as satisfying the criteria if one of them was marginally not met, but two or more were amply met."

20 Annex C, Part V, B, chapeau. Compare, in this context, ECE Regional Protocol, Annex V, *International Legal Materials* 37 (1998): 505, referring in paragraph 2 to the "principles of precaution and prevention."

21 Although the obligations for existing chemicals are weaker than those for new chemicals, the proposal for existing chemicals nearly did not make it into the final text. Draft text had been developed in Bonn but had been omitted from the chair's text on which INC-5 negotiations were based.

22 It would also have been fairly easy to refer here not just to Annex D but to the objective of the convention and thereby incorporate the precautionary idea in that manner. Nevertheless, it will still be possible to argue that the references to precaution in both the objective and the preamble should and will inform the proper interpretation of this article.

23 *Earth Negotiations Bulletin*, INC-5 para. 6; a member of the JUSCANZ group.

24 See for example INC-2, para. 38(a), a number of representatives referring to the Montreal Protocol, and the practice for conventions within the family of the International Maritime Organization (IMO).

25 Timing, however, was a continuing concern. Delegates were very much aware that the multi-stage committee procedure when combined with the CoP adoption procedure would make it difficult to add new chemicals expeditiously. See, for example, INC-3, paras 64–66.

26 It would follow from this that the committee itself cannot propose an amendment, notwithstanding the language of Article 8(8). Instead, an amendment must be proposed by a party, although it is conceivable that this might be taken to have been done when the party originally proposed the chemical to the committee for possible listing.

27 For the relevant ECE texts see the POPs Protocol, *International Legal Materials* 37 (1998): 505: Article 14(3), amendments to obligation annexes by consensus and entry into force after deposit of instruments of ratification for two-thirds of the parties; Article 14(4), amendments to recommended BAT and control measures adopted by consensus and automatic entry into force but reserving a right of opt-out; and Article 15(7), any decision to amend EB 1998/2 (the decision of the executive body that stipulates the criteria for consideration as POPs) shall be taken by consensus and shall enter into force automatically sixty days after adoption.

28 There is a partial precedent for this procedure in the Convention to Combat Desertification (CCD), articles 31 and 34(4). In the CCD these provisions apply to both new annexes and amendments to existing annexes.

29 To the same effect in the context of the Biosafety Protocol, see Peter-Tobias Stoll, "Controlling the Risks of Genetically Modified Organisms: The Cartagena Protocol on Biosafety and the SPS Agreement," *Yearbook of International Environmental Law* 10 (1999): 99. The Biosafety Protocol refers to precaution in the preamble and in Article 1 (Objective) and endeavours to operationalize the term in articles 10(6) and 11(8), which deal with the decision-making procedures of parties of import.

30 For a useful recent discussion, see Robin Churchill and Geir Ulfstein, "Autonomous Institutional Arrangements in Multilateral Environmental Agreements: A Little Noticed Phenomenon in International Law," *American Journal of International Law* 94 (2000): 636.

31 Montreal Ozone, Article 2, and CITES, Article XV.

32 Take Canada as an example. An instrument of ratification always requires a cabinet submission and, ultimately, adoption by an order in council. By contrast, instructions to a delegation attending a COP will be developed interdepartmentally and ultimately approved by the minister of Foreign Affairs or other relevant minister (for example, the minister of Transport in the case of IMO agreements). A cabinet submission will not be required and will be subsequently required in the opt-out process only if the amendment needs to be incorporated in domestic law by either a statutory amendment (which will require parliamentary approval) or an amendment to a set of regulations that requires the intervention of the governor in council rather than simply a responsible minister. Canada, in its instrument of ratification (23 May 2001), elected to avail itself of the opt-in procedure. Other arguments in favour of the opt-in procedure include the idea that one year may be too short a time to make a decision and, thus, that states may have to opt out simply to buy more time. It is argued that the optics of such a move would be unfortunate and might send some quite unintended messages.

33 Rio Declaration, principle 7.

"States shall cooperate in a spirit of global partnership to conserve, protect and restore the health and integrity of the Earth's ecosystem. In view of the different contributions to global environmental degradation, States have common but differentiated responsibilities. The developed countries acknowledge the responsibility that they bear in the international pursuit to sustainable development in view of the pressures their societies place on the global environment and of the technologies and financial resources they command."

Unlike some of the other principles discussed here, this principle has no direct precursor in the Stockholm Declaration, although there is a recognition of the special needs of developing countries for financial assistance in Principle 12.

34 There are two particularly useful articles on CDR: Duncan French, "Developing States and International Environmental Law: The Importance of Differentiated Responsibilities," *International and Comparative Law Quarterly* 49 (2000): 35; and Daniel Barstaw Magraw, "Legal Treatment of Developing Countries: Differential, Contextual and Absolute Norms," *Colorado Journal of International Environmental Law and Policy* 1 (1990): 69.

35 Although there are similarities, there are also important distinctions. The actual label "precautionary principle" carries huge emotive weight. It became very important to some of the negotiators (especially to the EU negotiators) to use that particular term because of its rhetorical and political sigNIficance. As a result, it becomes harder to proceed pragmatically by asking just what it is that we want precautionary language to achieve and then thinking of different methods of achieving that objective. By contrast, negotiations on the CDR principle will always be more pragmatic.

36 See, for example, INC-2, paras 38(a) and 66 (referring to the conditional issue) and Annex V, position of the G-77 and China, especially paras 2 and 3; INC-3, para. 28; INC-4, para. 22.

37 See INC-3, Article D, containing the phrase (or variations thereof): "[Subject to the accessibility of financial and technical assistance] each Party shall ..."

38 One of the more innovative proposals that almost dropped by the wayside was the Canadian proposal for a Capacity Assistance Network (CAN). The CAN was informed by two ideas: the first was that there was a need to coordinate the many possible sources (public, private, multilateral, bilateral, etc.) of support and assistance with the needs of individual countries. The second was the idea that the secretariat might perform this sort of function. While dropped from the formal text, the idea was retained in Resolution 3 of the Final Act, UNEP/POPs/CONF/4. The resolution calls on UNEP's executive director and the CEO of the GEF "to develop the modalities for" a CAN and to report accordingly to INC-6.

39 See *Earth Negotiations Bulletin*, INC-5, para. 5, noting that a proposal to delete the bracketed conditional text from the control paragraphs was ultimately only "deleted in the final Plenary as part of a package deal on the financial mechanism."

40 See also Annex D, para. 3.

41 It bears emphasizing that the qualifying words "successful" in Article 12 and "effectively" in Article 13 ultimately undermine any argument that obligations undertaken in articles 3 to 6 are conditional.

42 *Earth Negotiations Bulletin*, COP-5, para. 3.

43 For Basel, see Article 10 (general article on international co-operation) and Article 14 (financial aspects and calling for regional and sub-regional centres financed on a voluntary basis and consideration to be given to the possibility of establishing a revolving fund to help deal with emergency situations); for PIC, see Preamble para. 4, and Article 16 (parties shall co-operate to promote technical assistance and parties with advanced regulatory programmes should provide technical assistance to assist other parties in developing infrastructure and capacity to manage chemicals). Apart from the preamble, PIC is silent on the question of financial measures.

44 Convention on Biological Diversity (CBD), Article 20, Financial Resources. The preamble does not refer specifically to the CDR principle, although a number of the paragraphs do refer, of course, to the special position of developing countries.

45 See, for example, Framework Convention for Climate Change (FCCC), Preamble paragraphs 2 and 3, recognizing the disproportionate contributions of developed and developing countries to the problem of greenhouse gas emissions; para 5, referring explicitly to CDR; and, in the operative portion of the text, the reference to CDR in the lead-in to Article 4(1), dealing with commitments and to "new and additional" financing and "agreed full costs" in Article 4(3) and the quasi-conditional language of Article 4(7).

46 See INC-1, paras 55(a) and 60 and Annex, para. 7; INC-2, paras 36(a) and Annex III; INC-4, paras 22 and 100; and INC-5, para. 32. And see also *Earth Negotiations Bulletin* for INC-5, para. 3: "Colombia proposed inclusion of a new article stating that specific guidelines regarding liability, responsibility and compensation would be developed in future." Presumably, other countries must have been prepared to support Colombia on this, or the chair simply elected not to press the point; however, inclusion of the item in the COP work plan does raise the question of when one country or a very small group of countries should be able to claim priority for their special interest if it is not a widely shared interest.

47 UNEP/POPs/CONF/4.

48 In addition to the MEA experience, a full appreciation of the difficulties faced also needs to take account of the long-standing efforts of the International Law Commission (ILC) to develop a general liability regime. Work began in 1980 and still continues. The progress made has in fact worked backwards from liability to try to construct a procedural regime to deal with the prevention and avoidance of harm rather than the liability for harm should it occur. The ILC adopted its "Draft articles on prevention of transboundary harm from hazardous activities" in 2001 at its fifty-third session.

49 The need for a liability regime was identified in Article 12 of the Basel Convention. Negotiations commenced in 1993 and were completed in December 1999 after ten sessions of the ad hoc working group. The text was adopted by COP-5 of Basel. As of 9 April 2002, there were thirteen signatories and no ratifications.

50 Protocol on Environmental Protection to the Antarctic Treaty, Madrid, 4 October 1991, *International Legal Materials* 30 (1991): 1461. The need for a liability

regime was identified by Article 16 of the protocol, which envisaged that the parties would undertake to elaborate liability rules (not just to study the need for) to be included in an annex. Negotiations commenced shortly thereafter and continue.

51 Article 27 of the Biosafety protocol calls upon the first meeting of the parties to adopt a process for the elaboration of appropriate international rules and procedures and to complete that process within four years. The topic is being addressed by the Intergovernmental Committee for the Cartagena Protocol, UNEP/CBD/ICCP/2/1/Add.1, 19 July 2001, Agenda of the Second Meeting and Work Plan. See also Article 15 of the protocol to the London Dumping Convention (LDC), calling on the parties to "undertake to develop procedures regarding liability arising from the dumping or incineration at sea of wastes or other matters." In fact, no work has been undertaken on this topic to date.

52 To the extent that developing countries continue to use DDT for vector control or other POPs for a variety of exceptional uses, they may be more exposed to the implications of a liability regime than, say, northern hemispheric countries, which are the recipients of POPs generated and used in warmer climates through a process of atmospheric transport and cold condensation.

53 Scholars are paying increasing attention to this issue. In the context of polar regions see, for example, the collection of essays in Davor Vidas, ed., *Protecting the Polar Marine Environment: Law and Policy for Pollution Prevention* (Cambridge: Cambridge University Press 2000). The issue (although described differently) has also received the attention of the International Law Commission; see Annex to the ILC's Annual Report for 2000, Gerhard Hafner, "Risks Ensuing from the Fragmentation of International Law," 321–39.

54 See, in particular, Nancy Kontou, *The Termination and Revision of Treaties in the Light of New Customary International Law* (Oxford: Oxford University Press 1994).

55 See, for example, the North Sea Continental Shelf Cases [1969] ICJ Reports 43.

56 See, for example, the Cartagena Protocol on Biosafety and the savings clauses in paragraphs 3 and 4 of Article 2 and paragraph 12 of Article 4 of the Basel Convention.

57 See the *sic utere tuo* principle enshrined in Principle 2 of the Rio Declaration and Principle 21 of the Stockholm Declaration.

58 Vienna, 23 May 1969 (hereafter VCLT).

59 For example, it glosses over the more basic obligation of parties to observe all agreements to which they are parties. It does not tell us what amounts to a conflict or deal with other priority rules such as the idea that the *lex specialis* prevails over the *lex generalis*.

60 See, for example, the CBD Article 22, CITES Convention Article XIV, CCD Article 8.

61 In my view, either technique is effective if the language is clear because a preamble, while incapable of creating legal obligations, can serve as an authoritative statement of the intent of the negotiators. See the PIC Convention and the Cartagena

Protocol. However, most commentators take the view that the specific compromise language of Cartagena creates confusion and ambiguity and is unhelpful. See, for example, Stoll, "Controlling the Risks," 118–9.

62 *Earth Negotiations Bulletin* (ENB), INC-5, para. 12, noting the last-minute deletion of the Nbis provision but on the grounds that the matter had been addressed in the preamble. This seems to be an optimistic interpretation. Only paragraph 9 of the preamble addresses the matter, recognizing that "this Convention and other international agreements in the field of trade and the environment are mutually supportive." Clearly, this does not tell us what happens in the event of conflict; it merely urges states to do what they are already required to do by the VCLT, which is to fulfil their obligations under all treaties to which they are party. The clause is quite different from the two preambular paragraphs of the PIC Convention that address successive agreements, and, while modelled on the similar clause in the Biosafety agreement, it lacks that agreement's further stipulation that rights and obligations under existing agreements are not to be taken to have been changed. Australian efforts to introduce PIC-style language were defeated (*ENB*, INC-5, para. 4).

63 Whereas they might well be appropriate to desertification, for example, and are explicitly contemplated in Article 15 of the CCD.

64 For example, under Article 3(5) a party to the protocol is required to assume the obligation to reduce its total annual emissions of polycyclic aromatic hydrocarbons (PAHS), dioxins and furans, and hexachlorobenzene.

65 Ratification of Basel is widespread. Adopted in 1989, Basel entered into force in 1992. As of 9 April 2002, 150 parties have ratified; the United States, Afghanistan, and Haiti have signed the treaty but have not yet ratified it.

66 UNEP/POPS/INC.5/3 and INF/3 and INC.4/INF/7.

67 INC-4, Annex II, Draft Text, note 10 commentary on Article D.4.

68 The focus of the Basel Convention is waste (and not stockpiles of chemicals that may still have a value and a use) and the transboundary movement of that waste. But that is simply a question of focus. Parties to the Basel Convention also undertake obligations to reduce the generation of waste and to provide for the environmentally sound management of wastes (Article 4(2)).

69 The Basel technical guidelines on the environmentally sound handling of various streams of waste. The precise legal status of the guidelines depends on the view that one takes of (1) the definition of the term "environmentally sound management" of wastes and the role of the technical guidelines in elucidating the meaning of this term for particular waste streams; (2) the measures required by Article 4(2); and (3) the role of the COP in further defining the measures required of contracting parties.

70 Bamako Convention on the Ban of the Import into Africa and the Control of Transboundary Movement and Management of Hazardous Wastes within Africa, 29 January 1991, reproduced in *International Legal Materials* 30 (1991): 773.

71 INC-5, para. 43.

72 VCLT, articles 26 and 27; International Law Commission, draft articles on state responsibility, as adopted on second reading, articles 1 and 12, A/CN.4/L.602/Rev.1, 26 July 2001.

73 See also the resolutions adopted by the diplomatic conference that are frequently designed to bridge the period between adoption of the text and entry into force, and, in the meantime, to assist parties in getting into a position from which they will be able to fully implement the convention from the date of entry into force: Resolution 1 on interim arrangements; Resolution 2 on interim financial arrangements; and Resolution 3 on capacity-building and Capacity Assistance Network, UNEP/POPs/CONF/4.

74 Clearly a very contentious issue in the context of some MEA negotiations, e.g., climate change.

75 See also Article 4(6), dealing with the review of specific exemptions. While the article is silent on the criteria to be applied by the COP, it does call on the applicant party to justify its continuing need, and presumably the COP will have to balance that need against the overall objective of the convention and in light of the data demonstrating the convention's success or lack of success.

76 A partial exception to this statement is the coordinating role accorded to the secretariat for compiling reports: Article 20(2)(d).

77 The POPs convention has the standard formulation in Article 18: a party may elect a compulsory form of dispute resolution. If parties elect different forms of binding dispute settlement or make no election, parties shall at the request of either submit the dispute to conciliation. While MEA dispute-settlement provisions are not (apart from conciliation) in any sense "compulsory," since a party may not elect any binding dispute-resolution mechanism (arbitration or judicial settlement), we are seeing increasing reliance on compulsory and binding dispute settlements in two other areas of international law: trade relations and the law of the sea.

78 There is a large body of literature on compliance, including Kamen Sachariew, "Promoting Compliance with International Environmental Standards: Reflections on Monitoring and Reporting Mechanisms," *Yearbook of International Environmental Law* (*YBIEL*) 2 (1991): 31; Maas Goote, "Non-compliance Procedures," *YBIEL* 9 (1998): 146; Jutta Brunnee, "A Fine Balance: Facilitation and Enforcement in the Design of a Compliance Regime for the Kyoto Protocol," *Tulane Environmental Law Journal* 13 (2000): 223 (excellent review of the complexities associated with the development of a compliance scheme for Kyoto and noting the need to have some "hard" compliance measures so as to accord the Kyoto regime the necessary credibility).

79 See, for example, Article 4(5) (review process for entries on the register); Article 13(7) (guidance for the financial mechanism); Article 15(3) (format for reporting); Article 16(2) arrangements for monitoring; Article 19(4) (rules of procedure and financial rules), (6) appointment of the committee; as well as the resolutions of the diplomatic conference UNEP/POPs/CONF/4, resolution 2(6), review of availability of financial resources, and resolution 4, review of liability and redress issues.

80 For example, the following agreements contain no reference to a compliance mechanism: Basel, the CBD, the Vienna Ozone Convention, and the CCD. More recent agreements generally provide for a compliance mechanism but stipulate that it should be developed by the COP; e.g., PIC, Article 17; Biosafety Protocol, Article 34; 1996 Protocol to the London Dumping Convention, Article 11 (compliance procedures to be developed within two years of entry into force).

81 See Convention on Access to Information, Public Participation in Decision-making and Access to Justice in Environmental Matters, Århus, 25 June 1998. Article 13 contemplates that the parties will establish "optional arrangements of a non-confrontational, non-judicial and consultative nature for reviewing compliance with the provisions of this Convention. These arrangements shall allow for appropriate public involvement and may include the option of considering communications from members of the public on matters related to this Convention." Provisions for public participation are not yet the norm in MEAs. Others take the view that the public may participate indirectly through submissions to their own governments and to the secretariat and suggest that analogies to the citizen petition procedure of the NAFTA CEC are inapt.

82 Adopted as an interim procedure at the second meeting of the parties II/5 and subsequently revised at the fourth (IV/5) and tenth meetings (X/10). See also the compliance procedure adopted under the CLRTAP by EB 1997/2 and Decision II/4 of the Meeting of the Parties to the Espoo Convention, ECE/MP.EIA/4, 7 August 2001.

83 Efforts to develop a compliance mechanism began some years ago. Those efforts continue today within the Legal Working Group and have yet to reach fruition. See Report of the Fourth Session of the Legal Working Group, 18–19 January 2002, UNEP/CHW/LWG/5. Article 19 of Basel contains reference to a procedure (euphemistically called "verification") for parties to make complaints about another party to the secretariat.

84 Other MEAs only touch on the subject; e.g., Basel, Article 15(7), but this type of provision is becoming more common. See Article 35 of the Biosafety Protocol.

9
POPs and Inuit: Influencing the Global Agenda

TERRY FENGE

INTRODUCTION

Representatives of thirty-six nations gathered in Århus, Denmark, on 28 June 1998 to sign a protocol on persistent organic pollutants (POPs) to the 1979 United Nations Economic Commission for Europe (UNECE) Convention on Long-range Transboundary Air Pollution (CLRTAP). Nearly three years later, representatives of 111 nations gathered in Stockholm, Sweden, to sign a global convention on POPs. Both conventions aim to significantly reduce emissions to the environment of certain POPs and both identify environmental and public health concerns in the Arctic as reasons for concluding these legally binding agreements.

While yet to be fully ratified and implemented, these agreements effectively respond to the deep concerns of Inuit and other northern Indigenous peoples about their cultural survival and the health of their fragile and vulnerable natural environment. During negotiations, Arctic Indigenous peoples did not quietly play the role of victims of change powerless to influence events. Rather, they evocatively and successfully pressed for international action. Their objectives were to protect themselves and their immediate environment from a global threat and to promote the health of the global commons. Major media in western Europe and North America reported their successful advocacy.

While Arctic Indigenous peoples can justly claim to have had an impact on international negotiations, it is important not to overstate their influence. Arctic Indigenous peoples were absent from many of the international meetings in the mid-1990s that put POPs on the international agenda but were active participants in key Canadian and circumpolar research programs that convinced Arctic states of the need for international remedial action.

THE NORTHERN CONTAMINANTS PROGRAM AND THE ARCTIC MONITORING AND ASSESSMENT PROGRAMME

Blood and fatty-tissue samples taken from Inuit in southern Baffin Island and northern Quebec in the late 1980s showed surprisingly high levels of certain POPs, including PCBs and DDT.[1] These unexpected results raised red flags among Inuit organizations and the research, public health, and policy communities in Canada and elsewhere. The presence of these toxins was suspected to have been a result of their long-range transport from tropical and temperate countries to the Arctic and their bioaccumulation in the food web – ultimately in the fat of marine mammals consumed by Inuit.

Although the revelation of POPs in the Arctic was unwelcome, its timing for decision-making purposes was opportune. The findings were highlighted in Canada's 1991 national state of the environment report[2] and had earlier been raised by northerners in the consultations that shaped Canada's 1990 *Green Plan for a Healthy Environment.*[3] The Green Plan committed $100 million over five years to the Arctic Environmental Strategy (AES),[4] a component of which, the Northern Contaminants Program (NCP), was funded $5 million annually for five years with a subsequent renewal for a further five years.

The NCP was designed to determine the types and levels of contaminants in the Arctic, the extent to which people are exposed and the effects of such exposure, and the impacts of strategies to reduce or eliminate contamination and human exposure to contaminants. In 1997 the NCP published the *Canadian Arctic Contaminants Assessment Report* (CACAR) to document the state of contamination in the Canadian Arctic and to compare findings there with findings from other areas of the world, particularly the circumpolar Arctic.[5] Five Indigenous peoples' organizations – Inuit Circumpolar Conference Canada, Inuit Tapirisat of Canada [now the Inuit Tapiriit Kanatami (ITK)], Dene Nation, Métis Nation–NWT, and the Council of Yukon First Nations – joined four federal agencies and three territorial governments to direct and manage the program. Participation in the NCP put the Indigenous peoples' organizations on a steep learning curve, enabling them to intervene in international negotiations some years later with a good understanding of the science of contaminants.

Worrisome environmental contamination and public health data were emerging from behind the rusting Iron Curtain when eight Arctic states – Canada, United States, Russia, Sweden, Norway, Finland, Denmark, and Iceland – began discussions in 1989 to increase environmental co-operation. They easily factored transboundary contamination, including that from sources far to the south, into these discussions and, in 1990, prepared a paper summarizing all that was then known about POPs in the circumpolar Arctic.[6]

Following negotiations lasting nearly two years, political representatives of the eight Arctic states signed a declaration on 14 June 1991 in Rovaniemi, northern Finland, and adopted the forty-five-page Arctic Environmental Protection Strategy (AEPS) to implement the declaration.[7] The AEPS mandated the Arctic Monitoring and Assessment Programme (AMAP), chaired by Canada, with a vice-chair from Sweden and a secretariat from Norway, to examine six pollution issues, including POPs.

The 1991 Rovaniemi Declaration was a milestone. It committed all eight Arctic states to an expansive program of work and promised to promote the results of that work. Importantly, it acknowledged

> growing national and international appreciation of the importance of Arctic ecosystems and an increasing knowledge of global pollution and resulting environmental threats.

and resolved

> to pursue together in other international environmental fora those issues affecting the Arctic environment which require broad international cooperation.[8]

AEPS ministers, in political declarations issued every two years, continued to recognize the need for international action to protect the Arctic. At their 1993 meeting in Nuuk, Greenland, they undertook to consider the development of "regional instruments" to protect the Arctic, a reference to the then-proposed POPs protocol under the CLRTAP.[9]

Three international Indigenous peoples' organizations – Inuit Circumpolar Conference (ICC), Saami Council, and the Russian Association of Indigenous Peoples of the North (RAIPON) – were accorded observer status in the AEPS, enabling them to participate actively in AEPS programs. AMAP quickly became their priority and ICC Canada, with its experience with the NCP, was particularly well placed to contribute to AMAP. (In 2000, other Canadian Indigenous NCP participants organized themselves as the Arctic Athabaskan Council and the Gwich'in Council International to participate in the Arctic Council, which had subsumed the AEPS and its continuing programs in 1996. Indigenous peoples' organizations were accorded permanent participant status to the Arctic Council, increasing their involvement in the council's programs.)

Instructed to monitor the levels and assess the effects of selected anthropogenic pollutants, AMAP produced the world's most detailed and comprehensive regional contaminants assessment report, the *AMAP Assessment Report: Arctic Pollution Issues*, for the ministers at their 1997 meeting.[10] Almost 1,000 pages in length and the result of the collaboration of 400 scientists, the assessment recommended:

The AMAP countries, all being parties to the Convention on Long-range Transboundary Air Pollution (LRTAP), should work vigorously for the expeditious completion of negotiations for the three protocols [including POPs] presently being prepared. ... The protocols should apply throughout the full extent of the geographic area covered by the Convention [North America, Europe, and the countries of the former Soviet Union], ... In addition, the AMAP countries should strongly support the work of the international negotiating committee, to be established early in 1998 following a decision of the Governing Council of the United Nations Environment Programme (UNEP), to prepare an international, legally-binding global agreement on controls for twelve specified POPs.[11]

In response, AEPS ministers committed to

take [AMAP's] findings and recommendations into consideration in our policies and programmes. **We agree** to increase our efforts to limit and reduce emissions of contaminants into the environment and to promote international co-operation in order to address the serious pollution risks reported by AMAP. **We will** draw the attention of the global community to the content of the AMAP reports in all relevant international fora, particularly at the forthcoming Special Session of the General Assembly, and **we will** make a determined effort to secure support for international action which will reduce Arctic contamination.[12]

TURNING SCIENCE INTO POLICY: ARCTIC INDIGENOUS PEOPLES AND THE CLRTAP POPs PROTOCOL

Henrik Selin's chapter in this book addresses the importance of Arctic data in persuading UNECE nations, including Canada, to negotiate a POPs protocol to the 1979 CLRTAP. Although immersed in the NCP- and AMAP-sponsored assessments, Inuit did not participate in the early CLRTAP POPs negotiations, all conducted in Geneva. The Department of the Environment (DOE), initially the lead Canadian agency in the negotiations, sponsored conference calls with representatives of industry, governments, Aboriginal peoples, environmental groups, and others to keep interested parties informed, but the calls proved unwieldy and did little to equip these interests to influence events. Moreover, the Aboriginal peoples were uneasy participating simply as one interest group among many in the conference calls. They were acutely aware of their Aboriginal and treaty rights and of the Crown's fiduciary obligations that placed significant consultative responsibilities upon the federal government before engaging in international negotiations that might affect their rights. At no stage were Inuit or other Aboriginal peoples' organizations invited to assist the federal government in developing its formal negotiating position.

As concerns grew over the public health implications of POPs in the Arctic, the five Aboriginal peoples' organizations involved in the NCP formed a coalition – Northern Aboriginal Peoples' Coordinating Committee on POPs – in March 1997 to participate in and influence Canada's position in both the CLRTAP and the proposed global POPs negotiations. DOE encouraged the Aboriginal peoples' organizations to form the coalition, and modest funding was obtained from the NCP and DOE. The coalition hired a part-time technical advisor and used left-over funds to pay travel costs for a small delegation to attend the remaining CLRTAP POPs negotiations.

Some weeks later, the coalition members obtained a March 1997 letter written to the Department of Foreign Affairs and International Trade (DFAIT) on behalf of the federal departments of Industry, Natural Resources, and Agriculture and Agri-food and the Pest Management Regulatory Agency (PMRA). The letter suggested that all was not well in the Canadian approach to the issue or on the Canadian negotiating team. Citing the "urgent need for a Federal Strategy on POPs," including Cabinet instructions to the negotiating team and the need for the DFAIT rather than DOE to lead and co-ordinate Canada's efforts, the letter stated:

> Actions on these substances can be expected to have significant economic impacts, by virtue of reducing or eliminating markets for specific products and associated industries, and by requiring substantial investment in technology designed to reduce or eliminate byproducts of a wide range of manufacturing processes.[13]

The coalition reacted immediately to this letter's characterization of the POPs issue as an economic matter, rather than as a matter of public health. This was particularly puzzling in light of the reporting relationship of PMRA to the minister of Health, who it was assumed should automatically see this as a health issue. In addition, the coalition was surprised to learn that Canada's position in Geneva was the result only of agreements among federal civil servants. The coalition felt this did not reflect northerners' concerns about POPs and public health. Moreover, it was not consistent with the significant investment in the NCP or with Canada's considerable efforts in the early to mid-1990s to persuade other countries to negotiate a POPs protocol.

The letter revealed disagreement among federal agencies on Canada's position and on the make-up and leadership of the negotiating team, and the coalition feared that the proposed Cabinet process could weaken Canada's commitments to reduce emissions of POPs. These concerns prompted the coalition to write to the minister of Foreign Affairs and International Trade pointing out that the proposed POPs protocol would, in fact, have few economic implications: none of the initial eighteen named substances were manufactured in Canada and only four were used in the country. The coalition noted that the

International Council of Chemical Associations, of which the Canadian Chemical Producers Association was a member, supported the proposed protocol. In a direct appeal to the minister the coalition said:

For more than five years we have observed and applauded an aggressive Canadian commitment to dealing with long-range transport of POPs. We are puzzled by this apparent change of heart on an issue the federal government has so effectively convinced us to be of immediate concern and importance to our environment and our health. We ask you to give this issue your personal attention and to encourage your Cabinet colleagues to define a negotiating position that fully takes into account the environmental and public health implications of POPs to northern aboriginal peoples.[14]

That political leaders of all northern Aboriginal peoples would jointly sign a letter well illustrated the political importance of the POPs issue.

Further evidence to support the appeal to the minister was revealed in a May 1997 letter from Indian and Northern Affairs Canada (INAC) to PMRA concerning the inclusion in the protocol of the pesticide lindane. The letter acknowledged that the Canadian team had

decided to recommend that Canada will not support the inclusion of any controls on Lindane including restrictions which may be compatible with current Canadian regulations [as] the presently valid Canadian risk assessment for this substance would not justify such action.[15]

Having noted that INAC would "respect" this decision, the letter stated that significant, if preliminary, evidence in the *CACAR* and the AMAP assessment showed lindane to be a real public health concern in the North. In the Baffin region, dietary surveys had shown that nearly fifteen per cent of Inuit women studied ingested more than the advised tolerable daily intake (TDI) of hexachlorocyclohexane (HCH) – a component of lindane.[16] When PMRA, citing proprietary concerns, declined to provide INAC with copies of the risk assessment and supporting data for lindane, obvious questions about the methods, data, and current validity of the assessment were raised. PMRA's actions in denying requested information to a sister agency on the same negotiating team were questioned by the coalition. In December 1997, ICC Canada formally asked PMRA for this information[17] and was similarly refused.[18] To the consternation of the coalition, the minister of Health supported PMRA and also refused a request by ICC Canada[19] to release information on the health assessment of lindane.[20]

Now thoroughly concerned about Canada's position, the coalition used ICC's consultative status to the United Nations Economic and Social Council (ECOSOC) to send an observer to the June 1997 POPs negotiations in Geneva. The observer reported that Canada was taking a more cautious and conservative position than virtually any other nation on the inclusion of substances and on import and export controls, a position curiously at odds with Canada's successful attempts to persuade the CLRTAP Executive Body to negotiate a POPs protocol. The observer also reported that representatives of industry observing the events were in close contact with the U.S. delegation, but that no environmental, public-interest, or other non-government groups were in attendance. National and international media either did not know about the negotiations or had chosen not to cover them. Of particular interest, in the context of the interdepartmental correspondence noted earlier, the Canadian delegation was now co-chaired by the departments of Environment and Foreign Affairs and International Trade.

These changes in personnel and attitude did not go unnoticed: representatives of two other Arctic delegations approached the ICC observer asking whether Canada had changed its view on the need for a POPs protocol. The coalition suspected that lobbying by the economic development agencies in Ottawa was affecting Canada's negotiating position and posture; this conclusion was reinforced in a letter to the coalition from the minister of Foreign Affairs and International Trade:

> You express concern that consideration of economic interests could delay and weaken potential control mechanisms for POPs. I wish to assure you that the final Canadian mandate for the negotiations will be based on principles of sustainable development, which will involve a thorough consideration of all environmental, social, and economic concerns. Our position will also take into account the views expressed by Northern Aboriginal organizations during the consultations that preceded the launch of negotiations.[21]

The final sentence left the coalition wondering what "consultations" the minister assumed had taken place. That the minister had characterized POPs as a sustainable development issue and not as an issue of public health confirmed suspicions that important federal agencies, with the notable exception of INAC, saw the issue differently from those it directly affected. The coalition feared that the public health of Aboriginal peoples might be traded off in the rough and tumble of international politics.

Based on observations in Geneva and taking the minister's commitment that Canada would soon provide a "final mandate," the coalition wrote to him in August that Canada's current position was "not acceptable." The letter cited Canada's refusal to include short-chain chlorinated paraffins (SCCPs), lindane, or pentachlorophenol (PCP) in the protocol; to support inclusion of trade-restrictive measures (import/export controls); and to invest authority in the convention's executive body to add substances to the protocol as and when science showed them to be of concern. The coalition concluded:

> To Inuit, Dene and Métis, and First Nations in northern Canada, contamination of the wildlife we hunt and eat is a matter of public health, and as such is an issue we take

seriously above all others. A significant percentage of aboriginal people in the North have levels of certain POPs in their bodies which greatly exceed Health Canada's "level of concern." We recommend to you that the Canadian position be reassessed. Canada should join the majority of other LRTAP Convention countries in supporting an effective Protocol that lays a strong foundation for negotiation of the UNEP legally binding POPs instrument.[22]

The federal Cabinet approved a final mandate in October 1997. The minister of the Environment tabled the correspondence from the coalition to the minister of Foreign Affairs and International Trade with the Cabinet submission to illustrate the depth of opinion on the issue. Several members of Parliament with northern interests contacted ministers about this issue, characterizing it as "nationally important," and were influential in raising its profile in Ottawa. The summer 1997 report of the Standing Committee on Foreign Affairs and International Trade on Canada's Arctic policy called upon the federal government to "redouble" its efforts to conclude an expansive POPs protocol.[23] Also influential was the lobbying of the Canadian Polar Commission (CPC) and the Canadian Arctic Resources Committee (CARC).

Two representatives of the coalition observing proceedings at the October 1997 POPs negotiations received help from the Danish, Norwegian, and Canadian delegations in facilitating their hallway involvement and advocacy. Canada's position softened slightly from where it had been in June. Canada now accepted that lindane could be included in the protocol as a restricted substance but insisted that all existing uses in Canada be allowed to continue and remained firmly against import and export controls. While refusing to accept a phase-out date, Canada agreed to a future review of lindane's uses.

With an eye towards future proposed negotiations for a global agreement on POPs, the coalition tabled, through the very supportive Swedish chair, five clauses that, if accepted, would establish the protocol's context in Arctic, Aboriginal, and public health concerns:

AWARE that persistent organic pollutants resist degradation under natural conditions, particularly in cold climates, and that certain persistent organic pollutants have been associated with adverse effects on human health and the environment, and that this is an immediate public health issue for Arctic indigenous peoples;

RECOGNIZING that many persistent organic pollutants migrate to the Arctic where they deposit and accumulate in terrestrial and aquatic ecosystems;

ACKNOWLEDGING that Arctic ecosystems are especially vulnerable to the serious threat posed by persistent organic pollutants which have been shown to bioaccumulate in the lipid-rich tissues of Arctic organisms;

COGNIZANT of the particular and immediate threat posed by persistent organic pollutants to the physical and cultural well being of indigenous peoples and others who are dependent on the harvest of country foods;

The ultimate objective of this protocol is to protect human health and the environment from the adverse effects of persistent organic pollutants subject to long-range transboundary atmospheric transport by taking measures, consistent with the precautionary principle, to control, reduce or eliminate their discharge, emission, and loss.[24]

Privately, the American delegation and representatives of Scandinavian countries warmly welcomed this language and proposed to support it. The Canadian delegation did not inform the coalition of its views. In subsequent meetings of heads of delegations, from which observers were excluded, the coalition's preambular language was, in part, accepted. The objective clause detailing the protocol's scope was not.

THE COALITION'S CONCLUSIONS FROM THE CLRTAP POPs PROCESS

The coalition welcomed the POPs protocol to the CLRTAP, but was not impressed with Canada's performance in the negotiations or with the manner in which the federal team approached the coalition. It seemed obvious to the coalition that the federal government's delegation should seek to protect the rights, interests, health, economy, and culture of Inuit and other northern Indigenous peoples. The coalition assumed, perhaps naively, that Canada's approach, posture, and position in the CLRTAP negotiations would remain consistent with the spirit of co-operation between federal agencies and northern Aboriginal peoples in the NCP and reflect the excellent work of federal agencies in the AEPS and Arctic Council. That the public health of northerners did not trump interdepartmental wrangling and economic concerns among federal agencies came as a disappointment.

Most federal representatives on the delegation seemed to have little knowledge of the political, constitutional, environmental, and socio-cultural circumstances that distinguish the territorial North from the provincial South. Neither did they fully appreciate federal obligations to northern Aboriginal peoples under comprehensive land-claims or self-government agreements and Aboriginal rights or as a result of the Crown's fiduciary relationship with Inuit, Dene, Métis, and First Nations. At the very least, these obligations suggest that the departments of Environment and Foreign Affairs and International Trade should have consulted with and directly involved northern Aboriginal peoples in defining Canada's negotiating position. The coalition interpreted the delegation's stand-offish approach and demeanour to be serving federal rather than national interests. It also noted that foundations, environmental groups, and the media based in southern Canada had not responded to the CACAR and the AMAP assessments and had not expressed interest in the CLRTAP POPs process. Coalition members asked themselves whether silence would similarly shroud this issue if the levels of POPs in Inuit women were being discovered in mothers in southern Canada.

The coalition strongly believed that Canada should adopt a more open and participatory approach in preparing to negotiate a global POPs agreement. Moreover, they felt that defending the interests of northerners should be a central objective and the delegation should be coordinated by representatives with real knowledge of the North. The coalition also hoped that the environmental movement, particularly in Canada, the United States, and western Europe, would intervene in the global POPs process and, in so doing, help attract media attention.

In early 1998 the coalition began looking for allies to help promote a global POPs convention that would reflect Arctic and Aboriginal perspectives. This search led to discussions between the coalition and environmental and public health organizations in tropical and temperate lands. Soon, this quest came to consider how Aboriginal peoples from the Arctic could make common cause with residents of tropical and temperate lands – the origin of many Arctic POPs – bridging north and south through mutually supportive advocacy.

TURNING SCIENCE INTO POLICY: ARCTIC INDIGENOUS PEOPLES AND THE GLOBAL POPS CONVENTION

UNEP's governing council met in 1995 and 1997 and, with a key role played by Canada, decided to assess twelve POPs – the "dirty dozen" as they came to be known – to make the case for global action. The governing council requested its executive director to convene an Intergovernmental Negotiating Committee (INC) to prepare an international legally binding instrument to address the identified POPs and to identify additional candidate substances for future action. Negotiations began in summer 1998 in Montreal and continued later in Nairobi, Geneva, Bonn, and Johannesburg, with the resulting convention being signed in Stockholm on 23 May 2001. The coalition, by then renamed the Canadian Arctic Indigenous Peoples Against POPs (CAIPAP), participated actively in all negotiations and at the signing ceremony.

At the beginning of this process, CAIPAP developed a basic position from which it never wavered: it sought a comprehensive, verifiable, and rigorously implemented convention to protect the health and way of life of northern Indigenous peoples. This position was supported by the coalition's technical analyses that the convention should commit to the elimination of POPs rather than to perpetual management of them and that generous financial and technical assistance be provided to developing countries and those with "economies in transition" to enable them to meet obligations and duties under the convention. The coalition also developed positions on destruction of stockpiles and import and export controls and proposed other features of a "model" convention.

Concerned that states in the developing world would sign the convention with a political flourish but, lacking capacity, would fail to implement it, the coalition received advice from faculty members and graduate students at the Faculty of Law, University of Calgary, suggesting the convention include language similar to arms-control treaties in promoting monitoring and verification. The coalition worked hard to make its position and background material known to the Canadian delegation and representatives of other Arctic states and prepared a detailed "wish list" addressing each draft article, which it presented to the Canadian delegation before the Nairobi INC.

The chair of the negotiations, a Canadian with the federal Department of Environment, proved highly skilled: His sure-footed ability and calm and inclusive approach convinced states and non-government organizations that they would be heard and their views considered.[25] With CAIPAP again operating from the back of the room, the chair generously acceded to its requests to intervene at strategically important moments. Sheila Watt-Cloutier, president of ICC Canada and vice-president of ICC, attended and spoke at all negotiations.[26] Ms Watt-Cloutier, an Inuk from Kuujjuaq in Nunavik (northern Quebec), proved to be a gifted public speaker able to convey technical information to a large audience and to do so from the heart. Political representatives of the Council of Yukon First Nations (CYFN) also attended and intervened at certain sessions.

CAIPAP's first plenary intervention at the Montreal session proved to be highly effective and of long-term importance. It illustrates well the high moral ground the coalition came to occupy and its style of delivery. Beginning in Inuktitut, with no translation provided, and then switching to English, Ms Watt-Cloutier said:

> [I]magine for a moment if you will the emotions we now feel – shock, panic, rage, grief, despair – as we discover that the food which for generations has nourished us and keeps us whole physically and spiritually is now poisoning us. You go to the supermarket for food. We go out on the land to hunt, fish, trap and gather. The environment is our supermarket ... Our living land itself, whose healing energy is so strong it can be palpably felt by anyone, is being made to quietly absorb layer upon layer of contaminants ...
> Many of these contaminants are passed from one generation to the next through the placenta and breast milk. As we put our babies to our breasts we feed them a noxious chemical cocktail that foreshadows neurological disorders, cancer, kidney failure, reproductive dysfunction, etc. This is truly worrying. The bond between mother and child

lies at the core of our culture. I expect this is the same for all peoples around the world. That Inuit mothers – far from areas where POPs are manufactured and used – have to think twice before breastfeeding their infants is surely a wake-up call to the world... . [W]e Inuit are few in numbers and don't constitute a major lobby group. Until recent years, we have not been influential in world affairs as we were still reeling from tumultuous change. But we are back now and wish to speak out on behalf of the land that has sustained us for hundreds of generations. We are the land and the land is us. We cannot stand by, waiting for slow moving governments to step in and make everything right, rather we must try to effect what change we can. ...
If we can help people to see that a poisoned Inuk child, a poisoned Arctic and a poisoned planet are one in the same, then we will have effected a shift in peoples' awareness that will result without doubt in positive change.[27]

The coalition's position was reasonable, technically well reasoned, and consistently advocated. That Inuit and other Arctic Indigenous peoples were "exotic" to most of the participating delegates strengthened the influence wielded by the coalition. At all negotiations, the coalition organized and staffed a display of country food, furs, and hunting paraphernalia to remind delegates of the reality of living in the Arctic and the threat posed by POPs to this way of life. An important bridge-builder between the coalition and the environmental movement occurred at the Geneva session when the Greenpeace delegates allowed themselves to be photographed wearing sealskins. Inuit communities had been economically devastated in the 1980s and 1990s after Greenpeace had successfully lobbied European governments to ban the importation of furs and staunchly supported the U.S. Marine Mammal Protection Act, which prohibits the importation of marine mammal products, including sealskins, to the United States.

The global POPs negotiations attracted the attention of many organizations, including the World Wide Fund for Nature from the United States and CARC from Canada as well as Greenpeace. At the lead of the Washington, D.C.-based Physicians for Social Responsibility, many non-government groups, including Indigenous peoples from the United States, came together in the International POPs Elimination Network (IPEN). CAIPAP agreed with IPEN's position but chose to maintain its independence and to participate in selected IPEN conferences and workshops held before and during negotiations.

IPEN brought significant intellectual and financial resources and media savvy to the POPs negotiations. Events such as street theatre and enthusiastic but always peaceful protests galvanized media attention. IPEN, WWF, and Greenpeace generously encouraged CAIPAP to share their press resources and to participate in their press conferences. The media gravitated to the Arctic and Aboriginal peoples' dimensions of the issue; Arctic Aboriginal peoples featured prominently in wire stories filed from all negotiating venues.

The coalition learned a great deal from the Montreal negotiations; in the POPs context, Arctic Indigenous peoples are newsworthy in European capitals. In subsequent negotiations CAIPAP sought to push the envelope. The Government of Germany's invitation to Indigenous peoples' dance and song troupes from Nunavut and Yukon to perform at an evening social event in Bonn generated an opportunity for political representatives of CYFN and ICC Canada to speak to the negotiators about POPs and culture.

Larissa Abroutina, a Chukchi medical doctor from Chukotka in the Russian Far East and vice-president of RAIPON, joined the coalition for the negotiating sessions in Nairobi, Geneva, Bonn, and Johannesburg and spoke convincingly of the POPs-related health concerns of the 200,000 Indigenous people in the Russian Arctic. The trust built through a joint ICC Canada/RAIPON democracy- and institution-building project in northern Russia and solidarity developed in AEPS and Arctic Council meetings contributed to Dr Abroutina's presence. As negotiations progressed, CAIPAP assumed the mantle of a circumpolar coalition of Indigenous peoples, and at its request the Canadian delegation included among its members an Aboriginal woman from Yellowknife well versed in contaminant issues.

CAIPAP's informal moral influence strengthened through the negotiations. At Nairobi, UNEP sponsored an evening reception where Ms Watt-Cloutier spoke and presented Dr Klaus Töpfer, executive director of UNEP and ex-minister of the environment for Germany, with an Inuit carving of a mother and child. Dr Töpfer immediately passed the carving to the chair of the POPs negotiations and gave an impromptu speech suggesting that Indigenous peoples were the "conscience" of the negotiations and that the world was obliged to take their concerns seriously. During all subsequent negotiations the carving sat on the chair's table – a symbolic reminder of that conscience. Dr Töpfer referred to it, sometimes wistfully, in every speech he delivered at meetings of the INC, urging negotiators to conclude a strong convention. A photo of the carving was posted on the UNEP POPs website, and Dr Töpfer invited Ms Watt-Cloutier to join him in hosting the official press conference at the conclusion of the Nairobi negotiations, where both spoke of connections between the Arctic and the wider world. The coalition had made an important and influential friend.

FINANCING THE CONVENTION

As the process continued, several key issues emerged: financing the implementation of state obligations; providing for technical assistance for developing countries and those with "economies in transition"; and placing DDT in the convention. Canada played a positive role in these debates. Authorized by the February 2000 federal budget, Canada's co-chief negotiator from DFAIT announced in Bonn, Germany, a contribution of CAN$20 million to assist convention implementation. This was the first financial contribution announced during negotiations, and the funds were almost immediately transferred to the World Bank for distribution. Canada also organized informal meetings of donor countries to persuade them to commit to funding support.

The role of the Global Environment Facility (GEF) in financing convention implementation was central in debates in the final three sessions. Established as a result of the 1992 Earth Summit, the Washington, D.C.-based GEF finances the "incremental" costs of delivering "global benefits" through national projects related to international conventions (including those on climate change and conservation of biological diversity). Funding for the GEF comes from its 171 member states and a range of corporate and academic partners. A GEF offer to establish a program to fund POPs projects was not immediately accepted by the developing world, which claimed the GEF to be overly bureaucratic, difficult to access, and dominated by donor countries. Instead, developing countries suggested a new multilateral fund for POPs projects similar to that included in the Montreal Protocol on Ozone Depletion. While sympathetic to the developing world but mindful that western Europe, Japan, and North America would pay the piper and call the tune, the coalition spoke of the need for substantial and stable funding and transparent processes to allow timely access.[28]

ICC Canada was heavily involved in two initiatives regarding the potential role of the GEF in implementing a global POPs convention. Following Sheila Watt-Cloutier's remarks at the Montreal negotiations, the coalition was approached by two World Bank-based GEF secretariat officers who asked whether, how, and by whom a POPs-related Indigenous peoples' project could be mounted in northern Russia. With the GEF providing US$25,000, ICC Canada, RAIPON, and the AMAP Secretariat jointly prepared a US$2.5 million proposal to conduct in northern Russia the type of research undertaken through the NCP in northern Canada. The GEF subsequently contributed US$750,000 to this three-year project, and additional funds were obtained from six Arctic states and a private Toronto-based foundation. GEF personnel attending the POPs negotiations used this project, initiated in late 1999, to demonstrate their reinforced commitment to fund implementation of the POPs convention and to work with non-government organizations.

At the August 2000 meeting of Arctic Parliamentarians in northern Finland, Ms Watt-Cloutier shared the podium with Mr Mohamed El-Ashry, chief executive officer of the GEF – a meeting carefully arranged by Clifford Lincoln, a Canadian member of Parliament and chair of the Arctic Parliamentarians. Ms Watt-Cloutier spoke of the central issue of financing the convention and urged Mr El-Ashry to attend the final negotiations in Johannesburg to put forward the GEF case in person and to assure all delegates that the GEF was willing and able to reform its granting procedures. This is precisely what Mr El-Ashry did, and delegates eventually agreed that the GEF would be a key mechanism to implement the convention.

OBSERVATIONS ON THE ROLE OF THE UNITED STATES

Highly influential in any global negotiations, the United States is also an Arctic state many thought would play a leadership role in promoting a strong and effective global POPs convention. Expectations were high because of its accession to AEPS political declarations. That it did not do so created difficulties both for the coalition and for Canada, which invariably seeks positions in close accord with its giant neighbour.

Following three negotiating sessions, the United States sent a diplomatic note to the European Union about the state of the negotiations. Leaked in Europe, this note said:

> [T]he POPs issue is not a global commons issue to the same degree as ozone depletion or climate change. The most serious deleterious effects of POPs are often felt near to where they are produced and/or used. Thus, the G77 [developing world] needs to accept that the OECD [developed world] is not prepared to bear all the costs of implementing POPs.[29]

Missing from this analysis was an Arctic or even an Alaskan dimension acknowledging the long-range transport phenomenon and, therefore, the global nature of the POPs problem. The position seemed to reflect the difficulties the United States had in committing funding in advance of Congressional consideration of the POPs convention and the fact that it was already in arrears in its contribution to the GEF.

When this position was reiterated orally in Geneva, the president of ICC Alaska wrote to the U.S. secretary of state to seek clarification and to remind her of Arctic and Inuit concerns.[30] The reply, curiously sent to Alaskan Senator Ted Stevens rather than to ICC Alaska, essentially restated the diplomatic note.[31] ICC felt that the U.S. Department of State did not understand the Arctic dimension to the issue or appreciate the depth of concern about POPs held by Indigenous peoples in Alaska.

Fortunately, the Arctic dimension to POPs did find its way onto the agenda of the United States. As chair of the Arctic Council from 1998 to 2000, the United States hosted a meeting of council ministers in Barrow, Alaska, in autumn 2000. This provided an important platform for Alaskan interests, including the governor, to press visiting senior State Department officials to adopt a more positive position in the negotiations. The governor urged the U.S. government to support a strong POPs convention and to sponsor badly needed

POPs research in Alaska.[32] At the same time, Alaskan Indigenous peoples began a campaign to establish a POPs research and monitoring program modelled partly on Canada's NCP.[33] Sensitive to Alaskan concerns, the Arctic Council ministers issued a political declaration that further committed all Arctic states to press for a POPs convention that would benefit the Arctic.[34]

Almost simultaneously, the Montreal-based Commission for Environmental Co-operation, established through the North American Free Trade Agreement, released a theoretical and computer analysis of long-range transfer to eight communities in Nunavut of dioxins released to the environment by industrial and waste incineration facilities.[35]

The study concluded that the vast majority of dioxins in Nunavut came from the mid-west and the eastern seaboard of the United States. It was released in New York City, and once more the media distributed the Inuit and Arctic dimensions to the issue worldwide. All of this heightened pressure on the U.S. Department of State to take a more forward-looking position in the global POPs negotiations. At the final negotiations in Johannesburg, the delegation from the United States included Indigenous Alaskans and senior representatives of the state government. These representatives worked with CAIPAP to draft a preambular clause singling out Arctic concerns that was tabled by the United States and accepted in plenary.

CAIPAP AND CANADA

Unimpressed with the less-than-stellar performance of the Canadian delegation in the CLRTAP negotiations, CAIPAP pressed hard for improvements. Coalition members engaged in many hours of debate with the delegation and met with ministers, members of Parliament, parliamentary committees, senior civil servants, and selected media to press the case. In March 2000, immediately before the Bonn sessions, a national Canadian newspaper, *The Globe and Mail*, published an opinion-editorial by ICC Canada noting the decline of Canada's international environmental standing caused by its lacklustre performance in meeting commitments from the 1992 Earth Summit. The article challenged the delegation:

> Next week in Bonn, we'll find out much about Canada's foreign-policy priorities – and whether Canada still has what it takes to lead the world on a public health and environmental security issue of compelling importance to its own people.[36]

The nadir had been reached about one year earlier, a few weeks after the Nairobi session. In a March 1999 conference call sponsored by DOE to inform interests about the Nairobi meeting, the ICC Canada participant cut short complaints of the co-chief negotiator from DOE about the positions of developing countries. When asked about Canada's position on key substantive issues he said that Canada had yet to formulate positions. This response was clearly unacceptable and prompted the ICC Canada representative to write privately to the DOE negotiator recommending that he resign. In a lengthy and very difficult meeting between ICC Canada and both DOE and DFAIT co-chief negotiators, ICC repeated demands for effective, committed, and skilful leadership, leaving no doubt that the coalition meant business.

Ms Watt-Cloutier repeated the demand for effective leadership in a meeting with the minister of the Environment; political leaders of all coalition members also informed the minister in a five-page letter of their "deep concern" about Canada's performance:

> We observe that Canada is participating in these global negotiations with neither a "top line" position to eliminate POPs nor clear strategies and tactics to achieve this goal ... We expect Canada to lead by example in these negotiations, yet we find it seeking the most politically acceptable positions supported by our neighbours to the south ... Our continued requests for information on Canada's position and negotiating strategy have yielded only general, uninformative, and unhelpful responses ... In sum, we believe Canada's negotiating position and its posture are not appropriate to the seriousness of the issue. Furthermore, Canada's position – or lack thereof – does not reflect what we have articulated and provided to the negotiating team during the consultative process. Nor does it reflect the Canadian statement made in Nairobi that Canada favours "aggressive action on POPs."[37]

CAIPAP recommended that the minister seek the authority of Cabinet to renew the vacant Office of Environment Ambassador, and that the person chosen to occupy this office lead Canada's delegation to global POPs negotiations. Canada's delegation became very much more co-operative, collegial, and engaged with the coalition after these interventions.

DDT AND BRIDGING NORTH AND SOUTH

Although banned in Canada, DDT is used in tropical countries as a vector control for malaria, saving the lives of thousands of people every year. At negotiations in Geneva some developing countries, aided by a public-interest organization, balked at DDT being included in the convention. The issue was threatening to split negotiations along north–south lines when Ms Watt-Cloutier announced that Inuit would refuse to be party to a convention that threatened the health of others notwithstanding the threat of POPs to their own health. The coalition wanted only a win-win solution and such selfless remarks bridged the north–south divide.

This issue arose again at the final negotiations in Johannesburg. An opinion-editorial written by Amir Attaran, a Canadian apparently associated with Harvard University and an active lobbyist at the negotiations, and published in December 2000 in *The Globe and Mail* bitterly criticized the Canadian delegation:

> [N]ot only is Environment Canada arguing in the Johannesburg treaty negotiations that DDT should be eliminated once and for all, it has proposed that the treaty not include a financial aid mechanism to help poor countries finance the alternatives. Canada is alone among wealthy countries in advocating this parsimony. Such policies literally kill.[38]

Telephone lines buzzed between Ottawa and Johannesburg as ministers in Ottawa demanded to know why Canada was getting such bad press. Canada's delegation was temporarily thrown off balance at a critical juncture in the negotiations. The coalition knew these criticisms to be outrageous and wrong. Urging the Canadian delegation to remain focused on negotiations, CAIPAP replied a few days later through *The Globe and Mail*:

> [N]obody supports a ban [of DDT] that puts lives at risk. When this issue surfaced in 1999, Canadian Indigenous peoples said they would refuse to be party to an agreement that threatened the health of others, notwithstanding the threat of POPs to their own health. [This] is also the fundamental position of all countries participating in the negotiations and all non-government organizations observing the debate ... Mr Attaran's contention that Canada refuses to help finance alternatives to DDT is demonstrably untrue. The Minister of Finance announced $20 million in his February budget for exactly this purpose. Canada was the first nation to provide such support and is effectively advocating additional financial and technical assistance to developing countries and "economies in transition."[39]

Canadian Indigenous peoples publicly defended the honour of the Crown in a delicate and embarrassing situation, demonstrating the trust and openness that had developed between them and the Canadian delegation in the year or so following the second INC. The coalition realized that the Canadian delegation was best able to advance Arctic views; Canada had the largest delegations at negotiations and its position on many issues of importance to the coalition had improved significantly. For example, the capacity-assistance network Canada pressed upon delegates bore measurable resemblance to the monitoring and assistance suggestions made earlier by the coalition. Canada had put money on the table in Bonn precisely to encourage others to do likewise as recommended by the coalition. The Canadian delegation perhaps realized that the coalition had some good ideas and was right to fight hard and to make waves. And, it was indisputable that the coalition had influential colleagues in Ottawa who were scrutinizing Canada's performance.

SIGNING AND CELEBRATION

On behalf of CAIPAP, Ms Watt-Cloutier joined the Hon. David Anderson, Minister of the Environment for Canada and Chair of UNEP Governing Council, in Stockholm at the May 2001 global POPs convention signing. UNEP's magazine, *Our Planet*, subsequently printed key points she raised at the signing:

> Klaus Töpfer thanked us for our efforts in support of a global POPs Convention. He acknowledged that we had helped. This was important and uplifting. Inuit have come a long way in a short time. Robert Peary, the famous American polar explorer once said of Inuit: "Of what value to the world are these people? They have no culture to speak of, no written language. They value life only as a fox or a wolf." We have shown the world our value and we hope and intend to do so again on other global issues of importance in the Arctic. We are able to do this because we can draw upon our rich and diverse cultural heritage. Hunting, fishing, and trapping continue to provide us with lessons of great relevance. We remain guardians of the natural environment. As we continue to navigate rapid social change it seems highly appropriate that Inuit provide advice to the world on issues that affect the health of our planet.[40]

Some weeks later in Ottawa, at a reception for about 100 people who had worked on the issue, the Rt. Hon. Herb Gray, Deputy Prime Minister, noted that Indigenous peoples had exerted influence in the negotiations out of all proportion to their numbers. And in Washington, President Bush, supported by the secretary of state and the administrator of the Environmental Protection Agency, announced to the press that his administration would sign the convention.

LESSONS AND CONCLUSIONS

From the perspective of northern Indigenous peoples, what lessons or conclusions from the international POPs debate can be usefully drawn?

1. The Northern Contaminants Program (NCP) was of central importance to making the scientific case for international action and in persuading the government of Canada to argue for international agreements to control, reduce, and eventually eliminate key POPs. But more than this, the NCP educated and helped finance northern Indigenous peoples, allowing them to act

effectively on the circumpolar and global stages, defending their interests and promoting their ways of life. In times of fiscal restraint it is important to note that well designed programs such as the NCP make a difference.

2. The AEPS and the Arctic Council proved important in promoting international action; they generated and aggregated AMAP's scientific information and presented it to decision makers. Arctic Indigenous peoples used their "permanent participant" status in these bodies to prod the Arctic states to international action. The United States, although not, perhaps, psychologically an Arctic state, joined the POPs fight in large part because of lobbying by Alaskan interests and the fortuitous meeting of the Arctic Council in Alaska in 2000.
3. That Indigenous peoples from northern Canada worked together throughout the negotiations, and did so within domestic and international rules, exerting real and acknowledged influence, is precedent setting. Having an articulate political spokesperson committed to the global POPs process throughout the negotiations was of immense value. That the coalition was not established before the beginning of CLRTAP POPs negotiations was unfortunate and invites the question of what might have been.
4. Climate change, arguably the defining global issue of coming decades, is already influencing polar regions as computer models have predicted. The popular image of climate change in the Arctic consists of melting glaciers and fewer and thinner polar bears. Might the Arctic Council, aided by Arctic Indigenous peoples, draw upon the POPs experience to re-balance popular perceptions by giving global climate change a human face – perhaps an Inuk mother feeding her child? Climate change is also an issue of cultural survival to Inuit. Such action might stimulate more aggressive action by national governments to reduce emission of greenhouse gases that threaten the world and the Arctic in particular.

NOTES

1 E. Dewailly et al., "High Levels of PCBs in Breast Milk of Inuit Women from Arctic Quebec," *Bulletin of Environmental Contamination and Toxicology* 43, no. 1 (1989): 641–6; D. Kinloch, H.V. Kuhnlein, and D. Muir, "Assessment of PCBs in Arctic Foods and Diet: A Pilot Study in Broughton Island, NWT, Canada," *Arctic Medical Research* 47 (supplement) 1998. See also Henrik Selin, "Regional POPs Policy: The UNECE CLRTAP POPs Protocol," in this volume.

2 Canada, *The State of Canada's Environment* (Ottawa: Minister of Supply and Services Canada 1991).

3 Canada, *Green Plan for a Healthy Environment* (Ottawa: Minister of Supply and Services Canada 1990).

4 Arctic Environmental Strategy (Ottawa: INAC 1991).
5 J. Jensen, K. Adare, and R. Shearer, eds, *Canadian Arctic Contaminants Assessment Report* (Ottawa: Minister of Public Works and Government Services Canada 1997).
6 J. Jensen, "Report on Organochlorines," in *The State of the Arctic Environment Reports* (Rovaniemi, Finland: Arctic Centre, University of Lapland 1991), 335–84.
7 http://www.arctic-council.org/pdf/arctic_environment.PDF [Editors' note: To link to the PDF, you must key in the second arctic as "artic."]
8 Ibid, p. 6.
9 http://www.arctic-council.org/nuuk.asp.
10 *AMAP Assessment Report: Arctic Pollution Issues* (Oslo, Norway: Arctic Monitoring and Assessment Programme 1998).
11 AMAP, *Arctic Pollution Issues: A State of the Arctic Environment Report* (Oslo, Norway: Arctic Monitoring and Assessment Programme 1997), p. xii.
12 http://www.grida.no/prog/polar/aeps/alta.htm.
13 L. Bradet, Industry Canada, letter to R. Archibald, Department of Foreign Affairs and International Trade, 21 March 1997.
14 Northern Aboriginal Peoples Coordinating Committee on POPs, letter to Hon. L. Axworthy, Minister of Foreign Affairs and International Trade, 1 May 1997.
15 L. Whitby, Indian and Northern Affairs Canada, letter to R. Taylor, Pest Management Regulatory Agency, 15 May 1997.
16 H.V. Kuhnlein et al., "Arctic Indigenous Women Consume Greater than Acceptable Levels of Organochlorines," *Journal of Nutrition* 125 (1995): 2501–10.
17 T. Fenge, Inuit Circumpolar Conference Canada, letter to L. R. Leblanc, Pest Management Regulatory Agency, 18 December 1997.
18 L.R. Leblanc, Pest Management Regulatory Agency, letter to T. Fenge, Inuit Circumpolar Conference Canada, 14 January 1998.
19 S. Watt-Cloutier, President, ICC Canada, letter to Hon. A. Rock, Minister of Health, 20 January 2000.
20 Hon. A. Rock, Minister of Health, letter to S. Watt-Cloutier, President, ICC Canada, 29 March 2000.
21 Hon. L. Axworthy, Minister of Foreign Affairs and International Trade, letter to political leaders of the Northern Aboriginal Peoples Coordinating Committee on POPs, 22 July 1997.
22 Northern Aboriginal Peoples Coordinating Committee on POPs, letter to Hon. L. Axworthy, Minister of Foreign Affairs and International Trade, 8 August 1997.
23 "Canada and the Circumpolar World," Report of the House of Commons Standing Committee on Foreign Affairs and International Trade, 22 April 1997.
24 T. Fenge, "POPs in the Arctic: Turning Science into Policy," *Northern Perspectives* 25, no. 2 (1998): 8–14.
25 See J. Buccini, "The Long and Winding Road to Stockholm: The View from the Chair," in this volume.

26 See S. Watt-Cloutier, "The Inuit Journey towards a POPs-free World," in this volume.
27 S. Watt-Cloutier, "The Need for a Global Treaty on Persistent Organic Pollutants," *Silarjualiriniq* 2 (October–December 1999).
28 http://inuitcircumpolar.com/Index_of_Speeches/Index_of_Speeches.html.
29 United States, note to the European Union, undated and unsigned.
30 D. Tiepelman, President, ICC (Alaska), letter to M. Albright, U.S. Secretary of State, 20 April 2000.
31 B. Larkin, Assistant Secretary, Legislative Affairs, U.S. Department of State, letter to Senator T. Stevens, 8 August 2000.
32 Office of the Governor, State of Alaska, "Knowles Calls for National, International Action to Protect Arctic Environment," News Release 00–253, 11 October 2000.
33 J.E. Berner, "Alaska Native Traditional Food Monitoring Program," *Arctic Research of the United States* 15 (Spring/Summer, 2001): 32–5.
34 http://www.arctic-council.org/declarations.asp [Barrow Declaration].
35 See B. Commoner, "The Deposition of Airborne Dioxin Emitted by North American Sources on Ecologically Vulnerable Receptors in Nunavut," in this volume.
36 S. Watt-Cloutier and T. Fenge, "Poisoned by Progress," *The Globe and Mail*, 14 March 2000.
37 S. Watt-Cloutier et al., letter to Hon. Christine Stewart, Minister of the Environment, 20 April 1999.
38 A. Attaran, "DDT Saves Lives," *The Globe and Mail*, 5 December 2000.
39 S. Watt-Cloutier, R. Charlie, and J. Crump, "We Can All Win," *The Globe and Mail*, 11 December 2000.
40 S. Watt-Cloutier, "Wake-Up Call," *Our Planet* 12, no. 4 (2002): 12–14.

10
POPs in Alaska: Engaging the United States

HENRY P. HUNTINGTON
AND MICHELLE SPARCK

INTRODUCTION

Alaska is home to several distinct Indigenous groups, each of which continues traditional pursuits of hunting, fishing, and gathering. These practices account for a substantial portion of the diet of many Alaskans, both Native and non-Native, who stock their freezers with wild fish and game.[1] Native foods and their acquisition have cultural and spiritual value greater than their sustenance value; many Natives feel that, without them, their existence as a people would be in danger. The discovery of contaminants, including persistent organic pollutants (POPs), in Native foods was a shock: their existence is seen as a major threat. This chapter describes this discovery and the subsequent government action and inaction.

Alaska has been inhabited for at least 11,000 years, and probably for far longer. Some communities have existed for millennia, and the Indigenous peoples' sense of connection to and dependence on the land and sea reflect this depth in time. "Outsiders" from Russia arrived relatively recently, in the 1700s, and quickly exerted their influence along the southern coast, but large-scale immigration did not begin until after the United States purchased Alaska in 1867. Today, some 650,000 people inhabit the state and Alaska Natives account for one-sixth of the population. In rural areas, the proportion of Alaska Natives can reach ninety per cent. Alaska has about 220 tribes that are recognized by the federal government; several more communities consider themselves tribes, although they do not benefit from official recognition.

Alaska's tribes perpetuate many lifestyle traditions, but perhaps the most important aspect of cultural continuity is the harvesting and sharing of fish, animals, and plants. Collectively, these activities are known in Alaska as "subsistence." The English-language definition fails to fully convey the richness and inherent value in the manner in which this lifestyle sustains and binds families and communities. Nevertheless, the term has legal standing, and no preferable alternatives have been put forward.

Today, subsistence is under attack from a variety of sources, mostly political. The Alaska Native Claims Settlement Act (ANCSA) extinguished Aboriginal claims to hunting and fishing rights. Alaska Native subsistence is therefore vulnerable to the demands of "public process," a legal tool frequently used by the urban fishing and hunting groups that oppose the federally legislated preference given to rural residents in times of shortage. Many Natives feel this preference is inadequate because it does not grant specific recognition to Natives and fails to protect those who have moved to cities. The designation, for conservation or other purposes, of enormous tracts of federal and public land has also impinged on Native subsistence practices. Many laws designed to protect the environment or conserve wildlife tend to dismiss traditional knowledge and customs, which are rather more opportunistic and communal than regulated and individual. At the same time, modern conveniences have contributed to undermining the rigours of life on the land and sea and pose a threat to the pursuit of traditional activities and the perpetuation of the knowledge they require.[2]

Into this already potent mix come contaminants, posing an even more insidious threat to the Native way of life by invading the one aspect that so firmly binds people to their cultures and their legacy – their food. Native foods are usually regarded as a source of strength, well-being, and happiness. If they are also a source of environmental contaminants, what can replace them in the lives of Alaska Natives? Store-bought processed foods are packed with a variety of chemicals, are expensive, have no cultural value, and offer nothing to the soul. And, as many of these foods are linked to health problems such as obesity, diabetes, and heart disease, they constitute a public health concern in their own right. It is no wonder that Native communities have reacted to contamination of their food supply with frustration, anger, resentment, and fear.

History provides at least a partial explanation for the social and political reaction to contaminants. The United States is not known for taking an internationalist approach to solving its problems. Alaska, in particular, often displays a strong parochial tendency that can be summed up in the oft-repeated motto "We don't care how they do it Outside," referring to the other forty-nine states and elsewhere. Confronted with the spectre of contaminants, the state tended to ignore the resources and experiences available in Canada and other places that faced similar problems. Indeed, many of the experts available within Alaska were not used; instead, others stepped forward, with both good and bad results.

DISCOVERING THE PROBLEM

In the 1970s some studies, many of which were spurred by oil development on the North Slope, documented contaminant levels in Alaska but received little public attention. Documenting background levels became an important task. Much of the early work on contaminants in wildlife focused on heavy metals, such as cadmium and mercury. In some species, such as walrus, levels of these elements can be relatively high, leading those Alaska Native coastal inhabitants who depend on walrus to become rightfully concerned. As little was done to analyze these findings from a public health perspective, advice on consumption was either absent or given by persons with little expertise in human health matters.

The largest continuous monitoring effort is the Alaska Marine Mammal Tissue Archive Project (AMMTAP), begun in 1987. Samples are taken under a rigorous protocol to avoid the introduction of foreign materials after the animal is taken. Some analyses are done immediately, but the main goal of the program is to accumulate a collection of samples for later analysis. Analyzing trends is problematic because analytical techniques change so rapidly that older data often cannot be compared with new data. The AMMTAP circumvents this problem. If new contaminants are discovered or a particular event such as an oil spill raises concerns about contamination, the archived samples can be analyzed at the same time as new samples, using identical techniques.

Concurrently, Canadian researchers began discovering alarmingly high levels of POPs in Arctic Canada. As reports of the Canadian findings made their way to Alaska in the early 1990s, several researchers began pressing for more attention to the problem. A conference was held in Anchorage in 1993 to examine the pathways by which contaminants reached Alaska.[3] Neighbouring residents began to examine local sources such as old military sites more closely and with greater suspicion. Local observations by Elders, hunters, and gatherers reported subtle changes in food sources over the years, but their information was viewed as anecdotal and their concerns were dismissed as paranoid. Taking a cue from their Canadian neighbours, Alaska Natives began calling for further research and, perhaps more importantly, public outreach.

Meanwhile, public health officials, concerned about pronouncements by biologists and others with little health background, tried to put the findings in context, but were often criticized for downplaying the danger. Federal agencies, comparing POPs pollution in Alaska with contamination by local-point sources at sites in the lower forty-eight states, found that Alaska was simply not a priority. By contrast, radioactivity (i.e., anthropogenic radionuclides), in particular from Soviet sources in or flowing into the Arctic Ocean, was addressed in a major, but short-term, research program known as the Arctic Nuclear Waste Assessment Program (ANWAP), funded by the Office of Naval Research.[4] The State of Alaska occasionally expressed appropriate concern but provided little support for research or other activity.

INTERPRETING THE PROBLEM

As agencies cogitated and discussed, Alaska Natives recalled past battles. This was not the first time that pollution in the Arctic had been treated lightly; nor was the perceived inaction of federal agencies surprising.

In the late 1950s, the Atomic Energy Commission had sought to establish "peaceful" uses for nuclear weapons. One such effort, known as Project Chariot, was designed to excavate a harbour in northwestern Alaska by detonating a series of bombs to create a keyhole-shaped indentation in the coastline southeast of Point Hope. Officials either dismissed the potential effects of the resulting radioactivity as unimportant or, perhaps, viewed the project as a fascinating experiment. While local residents were not to be informed of the project, fear of radioactivity was already established in the community after atmospheric testing in Nevada resulted in deposition and accumulation of radionuclides in Alaska.

In the end, the project was abandoned, but not before some contaminated soils from the Nevada test site were brought north and deposited on the tundra near the site of the planned harbour. When the details of the project were made public by researcher Dan O'Neill in the late 1980s, community activists in Point Hope and elsewhere were irate about their perceived betrayal at the hands of the federal government.[5] The anxiety generated by this episode continues to complicate communication efforts to this day.

Project Chariot was not an isolated case. In a military research project on thyroid function, radioactive iodine was injected in North Slope Inupiat, who were not told they were the subjects of research. In both cases, many activists made a link to increasing cancer rates among Alaska Natives, insisting that the government acknowledge its responsibility and provide compensation. The low numbers of people involved make epidemiological proof or disproof unlikely, leaving both cases open to speculation. Similarly, the death of many caribou at or near the Project Chariot site in the past few years was attributed by many in Point Hope to radioactivity, even though necropsies indicated starvation as the likely cause of death and, as expected by researchers, levels of radionuclides in tissues were low.

However, recent developments may bolster the claims of those subjected to Project Chariot and the iodine injections. The Atomic Energy Commission set off three underground nuclear explosions on Amchitka Island in the Aleutians between 1965 and 1971. People who worked there subsequently developed radiation-related cancers (twenty-one documented varieties) or the lung diseases berylliosis or silicosis. They or their survivors are eligible for government compensation of $150,000 each.

Military sites throughout the state, oil development activities, and the underground detonation of nuclear bombs on Amchitka are interpreted as further evidence of disregard for the lives of Alaska Natives and the health of the environment upon which they depend. Contaminants, in the darkest view, were an intentional means of weaning Natives from their way of life, forcing them to adopt a Western lifestyle that had little relevance in remote villages. Perhaps an extreme view, the fact that it could be thought seriously shows the tone of the overall discourse. Even without an assumption of conscious intent, the mere presence of contaminants suggested at least malign neglect.

This sense of injustice was exacerbated by the fact that subsistence is the primary means by which many contaminants reach people. In light of other attacks on Native practices, the very idea that Native foods, following bioaccumulation in the food web, might no longer be healthy was a major blow. The public health community maintained that wild foods generally remained the best food available for Natives. This claim, however, was usually met with skepticism; contaminants were in the foods, contaminants were bad, and recommendations that people continue to eat contaminants were nonsensical at best. The lack of trust Alaska Natives had in most government agencies made it difficult to accept any reassurance. Those claiming the problem was minor were covering something up. Those who agreed that it was a major problem were not doing enough to solve it. But the question was, and continues to be, what exactly needs to be done? Most frustrating, and disconcerting from a public health perspective, is that Alaska Natives do not want an alternative to Native foods.

ACTIVISM AND INACTION

In the mid-1990s, the U.S. Environmental Protection Agency began to promote "environmental justice" as a priority policy. The emphasis was on helping poor and minority communities address the excessive pollution and poor environmental quality that were all too often their lot. The Indian General Assistance Program (IGAP) provided funds for Native Americans and Alaska Natives to hire environmental protection officers. In Alaska, many tribes participated, trying to document polluted sites, such as inadequate landfills or sewage outfalls, and similar problems common in remote villages. Long-range pollution was often raised as a topic of concern, but IGAP by itself could not tackle such a large-scale problem. Nevertheless, by raising general awareness of pollution problems, IGAP empowered Alaska's tribes to learn more about contaminants issues.

As contaminated sites were identified, tribes and others began calling for clean-up efforts and for compensation to the victims of pollution and other environmental injustice. The U.S. Department of Defense was often cast as the villain, as most of the contaminated sites had become so through military activities. Sites were studied, contaminants identified, and restoration plans prepared. Community activism found expression in organizations such as Alaska Community Action on Toxics (ACAT), formed specifically to assist in cleaning up polluted sites throughout the state. ACAT and others also paid close attention to long-range pollution, advocating more research and vigorous action to stop the problem at its source – the places where POPs and other contaminants were manufactured, used, or released.

Despite such calls for action on long-range POPs pollution, the United States was unable to prepare a coordinated approach. For federal agencies, the problems in Alaska were typically too small and too remote to be a priority. The various agencies were at times unwilling to divert the significant financial and human resources necessary to study contaminants. Coordination within and between agencies was difficult to arrange, as agencies battled over authority, responsibility, and money. The State of Alaska, occasionally calling for more research, was unwilling to allocate funds itself, arguing that Alaska did not create the problem of long-range POPs but was merely their unwilling recipient.

The North Slope Borough (a county-like government covering the northern fifth of the state) was among municipal governments that set up their own programs to look at contaminants. Funded by a variety of federal programs, the North Slope Borough's scientists measured levels of various contaminants in subsistence animals and reported the results to the region's communities and to various research journals and organizations such as the Arctic Monitoring and Assessment Programme (AMAP). Other groups, such as the Alaska Native Science Commission, a non-profit group funded largely by the National Science Foundation, undertook statewide projects. One such project documented community knowledge of and reaction to contaminants, creating a large database of statements and observations that indicate how seriously the problem is viewed in the Native community.[6]

Unfortunately, such programs and projects worked on their own, protective of their information, and with minimal effort at, or incentive for, coordination. Various results were sporadically reported in scientific journals, at conferences, and in community meetings. The need for more work was identified and repeated; this occasionally yielded more funding to extend an existing project or begin a new one, yet little was done to improve coordination of the various efforts under way. This approach produced two significant results: first, results of studies were often presented without adequate discussion of their importance; second, the United States was unable to present a coherent description of environmental contaminants in Alaska in a format acceptable to such international programs as AMAP.

Reporting only concentrations of POPs in animal tissues with no attempt to determine the effects in wildlife, or the level and rate of exposure or possible effects in humans, does little to contribute to our understanding of the threat posed by these contaminants. Much work was done in Alaska to measure POPs, heavy metals, and radioactivity in a variety of animals. Some of the measurements were duly reported to the communities that target these animals, but rarely were they put in any kind of context (e.g., reporting the benefits of traditional foods, mitigated by whatever threats are posed by contaminants, might help clarify the context). The biologists and chemists responsible for taking samples and analyzing them were usually reluctant to speculate on health consequences, either for the animals or for the humans who consumed them. Once a community is alarmed, however, communicating a reasonable risk assessment is many times more difficult, for the proof must be extremely convincing to overturn negative conclusions already reached.

The outcome proved damaging. Instead of a rational analysis conveying the seriousness of the contaminant levels found in Alaska wildlife, the public was left to speculate about the extent of the risk. Given the rise in certain cancer rates among Alaska Natives, contaminants seemed to provide a clear answer to many of the questions people had been asking about their health. An apparent weak association can quickly be perceived as a cause-and-effect link in the absence of sound information and reliable studies. Public health officials, concerned not only with contaminants but also with the documented high death rates from drowning, accidents, alcohol and drug abuse, and other unnatural causes were reluctant to spend much effort on the emotionally powerful but clinically undemonstrable problem of contaminants. Analyses of current levels of contaminants in animal and human tissues found little evidence to warrant public health advisories.[7]

The absence of a coherent process for collecting and reviewing such data tended to undermine the credibility of those who reported that there was little to be concerned about. The uncertainties inherent in assessing risks from contaminants provide ample room to challenge interpretations of the measured levels. That Alaska Natives were not substantively involved in much of the earlier process compounded the difficulty. People were typically unwilling to trust agency researchers they did not already know and placed little faith in the scientific basis upon which conclusions were founded. In addition, contaminants issues in Alaska are typically addressed in a crisis-management paradigm that does not allow for thorough scientific evaluation. Such an approach demonstrates that bringing attention to contaminants is to claim a crisis exists.

Canada's Northern Contaminants Program was held up as a model for the collaborative study of POPs, heavy metals, and radioactivity, but establishing such a program in Alaska has been difficult. For example, at a workshop held at the University of Alaska Anchorage, scientists and Native leaders discussed the results of the Northern Contaminants Program. They then attempted to judge whether each statement was applicable to Alaska. In the absence of data specifically from Alaska and enough appropriately qualified participants, this process proved of limited practical value.

The absence of Alaska data and dedicated personnel brings us to the second result of agency non-coordination. In the 1990s, AMAP was preparing a comprehensive assessment of environmental contaminants in the Arctic. Participating countries were expected to provide data to be used in a comparative analysis of contaminants around the region. Although each country faced its own obstacles in doing so, only the United States failed to provide data for large sections of the assessment report; U.S. POPs data were scarce. The void is apparent in many comparative maps in the AMAP report: Alaska is simply left blank.

Although little or no data were reported, much, in fact, had been collected and could have been included in the AMAP report. The agency in charge of U.S. participation in AMAP had allocated few funds to the task, and other agencies were unwilling to make up the difference.

By contrast, the United States is making a stronger contribution to the second phase of AMAP, in which updated reports are to be issued on POPs and other specific topics in autumn 2002. Some research and synthesis work has been funded, and many Alaskan researchers have started working with Canadian colleagues, establishing at least an informal basis for international collaboration.

The 1997 AMAP publication *Arctic Pollution Issues: A State of the Environment Report* was the first attempt to address the question of contaminants across the circumpolar Arctic. It examined the levels that have been found, their significance in the environment, and their implications for human health. The report, issued in scientific and plain-language versions, was the first accessible reference on the topic available to many Arctic residents. Its findings were largely reassuring in terms of human health and environmental effects of contaminants and helped to allay some of the fears expressed by Alaskans; however, the report also recommended further research in many areas, leaving open several questions regarding the threats posed by POPs and other contaminants. Without Alaskan data in the report that could be compared with those from other countries, there remains the question of whether the general conclusions are applicable to Alaska.

As AMAP completed its report, international concern about POPs had led to a call for negotiations seeking to ban their production and use worldwide. The presence of POPs in significant quantities in the Arctic was a key factor in demonstrating the tenacity and reach of these substances and the need for global action.

At this juncture, few Alaskan citizens, state officials, or their elected representatives in Washington, D.C., were aware of the concluding negotiations on a POPs Protocol to the UNECE Convention on Long-range Transboundary Air Pollution (CLRTAP) or the Intergovernmental Negotiating Committee sessions for a global POPs convention beginning in Montreal in June 1998. Influential in raising public awareness of the issue in Alaska and Washington were several advocacy organizations participating with the newly formed International POPs Elimination Network (IPEN), including ACAT, Indigenous Environmental Network, and the Circumpolar Conservation Union (CCU), an official observer to the Arctic Council.

During the period that the United States chaired the Arctic Council (1998–2000), the CCU, in partnership with ACAT, initiated a campaign to engage scientists, tribal leaders, Alaskan citizens, and the larger American public in actions to reduce or eliminate persistent contaminants that affect the environment and human health in the Arctic. The campaign saw Alaskans pressing the POPs issue at Arctic Council meetings in 2000 and lobbying successfully for a report and Senate briefing on POPs in Alaska.[9] The campaign also furthered the engagement of the Alaska Governor's Office; Governor Knowles wrote to President Clinton, urging the United States to take an aggressive leadership role in the negotiations and requesting that the state actively participate with the U.S. negotiating team in South Africa.[10]

A delegation of Alaska Natives, activists, and State of Alaska officials travelled to Johannesburg, South Africa, for the final negotiating session of the Stockholm Convention on POPs, and Alaska Natives participated in the convention signing ceremony in Stockholm. This action was late in coming but has led many of those involved to push anew for coordinated, internationalist action in Alaska to address environmental contaminants.

NEXT STEPS?

The most visible result of Alaska's renewed interest has been the attempt to create an Alaska Contaminants Program. As the Stockholm Convention entered its final diplomatic phases, the federal, state, tribal, and local agencies with an interest in or responsibility for some aspect of environmental contaminants worked together to issue a brochure describing the issue in Alaska and outlining the need for a coordinated program.[11] Unlike previous attempts, this effort, led largely by the Alaska Native Health Board, was successful in mobilizing the participation of a broad constituency. As before, questions of budget and control are among the obstacles to be surmounted, but there is hope that Alaska Natives may at last begin to understand and address POPs, heavy metals, and radioactivity in a comprehensive, coordinated fashion. Still required is the mobilization of scientific expertise in addition to the good intentions of agency staff and others.

Contaminants remain a significant concern in Alaska Native communities. When fish and wildlife management are discussed, the topic of contaminants is frequently raised. Suspicion, misunderstanding, fear, anger, and confusion still characterize much of the discussion, and it is unlikely they will ever be entirely absent. Nevertheless, the experience gained by Alaskans through involvement beyond their borders in international efforts such as AMAP and the Stockholm Convention negotiations has provided confidence that something can be done to combat the problem of contaminants. Alaska Natives have now become activists within the United States to press for national and global actions. Although the United States has yet to take a leadership role in pressing for international action on contaminants, the recognition that such actions can benefit Alaskans is a welcome step towards co-operation in addressing the threats posed by contaminants.

NOTES

1 AMAP, *AMAP Assessment Report: Arctic Pollution Issues* (Oslo, Norway: Arctic Monitoring and Assessment Programme 1998).
2 Henry P. Huntington, *Wildlife Management and Subsistence Hunting in Alaska* (London: Belhaven Press 1992).
3 Bruce F. Molnia and Kenneth B. Taylor, "Proceedings of the Interagency Arctic Research Policy Committee Workshop on Arctic Contamination, 2–7 May 1993, Anchorage, Alaska," *Arctic Research of the United States* 8 (Spring, 1994): 1–311.
4 Kathleen Crane and Jennifer Lee Galasso, *Arctic Environmental Atlas* (Washington, DC: Office of Naval Research 1999).
5 Dan O'Neill, *The Firecracker Boys* (New York: St. Martin's Press 1994).
6 Alaska Native Science Commission, http: //www.nativeknowledge.org.
7 Grace M. Egeland, Lori A. Feyk, and John P. Middaugh, "The Use of Traditional Foods in a Healthy Diet in Alaska: Risks in Perspective," *State of Alaska Epidemiology Bulletin* 2, no. 1 (1998).
8 AMAP, *AMAP Assessment Report.*
9 Lin Kaatz Chary, "Persistent Organic Pollutants (POPs) in Alaska: What Does Science Tell Us?" (Washington, DC: Circumpolar Conservation Union with Alaska Community Action on Toxics 2000).
10 Letter from Alaska Governor Tony Knowles to President Clinton, 20 October 2000.
11 Interagency Collaborative Paper, "Contaminants in Alaska: Is America's Arctic at Risk?" (Anchorage, Alaska: U.S. Department of the Interior and others 2000).

11
The Long and Winding Road to Stockholm: The View from the Chair

JOHN ANTHONY BUCCINI

PREFACE

Shortly after the adoption of the Stockholm Convention on Persistent Organic Pollutants (POPs) on 22 May 2001, I was asked to contribute a chapter to this book that would include my personal perspectives as chair of different intergovernmental groups that were involved in the process. I agreed to do so believing that the experience gained in developing this convention might be not only of interest to some but especially of assistance to others who may one day find themselves in a situation similar to mine, where they are presented with an opportunity to participate in a leadership role in an intergovernmental setting. However, I must caution at the start that, in my experience, such *opportunities* usually present themselves heavily disguised as work.

I will not address the details of the convention or the scientific information that justified its development because others do so elsewhere in this book. My contribution is limited to recollections of the major events and interactions that took place before and during the negotiations. This account is a personal one and the views are mine alone; I have attempted to recall events accurately and hope that I have not introduced any factual errors in telling this story.

GETTING STARTED

My involvement began in June 1995 with a telephone call from a colleague, Dr Michel Mercier, who worked with the World Health Organization (WHO) and also served as the secretariat to the Intergovernmental Forum on Chemical Safety (IFCS). The IFCS was formed in 1994 as a forum for stakeholders (i.e., representatives of governments, international associations, and intergovernmental and non-government organizations) to discuss issues related to chemical safety and coordinate global efforts to implement the commitments relating to the sound management of chemicals that were included in Chapter 19 of *Agenda 21*, agreed to at the 1992 UN Conference on Environment and Development in Rio de Janeiro, Brazil. We had both attended a meeting of the Organisation for Economic Co-operation and Development (OECD) Chemicals Group in Paris earlier in the month where I concluded a four-year stint as chair. Dr Mercier called to say that, as I was no longer chairing the Chemicals Group and he considered me now under-employed, he had an "opportunity" he thought I should consider.

I learned that on 25 May 1995 the United Nations Environment Programme (UNEP) Governing Council (GC) had adopted Decision 18/32, which invited the Inter-Organization Programme for the Sound Management of Chemicals (IOMC), working with the International Program on Chemical Safety (IPCS) and the IFCS, and assisted by an appropriate ad hoc working group, to initiate an expeditious assessment process, initially beginning with twelve specified POPs (aldrin, chlordane, DDT, dieldrin, dioxins, endrin, furans, heptachlor, hexachlorobenzene, mirex, PCBs, and toxaphene). Decision 18/32 specified that, taking into account the circumstances of developing countries and countries with economies in transition, the assessment process should:

a) consolidate existing information available from IPCS, United Nations Economic Commission for Europe (UNECE), and other relevant sources on the chemistry and toxicology of the substances concerned (particularly the impact on human, plant, and animal health);
b) analyze the relevant transport pathways and the origin, transport, and deposition of these substances on a global scale;
c) examine the sources, benefits, risks, and other considerations relevant to production and use;
d) evaluate the availability, including costs and effectiveness, of preferable substitutes, where applicable; and
e) assess realistic response strategies, policies, and mechanisms for reducing and/or eliminating emissions, discharges, and losses of POPs.

The IFCS was invited to take the results of this process, together with the outcome of the UNEP Intergovernmental Conference to Adopt a Global Programme of Action for the Protection of the Marine Environment from Land-based Activities (Washington, D.C., 23 October – 3 November 1995), and develop recommendations for and information on international action. The IFCS report was to include any information that would be needed for a possible decision on an appropriate international legal mechanism on POPs, to be considered at the respective 1997 sessions of the UNEP GC and the World Health Assembly (WHA), the policy body of the WHO.

The intergovernmental organizations of the IOMC (FAO, ILO, IPCS, OECD, UNEP, WHO) met on 29–30 June 1995 to discuss implementation of the GC decision and agreed that UNEP would convene, on behalf of the IOMC, an ad hoc working group with a proposed composition of IOMC members, non-government organizations (NGOs), and me on behalf of Canada as a lead country. It was proposed that the working group would convene for the first time on 28 October (during the Washington conference on marine protection) to prepare a work plan to implement the GC Decision for submission to the second intersessional meeting of the IFCS, scheduled for 5–8 March 1996, in Canberra, Australia.

I met with the IFCS and UNEP secretariats in Geneva on 6 July to discuss the invitation to chair the UNEP working group and to prepare the POPs discussion for the IFCS meeting in Canberra. The task was a complex and difficult one, with an especially short deadline. To be available in time for the January 1997 UNEP GC meeting, work would have to be completed and documents submitted by September 1996, which left only fourteen months to carry out the work.

I expressed my opinion that I should not be the only country representative on the working group. When the IFCS was formed in 1994, it had adopted a multi-stakeholder participation formula for its activities and, as the IFCS would be responsible for the development of the final report, I proposed that other country representatives should be invited to join the group from the very beginning to respect the need for balanced participation and for transparency in the process. This was agreed, and a stakeholder participation formula was developed.

The Canadian Network of Toxicology Centres was contracted to begin immediately with the preparation of an assessment report for IPCS on the twelve POPs (emphasizing points 'a' and 'b' in the foregoing mandate and adding any available information on points 'c' and 'd'). A first draft was needed for a preliminary review at the meeting of the working group planned for 28 October so that it could be completed by year end and distributed in time for consideration at the IFCS meeting in Canberra. The intention was to review the proposed work plan and the assessment report in Canberra and then ask the IFCS to adopt the working group and task it with the IFCS portion of the mandate. The latter task would require addressing, in depth, points 'c', 'd', and 'e' of the mandate and then developing recommendations for, and information on, international action and a possible decision on an appropriate international legal mechanism on POPs.

It was evident that it would be necessary to hold a meeting in the summer of 1996 to develop the IFCS report and recommendations and complete the requirements of the UNEP GC mandate. We recalled that in June 1995 more than 100 experts from forty countries attended a meeting in Vancouver, co-hosted

Persistent organic pollutants (POPs) are released to the environment through industrial processes, incineration of medical and municipal waste, and agricultural practices in tropical and temperate lands and are transported, mostly by air currents, to the Arctic, where they bioaccumulate in the land and marine food webs. In the food webs, they become toxic elements in the diet of Inuit and other northern Indigenous peoples.

Despite the significant social and economic changes of recent decades, Inuit remain a people whose way of life and economy reflect the reality of their culture: they are hunters and gatherers. Their spiritual and physical well-being is rooted in their traditional lifestyle and in their country food.

The chapters in this book portray well the impact of POPs on northern peoples - on their diets, their health, their very existence. These photos illustrate life on the land in the Arctic: gathering and hunting, preparing country food, eating marine mammal fats (maktak), and sharing food with family and community.

The presence of POPs in the Arctic undercuts the relationship between Inuit and the environment they know and use so well and threatens the continuation of the Inuit culture.

Photo credit: Eric Loring

Photo credit: Eric Loring

Photo credit: Eric Loring

Photo credit: Eric Loring

Photo credit: Eric Loring

Photo credit: Eric Loring

Photo credit: Hans Blohm

Photo credit: Hans Blohm

Photo credit: Hans Blohm

Photo credit: Hans Blohm

Photo credit: Hans Blohm

Photo credit: Eric Loring

by Canada and the Philippines, to address many of the scientific aspects of POPs. It had concluded that there was enough scientific information to warrant immediate national, regional, and international action, including bans and phase-outs of certain POPs, and proposed that a joint Canada-Philippines follow-up meeting be held in 1996 that would, in part, examine the global socio-economic aspects of POPs and their alternatives. I agreed to explore the possibility of an IFCS-sponsored meeting using the Canada–Philippines partnership that had been developed at the Vancouver meeting. Following my return to Ottawa, I approached my colleague Dr Harvey Lerer, who had served as the Canadian lead in the Vancouver meeting. He was interested and arranged for me to contact his Philippines counterpart, Mr Ramon Paje, Assistant Secretary for the Department of Environment and Natural Resources. Mr Paje was also interested, and we communicated over the summer and arranged to meet in person at the Washington conference in October to pursue arrangements for the meeting.

The next step in the planning process was a meeting on 18–19 October 1995 at the Paris office of UNEP with an expert advisory group on POPs comprising invited stakeholders. It was called to assist UNEP in developing its strategy and work program in response to recent governing council decisions, including the one on POPs, and to plan for the first meeting of the working group on 28 October. I introduced the draft plan of action to implement the GC Decision and the group agreed on a draft agenda, proposed terms of reference, and a formula for stakeholder participation for consideration at the working group meeting. The draft IPCS assessment report (*Persistent Organic Pollutants, An Assessment Report on DDT, Aldrin, Dieldrin, Endrin, Chlordane, Heptachlor, Hexachlorobenzene, Mirex, Toxaphene, Polychlorinated biphenyls, Dioxins and Furans)* was reviewed and there was consensus that the precautionary principle should be applied in the case of POPs and that the currently available information on the twelve POPs (with the possible exception of dioxins and furans) was sufficient grounds for action. This meeting also acknowledged that "the release of POPs is generally not a sustainable practice from an environmental viewpoint" and that "until socially and economically alternative processes and substitutes are available immediate cessation of releases of certain POPs may not be possible." My last recollection of this meeting was that, by chance, I happened to sit next to Jim Willis, the newly appointed director of the UNEP Chemicals program. Little did I know then how much time I would spend sitting next to him in the coming years!

THE AD HOC WORKING GROUP ON POPS

The 28 October meeting of the working group, chaired by UNEP, took place in Washington, D.C. I attended as the representative of the 18–19 October Paris meeting, and, following several background presentations on POPs, I facilitated a discussion on the proposed terms of reference and work plan for the working group. The group agreed on its terms of reference and decided that, while it would be open to observers, its core composition would include representatives of the six IOMC members; UNECE; four countries from each of the five UN regions (providing a global mix of countries producing, exporting, importing, using, or receiving POPs); four industry associations; four public-interest groups; and me as chair. The group was constituted as a UNEP ad hoc working group, with the intention that it would be adopted as an IFCS ad hoc working group at the Canberra meeting and it would complete its work with the consideration of POPs issues at the 1997 meetings of UNEP GC and WHA. The group agreed on its work plan and also agreed to forward the IPCS assessment report to the IFCS meeting in Canberra subject to some changes. While not required by its mandate, the group also noted the need to develop criteria for identifying POPs, in addition to the twelve specified in the GC Decision, as candidates for international action; to build on the POPs work of the UNECE; to take into account dispersion mechanisms for the atmosphere, the hydrosphere, and the biosphere; and to identify measures to evaluate the success of any implemented strategies. The prospect of a meeting in the Philippines was discussed at this time but could not be confirmed. Following the meeting, I consulted with Mr Paje on this matter and he agreed to seek approval upon his return to Manila.

It is noteworthy that shortly after the working group meeting, countries adopted a Global Programme of Action for the Protection of the Marine Environment that, in part, recognized the importance of controlling releases of POPs and specified actions that should be taken. The *Washington Declaration on Protection of the Marine Environment from Land-based Activities* (November 1995) called for "a global, legally binding instrument for the reduction and/or elimination of emissions, discharges and, where appropriate, the elimination of the manufacture and use of the persistent organic pollutants identified in decision 18/32..." In addition, in late November 1995, a decision was made to pursue the negotiation of a POPs protocol under the UNECE Convention on Long-range Transboundary Air Pollution. Thus there were indications that the timing was right to pursue a global agreement on POPs.

The final IPCS assessment report was submitted to the Canberra meeting along with a paper outlining the decisions that would be required of the meeting. The IPCS report noted that there was insufficient information available on the 'c' and 'd' elements of the GC mandate and, in January 1996, the IFCS president wrote to stakeholders and signalled the need to collect information in this area in preparation for Canberra and any subsequent meetings.

There were two separate meetings on POPs in Canberra. The first took place during the IFCS intersessional meeting (5–8 March 1996). I facilitated the

POPs session that began at 11:00 A.M., introducing the issue and highlighting the decisions that were required of the meeting. This was followed by thirty-four interventions from stakeholders from all sectors and the session concluded at 12:53 P.M. I remember this clearly because I was hoping that the president would call a lunch break so that I could collect my thoughts before attempting a summary; instead, he asked me to summarize in the remaining time before lunch. Another opportunity! This was a real challenge given the large number of interventions that had been made in slightly less than two hours. Although the majority supported the proposal, several had challenged the premise that there was enough scientific evidence available to warrant risk management action. I proposed that the IFCS agree that there was sufficient scientific evidence on the chemistry, toxicology, transport pathways, origin, transport, and deposition of the twelve POPs to demonstrate the need for international action and to provide a basis for moving forward on realistic response strategies. I paused to see if there were any objections to the proposal: there were none. At that moment I felt a distinct change in the room as everyone accepted the conclusion. This marked a turning point in the consultations from a discussion of *whether* action was warranted to a discussion of *what* action to take and *how* to proceed.

The IFCS meeting also agreed that additional information was needed to complete the tasks in paragraphs 'c', 'd', and 'e' of the mandate and to address socio-economic considerations as a basis for development of recommendations and information on international action. IFCS adopted the ad hoc working group and agreed to sponsor an IFCS meeting of experts in June 1996 in Manila to address tasks 'c' and 'd'. This was to be followed immediately by an open meeting of the IFCS working group to review the results of the experts meeting, to address task 'e', and to develop recommendations and information on international action. To ensure that the final report would be submitted in time for the 1997 meetings of UNEP and WHA, the IFCS agreed that the working group would submit its report based on the outcome of the June meetings to all IFCS participants for review and comments by 31 July. A summary of the comments would be then be submitted along with the report by 31 August 1996.

While IFCS would not develop criteria for adding POPs to the list, it agreed to build upon ongoing activities, such as those under the UNECE, and to propose a process for the development of science-based criteria in the final recommendations. The United Kingdom was invited to develop a paper on this for the June working group meeting. IFCS recommended that consideration be given to the need for monitoring networks to establish baseline levels of POPs in environmental and biological media and to assess trends following implementation of future measures. IFCS also noted that international action would have to take into account that the twelve POPs included pesticides, industrial chemicals, and unintentionally produced by-products and that different approaches may be needed for each category. It is noteworthy that all these ideas from the IFCS meeting would end up reflected in the final convention.

The IFCS meeting addressed all requested aspects relating to the mandate of the working group and now the latter was faced with considerable challenges in planning and organizing two meetings in Manila and finding the resources to fund them.

The second meeting in Canberra took place on 9 March, when the working group met to review ongoing POPs-related activities and the outcome of the IFCS discussions. A planning committee was formed to assist in making the arrangements for the Manila meetings and, in keeping with the IFCS approach, NGO representatives from Greenpeace and the International Council of Chemical Associations (ICCA) were included with government representatives (including Sweden, Philippines, Brazil, Mali, and Estonia) and the IFCS Secretariat. The decision to include the NGO representatives would prove very important.

The working group agreed that an Experts Meeting would be held on 17–19 June 1996 to acquire information and documentation on tasks 'c' and 'd' and that this would be followed by an open-ended working group meeting on 21–22 June to complete task 'e' and develop recommendations and the final report. The one-day break on 20 June was to allow time for the participants from the Experts Meeting to brief their delegates to the working group meeting. IFCS and UNEP would provide the secretariat, and all IFCS contact points would be requested to submit case studies for the Experts Meeting.

UNEP presented a trade survey that showed several POPs were still in use many years after most countries had banned them, suggesting a need to pursue action on the twelve POPs. UNEP also announced that earlier in 1996 it had set up a POPs information clearing house and that this would be further developed to coordinate future activities on POPs. Sweden made a presentation on work initiated on substitutes for POPs and invited other countries to participate in developing a report relevant to paragraph 'd' of the mandate for submission to the Experts Meeting. For the working group meeting, the Nordic countries offered to develop a report on optional models for strategies for development of international action on POPs, and other countries volunteered to assist in this work. The Montreal Protocol (on Ozone depletion) was identified as an important model to be considered during the June working group meeting, both for its control provisions and for its financial mechanism to support activities of developing countries in implementing the protocol. By the time the meeting concluded, we had a plan for the Manila meetings, a committee to oversee the preparations, and plenty of interest and enthusiasm for the upcoming events.

Immediately following the Canberra meetings, I went to Manila to discuss the arrangements for the meetings with Mr Paje and Ms Marinela Castillo, an assistant secretary for agriculture, who would co-chair the Experts Meeting with Dr Lerer. We advanced the plans for the meeting and I had the opportunity to meet many of the people who would be on the Philippines' team in June. On my return home, my attention turned to securing the funds needed for the Manila meetings.

Three related meetings were held before June. The first, sponsored by UNEP, took place from 20 to 22 May in Solna, Sweden. Representatives of twelve countries, Greenpeace, ICCA, UNEP, and WHO reviewed the progress made by Sweden in addressing key issues related to POPs alternatives; a report on this subject was provided to the June Experts Meeting. On 23 May a second meeting took place in Solna, attended by participants from fourteen countries, UNEP, Greenpeace, and ICCA, to review a draft paper on possible strategies for reducing or eliminating certain POPs that was developed by the Nordic Council of Ministers for submission to the June working group meeting. Among the issues addressed was the need to allow voluntary measures as an option for action in the future, notwithstanding eventual agreement on a binding instrument.

The third meeting, held in Ottawa, allowed the planning committee to consider a draft 'theme' paper developed as a springboard for the discussions at the Experts Meeting; many comments were received and it underwent a major revision before distribution. There was good interaction among the meeting participants and lots of 'creative tension' was evident. Numerous matters were addressed relating to agendas and planning and it became more evident to me just how difficult it would be to execute two international stakeholder meetings back-to-back and develop a final report in only two days.

As time passed, more countries and other stakeholders became interested in the POPs issue and registration for attendance at the meetings increased. Concerned about the prospect of a large working group meeting trying to develop a final report in only two days, I distributed two documents in advance of the meeting. One was a draft outline for the final report, with the historical background already filled in. This was to give meeting participants an idea of what the report could look like, recognizing that the important part (conclusions and recommendations) had to be developed at the meeting and would, in part, be based on the outcome of the Experts Meeting. The other document was a 'Chairman's Overview,' which laid out how I expected the working group meeting to proceed and what the participants should expect to happen over the two days.

The IFCS Experts Meeting on POPs, held in Manila from 17 to 19 June 1996, generated an excellent report, *Persistent Organic Pollutants: Considerations for Global Action*, with numerous conclusions and recommendations on the three groups of POPs (pesticides, industrial chemicals, and by-products). The participation of all stakeholders resulted in a report that enjoyed support from all sides as people literally toiled through the night to complete the meeting report. I remember going to bed after midnight on the second day while the IFCS Secretariat representative, Judy Stober, continued to work. The next morning I came into the same room to find her there and noticed that she was dressed in the same clothes as the day before. I quickly learned that she had worked through the night and that she was not the only one to have done so. This was typical of the dedicated effort made by many people during the week to ensure that our task would be completed on time.

The report of the Experts Meeting provided much of the substance for the working group meeting on 21–22 June and, indeed, for its final report. The working group met for two long days, during which I doubted, at times, that the meeting would find the necessary consensus to conclude its work. One of my favourite memories, a moment so unusual that I had never expected to see it, was when Greenpeace and ICCA, frustrated by governments' lack of ability to agree on an issue, stood before the meeting and made a joint proposal to the group. Despite both speakers asking that no photographs be taken of them, there were a number of delegates who could not restrain themselves; I suspect that these photos are still treasured by their owners. The spirit of compromise and consensus carried through the balance of the meeting and led to the *Final Report of the Intergovernmental Forum on Chemical Safety ad hoc Working Group on Persistent Organic Pollutants* (IFCS report) – unanimously supported by all stakeholders.

The key IFCS conclusion was that sufficient scientific information was available on the twelve POPs to demonstrate the need for immediate international action and to provide a basis for moving forward on realistic response strategies. The report recommended that immediate international action be initiated to protect human health and the environment through measures that would "reduce and/or eliminate the emissions and discharges of the 12 POPs" and "where appropriate, eliminate production and subsequently the remaining use of those POPs that are intentionally produced." The words captured in the IFCS report were very carefully chosen and parts of the report were so meticulously crafted that they would later be referred to as 'the covenant' and no changes to these would be acceptable. The strong stakeholder support for the IFCS report would prove to be an important factor in the coming months.

In July, the IFCS report was distributed to all IFCS contact points for comments. By prior agreement, the IFCS report was not changed after the Manila meeting and the comments received were provided with the report to UNEP and WHA for their 1997 meetings. The task assigned by UNEP GC Decision 18/32 had been completed using an open, transparent, and inclusive consultation process that achieved support for the consensus report from all involved stakeholders. This approach would continue to influence future processes in pursuit of the Stockholm Convention.

In looking back at the level and complexity of the activities in 1995–96, I realize that the combination of events, meetings, and mandates might seem confusing or somewhat haphazard. However, the planning for this phase of the POPs process was deliberate in many respects and required making best use of the opportunities that were available to develop support for and secure decisions that built sequentially to lead to the recommendation for international action on POPs. There were several countries (including those from the European Union, Sweden, Canada, the United States, and the Philippines) and NGOs (including Greenpeace, Pesticides Action Network, and the International Council of Chemical Associations) that were involved in generating the momentum that was needed to sustain the effort in this early period. These players were to be joined by many others in the following years as support for the development of the convention continued to grow. I am convinced, however, that the sustained participation of several governments and NGOs, in many cases of the same individuals, was one key to moving the agenda forward, and this would be demonstrated next at the nineteenth session of the UNEP Governing Council in 1997.

THE NEGOTIATION MANDATE

The next step in the process was the review of the IFCS report and recommendations at the nineteenth session of the UNEP GC in Nairobi in late January 1997. I attended the meeting as a member of the Canadian delegation and, as this was my first time at such a meeting, I was curious to see how the GC would respond to the report. There were several country and NGO representatives in attendance who were keen supporters of the process that had led to the development of the IFCS report and it was also the first time at a UNEP GC for many of them. It would soon become evident that they were attending to ensure that the delicate balance represented in the IFCS report would be preserved in any GC decision on POPs.

There turned out to be three other chemicals issues on the agenda, and the chair of the Committee of the Whole allowed a brief introduction of all four and then proposed that they be sent to a working group to develop draft GC decisions. I was "invited" to serve as the chair of this working group, and we spent the next three days hammering out draft GC decisions on all four topics. The discussion on POPs was interesting in that GC participants who had been part of the IFCS process adamantly refused to alter the wording in the IFCS report and insisted on an electronic 'cut and paste' approach to use the report to develop the draft decision, with editing kept to an absolute minimum while transforming the recommendations into decisions. The commitment to the IFCS report was absolute and the delicate balance developed in Manila was carefully preserved in the GC Decision drafted in Nairobi.

A package of four chemicals decisions was adopted on 3 February 1997 as GC Decision 19/13, consisting of four parts (A to D). In adopting the GC Decision on POPs (19/13C), all the conclusions and recommendations in the IFCS report were endorsed; a similar result was obtained at the WHA (Resolution WHA50.13) on 12 May 1997. The UNEP GC Decision 19/13C mandated action in four main areas.

The GC provided the negotiation mandate by agreeing to prepare for and convene "by early 1998," together with WHO and other relevant international organizations, an intergovernmental negotiation committee (INC) with a mandate to prepare, "preferably by the year 2000," an international legally binding instrument for implementing international action, initially beginning with the twelve specified POPs and taking into account the conclusions and recommendations in the IFCS report. The delay in starting the INC process was a result of the negotiations under way to develop the Rotterdam Convention on Prior Informed Consent (PIC) and the feeling of countries that they could support only one negotiation process on chemicals at a time. With the expectation that the PIC negotiations would be concluded in March 1998, a mid-1998 start date was accepted. The completion date of 2000 reflected the view of governments that the POPs issue required urgent attention by the international community. Participation in the INC would be open to governments and relevant NGOs and intergovernmental organizations (IGOs). Coordination among different regional and international initiatives on POPs was seen as essential to ensure harmonized environmental and health outcomes from mutually supportive and effective programs that result in the development of policies with complementary and non-conflicting objectives. In addition, consideration would be given to implementing voluntary measures as a complement to, or independently of, a legally binding instrument.

The second aspect of the GC decision dealt with development of science-based criteria and a procedure for identifying POPs not included in the initial twelve as candidates for future international action. This process was to incorporate criteria pertaining to persistence, bioaccumulation, toxicity, and exposure in different regions and take into account dispersion mechanisms for the atmosphere and the hydrosphere, migratory species, and the need to reflect possible influences of marine transport and tropical climates. An Experts Group would be established at the first meeting of the INC (INC-1) to carry out this work and it would include scientific and socio-economic expertise relevant to the POPs issue and representatives of countries in different stages of development and from different geographical regions, as well as relevant NGOs and IGOs. The Experts Group had to consider the criteria and procedure being considered by the UNECE in developing its POPs protocol and take full account of varied ecosystems and the circumstances of developing countries and countries with economies in transition, as well as the need to conserve biodiversity and protect endangered species. The principles set out in the Rio Declaration, especially the reference to the precautionary approach in Principle 15, and the provisions of Chapter 19 of *Agenda 21* were also to be taken into account.

In the third element of Decision 19/13C, governments were urged to initiate action on the recommendations in the IFCS report and to provide technical assistance, capacity-building, and funding to enable developing countries and countries with economies in transition to take appropriate action on POPs.

The final thrust of the decision requested UNEP to initiate immediate action on POPs in response to recommendations in the IFCS report, including general awareness-raising on the national, regional, and global aspects of POPs; information exchange, within and between countries and IGOs; promotion of information exchange on alternative products and processes to reduce or eliminate the generation, use, and release of POPs; assistance to countries in identifying and developing inventories of PCBS and in identifying worldwide capacity to destroy PCBS; development of inventories of information on dioxins and furans, including sources of releases and practices to manage releases; and collection of information to be used in the negotiations to assist the development of criteria and a process for identifying additional POPs.

Following acceptance of the GC decision, Sweden took the floor and offered to host the meeting at which the future POPs convention would be adopted and signed. According to custom, this meant that the treaty would be known as the Stockholm Convention on POPs. This was a surprise to all in the room, including me. I learned five years later how this offer came to be and that I had played an unwitting role in it. I was attending an OECD meeting in Paris in November 1996 (a few months after the submission of the IFCS report and two months before the GC meeting) when my colleague from Sweden (Bo Wahlstrom) invited me to a quiet dinner. I was unaware at the time that he was probing me for information on the intentions of the Canadian government with respect to the negotiations. I indicated that Canada felt it most important to get the negotiations started and that we were considering funding the first INC should the GC decide to proceed with the negotiations. On hearing this, Bo realized that there would be an opportunity for Sweden to host the final meeting. He pursued this on his return to Stockholm and secured approval for it.

Thus the mandate to negotiate a POPs treaty was in hand and time-bound and, even before we started, we knew that the treaty would be called the Stockholm Convention. However, the negotiation process was not funded and this would prove to be a challenge to completing the negotiations within the specified time frame.

THE UNEP/IFCS ROAD SHOW

On the heels of the UNEP GC meeting, the IFCS met in Ottawa from 10 to 14 February 1997, and I reported on the successful outcome of the work of the IFCS working group and proposed that, having fully met its terms of reference, the group could now be retired. Well, this was not to be. There was a lot of interest in continuing the group until the beginning of negotiations in 1998, and a revised terms of reference was adopted at the IFCS meeting to empower the group to assist in preparing for the UNEP negotiation process and to focus the efforts of governments to take action on POPs. Under its new mandate, the working group would promote implementation, by IGOs, NGOs, and governments, of the recommendations in the IFCS report and UNEP GC Decision 19/13C; facilitate information sharing, emphasizing scientific, technical, and socio-economic information, to prepare governments – especially those of developing countries – for taking action on POPs and for the INC process; provide, before INC-1, scientific and technical information pertaining to the criteria and processes to be developed in the INC process for selection of POPs in addition to the initial twelve; characterize the issues for each of the twelve specified POPs that may have to be addressed by IGOs, NGOs, and governments to design and implement action to reduce the risks to human health and the environment arising from the release of POPs; promote development of baseline data for sources, production, and uses of the twelve specified POPs; promote development of media, biota, and human monitoring data for the twelve specified POPs; and complete its activities with the commencement of INC-1.

From mid-1997, with financial support from several donor countries, UNEP Chemicals worked with the IFCS working group to assist countries in preparing for INC-1. In July 1997 a letter was sent to all IFCS and UNEP POPs contact points soliciting scientific information relevant to the development of criteria and a process for identifying future POPs candidates for international action. This information was analyzed and a summary submitted to INC-1.

Another joint UNEP/IFCS effort involved convening a series of regional and sub-regional workshops to raise awareness of POPs issues in developing countries and countries with economies in transition and of the need to identify and take action on POPs problems at the local, regional, and international levels. These workshops were intended to inform participants on the scientific information and policy development work that led to Decision 19/13C and to collect information from workshop participants on national and regional issues and opportunities that should be addressed during the negotiations. In total, eight workshops were held at the following locations: St. Petersburg, Russia (July 1997); Bangkok, Thailand (November 1997); Bamako, Mali (December

1997); Cartegena, Colombia (January 1998); Lusaka, Zambia (March 1998); Iguazú Falls, Argentina (April 1998); Kranjska Gora, Slovenia (May 1998); and Abu Dhabi, United Arab Emirates (June 1998). The workshops attracted representatives of governments, industry, academia, labour, and public-interest groups from 138 countries and provided an opportunity to gather the views and concerns of countries in each region with regard to the scientific, technical, social, and economic challenges that needed to be addressed during the development and implementation of a POPs convention.

The period between the IFCS meeting in February 1997 and June 1998 was a very active one, involving joint UNEP/IFCS information collection activities and workshops to prepare documents and participants for INC-1. The value of the workshops in the overall negotiation process must be fully appreciated. At the time of the nineteenth session of UNEP GC there was not a widespread understanding or acceptance in the developing world of the pervasiveness of the POPs problem in terms of environmental contamination and health and environmental impacts. Using the IFCS participation formula, stakeholders were invited to the workshops and asked to give papers on the POPs situation in their countries and regions. In so doing, it was shown time and again that there were local (as well as long-range) POPs environmental contamination concerns and that humans were carrying previously unrecognized body burdens of POPs. Recognition was also given to those countries that were taking action on POPs in advance of the negotiation process (e.g., Brazil and Mexico reduced the use of DDT in malaria control programs). Another result was that many of the participants in the workshops eventually took part in the negotiations and were aware of the background of the issues, the need for action, and the UNEP mandate for the negotiations. The environmental NGOs also became more involved and organized. A good example of this was the International POPs Elimination Network (IPEN) that was formed before INC-1: by the end of the negotiations it had grown to include more than 400 members. The involvement of Indigenous peoples' organizations was another notable development.

Another factor that assisted the POPs negotiations was the experience gained during the negotiations on the Rotterdam Convention (PIC) that concluded in March 1998. These negotiations established methodology and process, allowed the UNEP Secretariat to gain experience in supporting such a process, and established personal and working relationships among the negotiators and with the secretariat. These benefits carried over into the POPs negotiations and enabled a smoother start and finish than would otherwise have been the case.

The collective impact of all these developments was invaluable to the process and the benefits were evident from the very beginning.

THE NEGOTIATIONS

Canada hosted INC-1 in Montreal (29 June–3 July 1998), a few months after concluding the negotiations on the Rotterdam Convention, a few weeks after completing the last regional workshop in Abu Dhabi, and a few days after the signing of the CLRTAP POPs Protocol in Århus, Denmark. Thus, we began on a positive note. On the first morning of the meeting, as I approached the entrance to the conference centre, I saw Greenpeace putting on a silent demonstration in which women wore a plastic cover over their abdomens that made them look pregnant, to emphasize the intergenerational impact of exposure to POPs. The media were having a good time recording this, and Klaus Töpfer, Executive Director of UNEP, and Christine Stewart, then Canada's minister of the Environment, were interviewed in front of the demonstration. Such demonstrations would become a regular feature at all INC meetings and served to generate publicity and public interest in the POPs issue.

Representatives of ninety-four countries, sixteen IGOs, and fifty-six NGOs attended the meeting. It was generally understood that INC-1 would be a planning and organizing meeting, and the first order of business after the opening ceremony was the election of the bureau. I was elected to represent the Western European and Other Governments (WEOG) region and the following were elected for the other regions: Ephraim Buti Mathebula (South Africa) for the Africa region; Maria-Cristina Cardenas-Fisher (Colombia) for the Latin America and Caribbean region; M. A. Haque (India) for the Asia Pacific region; and Darka Hamel (Croatia) for the Central and Eastern Europe region. I was invited by the bureau to chair the meeting and Croatia agreed to serve as rapporteur.

The general tone of the meeting indicated solid support for developing the POPs convention by 2000, and there was a general expectation that it would be accomplished with five INC meetings (as was the case with PIC). I attributed a good deal of this spirit to the lasting influences of the IFCS process, the regional workshops, and the successful conclusion of the Rotterdam Convention. The rules of procedure were approved, without amendment, in a matter of hours (they were adapted from the PIC negotiations, where they had consumed most of the first INC). This approval was followed by a general discussion of the elements that should be included in the treaty, and the secretariat was tasked with developing an expanded outline to be considered at INC-2 as the basis for negotiations on the main provisions of the convention.

The meeting developed terms of reference for the Criteria Experts Group for POPs (CEG) to develop science-based criteria and a procedure for identifying future candidates for addition to the convention. It was agreed that the CEG would involve up to three meetings and meet in the period between INC meetings, with the first meeting scheduled for October 1998, and that its work would be completed before INC-4. It proved difficult to decide how many languages would be used in CEG meetings and we spent hours on this question, ultimately deciding on the six official UN languages. This issue frustrated many participants, as the CEG was originally foreseen as a relatively small body of specialists (forty to sixty persons). The interest in providing more languages was an indication that the group would become much larger and require more resources to fund its meetings even though none were yet available. I later concluded that we arrived at the right outcome, as accessibility to this and other aspects of INC work would become the hallmark of the negotiation process and prove to be what we needed to keep us on the right track.

In our attempts to discuss the control provisions of the convention, the issues relating to technical assistance and a financial mechanism repeatedly surfaced. It was agreed to separate this discussion from the negotiation of controls by establishing a subsidiary body – the Implementation Aspects Group (IAG) – to address measures for implementing the convention, including technical assistance, the transfer of technology, and financial aspects. The INC vice-chair, from Colombia, agreed to chair the group. The IAG would meet for the first time at INC-2, following a discussion in plenary of the issues to be addressed by the group, and its work would be concluded at INC-4, where its proposals would be taken up by the INC for inclusion in the convention.

The meeting concluded with plans to continue negotiations at INC-2 even though there was no indication of how the meeting would be funded. Fortunately, the United States had announced that it would fund the first meeting of the CEG, so we were able to maintain forward motion while fundraising continued.

Looking back on INC-1, I see there were a few surprises. The first was the reaction of the negotiators to an intervention by a representative of the Inuit Circumpolar Conference (ICC) who spoke so clearly and from the heart, putting a 'human face' on what many perhaps considered a scientific or abstract issue. She was the only person to receive a round of applause for an intervention, at this or any other INC. The second surprise was the time consumed in the language debate for the CEG meetings: it showed that there was more interest than I had expected in the matter of selecting future POPs and a strong desire to keep all aspects of the negotiation process open and transparent. The third surprise came mid-week as the session ended for the day. The INC was discussing how to organize the conduct of the negotiations at future meetings and I could not understand why the participants were not willing to accept the same model that had been used (successfully) during the PIC negotiations. There, a working group had been set up under the plenary to handle development of the technical aspects of the convention. Frustrated with not being able to get agreement on this, I adjourned the session, but before I could leave my chair I was approached by a few delegates who pointed out that I was probably the only one in the room who did not realize what was going on. They explained that the preference was for me to chair the discussions on the technical aspects of the convention and that I would not be allowed to delegate this responsibility. Another opportunity, I concluded!

One of the key factors in continuing the negotiations was the ability to fund the process through voluntary contributions. The idea of the "POPs Club" surfaced shortly after INC-1, and on 5 October 1998 the UNEP Secretariat mailed out a request for voluntary contributions, indicating the estimated cost for the remainder of the negotiations (about US$3.5 million) and announcing the creation of the POPs Club. The club would recognize annual contributions based on the amount, with different scales for developed countries, developing countries, least developed countries, academia, IGOs, and NGOs. This proved successful and, while the contributions were sometimes received after a session, we were able to schedule upcoming INC and CEG meetings within the necessary time frames to sustain the momentum in the negotiations.

The first meeting of the CEG was held in Bangkok from 26 to 30 October 1998. While 40 to 60 participants were expected, more than 100 participants from 57 countries, 4 IGOs, and 15 NGOs attended, confirming considerable interest in the issues to be addressed in establishing criteria and the process for selecting additional POPs for the future convention. The participants included experts in scientific and policy matters, a signal that the process for evaluating candidate POPs would not be purely technical. The bureau for the CEG included co-chairs from Germany (Reiner Arndt) and Gambia (Fatoumata Jallow Ndoye) and a rapporteur from Thailand (Jarupong Boon-Long). The CEG made good progress at this meeting as a result, in part at least, of the skill of the co-chairs (representing a developed and a developing country), and by the end of the meeting participants were already talking about completing the work in two sessions rather than the three originally planned.

INC-2 was held at UNEP headquarters in Nairobi (25 to 29 January 1999) and was attended by representatives of 103 countries, 10 IGOs, and 51 NGOs. In the absence of the vice-chair from India, Mir Jafar Ghaemieh (Iran) was elected to represent the Asia Pacific region; this would be the last change in regional representation on the bureau until the completion of the negotiations. This consistency was important to me, as I intended to engage the bureau more and more as the negotiations progressed, both before and during INC meetings. I felt doing so would ensure that I was getting needed input on issues and developments in the regional groups and that I would be able to effectively channel information concerning planning and other aspects of the negotiations back to the groups.

During the Monday opening formalities, and relevant to the IAG discussions that would begin later in the session, a representative of the Global Environment Facility (GEF) indicated its willingness to serve as the principal entity of the financial mechanism to be developed under the future convention. The co-chairs of the CEG reported on their first meeting, pointing out that socio-economics was viewed as an important element in the process and that the issue of precaution also had been raised. On the latter issue, it was recommended that the INC consider including in the convention a provision to prevent the introduction into commerce of new industrial chemicals or pesticides that have POPs properties. These issues would eventually become key pieces of the convention.

That evening, during the reception, the ICC presented Dr Töpfer with an Inuit carving of a mother and child in recognition of his personal commitment to the POPs process. He immediately grasped the significance of the carving with regard to the intergenerational nature of the POPs issue and then presented it to me with the request that it be displayed during the negotiations as a constant reminder to the delegates of the significance and importance of our task. The carving sat in front of the chair for the rest of the week, at press conferences, and throughout the remainder of the negotiations as a graphic reminder to all of the importance of completing our work.

The first item of business for INC-2 was to consider the secretariat's 'expanded outline' for the convention. The meeting quickly accepted this outline as the basis for its work in negotiating text – a big step forward as it led to a quick start for the negotiations and, by the end of the week, to a draft text that would be the basis for negotiation at INC-3. Following a general discussion of provisions, the IAG selected its bureau and the plenary split into two to allow the IAG to conduct its first meeting. The remainder of the INC participants met in what became known as the Negotiation Group, under my lead. Recognizing that some delegations were too small to follow what was going on in the different groups, the INC met in plenary each morning to receive reports from each of the two main groups and any others that were set up during the course of the meetings. It was important to ensure that all participants received daily briefings on progress so they would not lose touch with what was going on. This need would be pointed out to me several times during the process as some delegations struggled to keep up with the pace of the negotiations, particularly those with only one or a few members.

INC-2 had three key outcomes: the acceptance of the secretariat's 'expanded outline' as a basis for negotiating the convention text; the acceptance in principle that the long-term goal of the convention was to eliminate production and use of all the intentionally produced POPs; and the commencement of the IAG discussions. In the latter case, although the discussions were sometimes awkward, the 'ice was broken' and some issues and possible solutions were tabled. The separation of the negotiation and the implementation aspects seemed to be allowing progress to be made in each area, always with the knowledge that we would not get a convention unless we had a deal in both areas.

Following INC-2, I made some changes to the way we conducted business. It became apparent that regional groups were becoming more active and influential in the conduct of our work, and this reinforced my commitment to employing the bureau as a two-way communication mechanism. We expanded the bureau to include the IAG Bureau and, beginning with INC-3, the head of the Legal Drafting Group (LDG). I also sensed the need for firmer commitments for regional group meetings in advance of the INC sessions and raised this with the secretariat and the bureau. And I looked at how we used our time at INC meetings, in the context of our goal to conclude our work by INC-5. Recognizing that five-day meetings were too short, but not wanting to extend the meetings over two weeks, I felt that six-day meetings preceded by regional group meetings would improve our progress; this became the template for the balance of the negotiations.

It was at INC-2 that I recognized the value in keeping the INC process as open, inclusive, and transparent as possible. I was strongly influenced by the benefits of this approach on the previous IFCS working group activities and felt that it would be beneficial in the negotiations. I added another factor – accountability. I would be accountable to the bureau and INC for all my actions on behalf of the process and, in turn, I would expect the bureau and INC participants to be accountable to one another for their actions and decisions, especially in regard to how the INCs would be conducted. In committing to this I had to accept that, under its rules of procedure, there were limitations on how far the INC could go in reflecting the IFCS approach and that part of the GC Decision concerning stakeholder participation in the process. Finally, I felt that it would be easier for delegates to prepare for INC meetings if they could expect what would happen when they attended, and so I distributed scenario notes that explained my plans for each meeting, including the proposed timing for addressing certain issues. The main objectives of this approach were to manage participants' expectations, keep the process as open as possible, and keep surprises to a minimum.

About 130 participants from 63 countries, 6 IGOs, and 9 NGOs attended the second CEG meeting in Vienna from 14 to 18 June 1999. This increased participation was probably a result, at least in part, of the expectation that the CEG would complete its work at this meeting, which it did. The CEG developed a draft article on a procedure for evaluating POPs candidates for addition to the convention and proposed the criteria to be used in this procedure, although there was not complete agreement on the latter and a few options would remain for the INC to consider. The discussion on precaution surfaced again amid concerns that the procedure and criteria would be applied too rigidly in the future, thus preventing some candidates from being confirmed as POPs and, therefore, not including them in the convention. At times during this meeting, participants realized that they were attempting to negotiate matters that were the responsibility of the INC and they pulled back from this when it happened. This was a sure sign that the CEG participants had taken their discussions as far as they could and that the next step was to engage the INC.

INC-3, held in Geneva from 6 to 11 September 1999, attracted 420 participants from 115 countries, 17 IGOs, and 81 NGOs; interest in the negotiations continued to grow. The Legal Drafting Group was formed and Patrick Szell (United Kingdom) elected as its head. The report of the CEG was quickly adopted and became the basis for negotiating the relevant convention text on the review of candidate chemicals that would be proposed for future addition to the convention.

INC-3 proved to be a key meeting for the control provisions. Consensus was reached on policy approaches that would be refined over the remainder of the negotiations. For unintentionally produced POPs, there was agreement to move away from 'targets and timelines' towards an 'action plan' approach, based on the realization that not all countries could realistically establish baselines against which targeted reductions could be set. There was general agreement on the policy for intentionally produced POPs, with all but DDT slated for elimination. The use of DDT in malaria control programs had become a sensitive issue in the negotiations. WWF had earlier proposed that negotiators consider a target date of 2008 for the phase-out of DDT and some groups attacked this as a developed world proposal that would cost lives in the developing world. At this meeting, WWF withdrew its proposal. The negotiators agreed that there would be no phase-out of DDT for use in disease vector control programs until effective, feasible, and affordable alternatives were made available. This resolved a very delicate issue in the negotiations and though a few NGO participants tried to maintain this as an issue, they did not succeed.

While considerable progress was made in developing the articles on control provisions, the IAG was struggling to develop proposals for text, as there were basic differences of opinion in what was needed. For example, the developed countries were adamant that there would not be a new stand-alone fund created for this convention, whereas the Group of 77 (G-77)[1] and China were pushing for such a fund, indicating their dissatisfaction with the performance of the GEF. It was evident that the IAG issues would be among the last to be resolved and, given that the IAG was to conclude its deliberations at INC-4, I was beginning to get concerned about completing this part of the work on schedule.

Progress was made in developing text at this meeting, but the complexity of many issues was becoming increasingly apparent to all and the differences in points of view among delegations and regional groups were becoming numerous and more evident. The degree of difficulty was clearly increasing with each meeting and, just as I was beginning to wonder how we could finish by INC-5,

Switzerland tabled an offer to host the first meeting of the Conference of the Parties, preferably in a developing country. I took this as a good indication that delegates were still looking forward to a successful conclusion to the negotiations.

The German government hosted INC-4 in Bonn from 20 to 25 March 2000. The meeting was historic in that it was the first international meeting held in the former Bundeshaus building. In attendance were about 500 representatives of 121 governments, 17 IGOs, and 81 NGOs. The meeting started on a good note with Canada announcing that it would contribute CAN$20 million to the World Bank for capacity-building on POPs in developing countries. This appeared to be a good signal that donor countries would be willing to provide new and additional resources for POPs activities; however, no other country followed suit. One benefit from my point of view was that, as a Canadian chair, for a while I had the luxury of some 'moral high ground' in encouraging potential donor countries to be more forthcoming.

Before the meeting, I had identified three priorities for discussion (completing the controls provisions, determining the needs of developing countries to implement the convention, and defining how these needs would be met) and reminded delegates that no new issues could be introduced into the negotiations after INC-4. Challenging the negotiators to complete a draft treaty by the end of the meeting, I acknowledged that, while there would remain some unresolved issues, the policy choices had to be captured in the draft text to permit consultations to take place before INC-5. That would be the only way to set the stage for conclusion of negotiations by December 2000. I was in for both disappointments and surprises with the meeting outcomes.

By the end of INC-4, all areas of the draft convention had been addressed and the last new issues introduced, but many issues remained unresolved, including several related to the controls provisions. Other issues were also debated without making much progress in narrowing the differences between the various players. The interactions among the various groups became more noticeable, with the G-77 and China, the CEE group,[2] and the two groups of donor countries (the European Union and the JUSCANZ group[3]) divided on issues such as trade, the precautionary approach, and the convention objective for unintentionally produced POPs (minimization vs. elimination).

The most challenging part of the meeting related to the financial mechanism for the convention. The IAG met for the last time and, while it developed a draft article on technical assistance to address the needs of developing countries, it concluded its work without resolving the article on a financial mechanism. An impasse had arisen between the G-77/China group (which supported an independent multilateral fund under the convention) and the developed countries and countries with economies in transition (which wanted the GEF to serve as the financial mechanism for the convention). This issue came to a head in plenary late in the week and required an unscheduled session on Friday evening. The three groups spoke at length and with conviction on behalf of their positions and then there was an ominous silence, with no one willing to take the floor and offer a solution or path forward. The meeting looked to me to propose a solution to keep the negotiations from stalling on this key issue.

I had so far managed to preside over the negotiations without actively entering them. It had been my goal to complete the convention without doing so, although I fully expected to be drawn in at the last possible moment in the process. I wanted to preserve my position of neutrality for as long as possible. Faced with the circumstances, I offered two choices: I could defer the discussion of this issue until INC-5 or I could engage the three groups in a meeting to discuss the issues and determine if a document could be prepared for INC-5 that would bridge positions and conclude negotiations. After some discussion, the latter option was accepted. I would chair a group comprising representatives of ten countries from G-77 and China, two from Central and Eastern Europe, and seven from the WEOG group, which would meet not to negotiate or draft text, but to discuss the various options for a financial mechanism and to submit a report to INC-5 for consideration.

Immediately before the close of the meeting, in recognition that the current convention text was rather untidy, I was invited to develop a chair's draft text to be considered as the basis for commencing the negotiations at INC-5. I left Bonn with two more opportunities to occupy my time over the summer.

In retrospect, it was probably unreasonable to have expected more progress at INC-4. The complexities of some of the issues and the proposed solutions denied rapid progress. In addition, there was no real incentive or pressure to conclude key issues before INC-5. The normal practice is that agreements or concessions on the key parts of the convention are not made early in the process or one at a time. Typically, they are decided together as part of a larger package in the final stages of negotiations. But I was certainly uncomfortable with what was being left to the last meeting and, while I still felt that the task was achievable, the stakes seemed to have been raised for INC-5.

The intersessional meeting on the financial mechanism took place from 19 to 21 June 2000, in Vevey, Switzerland, in a quaint old hotel on a mountainside with a wonderful view of Lac Léman and the Alps. It was chosen for its isolated and pleasant location in an attempt to create an intimate atmosphere conducive to true dialogue and conversation, in contrast to the formal presentation of positions that takes place during an INC meeting. Before the meeting I distributed a scenario note to the participants outlining the purpose of the meeting: to analyze the proposals put forward on the financial mechanism; to identify common ground; and to develop proposals for a path forward that would help meet the needs and interests of the INC and contribute to the conclusion of negotiations. I reiterated that the Vevey meeting would not be a negotiation session. The main tasks of the group were to clarify issues related to the financial mechanism and to seek to build bridges among the various positions to identify solutions that could attract the support of the various regional groups at INC-5.

The setting did help to get the participants talking to one another, during both the working sessions and the meals, and I noted that the interpersonal communication developed quickly and well. The various proposals for a financial mechanism were critically evaluated to see how each would meet the demands arising from the obligations of the convention. The participants were careful not to actually engage in negotiations and, by the end of the meeting, I sensed we all had gained a deeper understanding of the various positions and, importantly, the reasoning behind the positions. I also received – and followed – some advice on how to introduce the issues at INC-5. At the end of the meeting, I undertook to prepare a report, circulate it to the participants, and, after receiving and responding to their comments, distribute it as a chair's report of the meeting for consideration at INC-5. Participants undertook to communicate on the meeting outcomes with their colleagues at the national and regional levels and inter-regionally in preparation for INC-5.

At the time, I did not fully appreciate the impacts that would result from the Vevey meeting. Key players came together in a way that previously had not been possible during the negotiation process. One result was a description of the attributes of a financial mechanism – many of which were viewed as common ground. Several of the attributes were intended to address shortcomings in the performance of the GEF from the point of view of recipient countries. Another outcome was a description of the options for a financial mechanism for the convention. This document would prove to be a good basis for moving forward on a number of fronts.

My other summer opportunity was the preparation of the chair's draft convention text. This was developed in London in July, where I met with the secretariat and the head of the LDG. The challenge, for me, was to respect the intent of the negotiators while cleaning up awkward, contradictory, or unnecessary language, all the while being careful not to introduce changes that might suggest I had entered the negotiations. In analyzing the convention text, I realized that it would not be possible to implement the proposed mechanism that would allow countries specific exemptions to produce or use intentionally produced POPs. This meant that much more work would be required at INC-5 to develop a fresh approach for a workable mechanism and it would have to be developed and completed during only one meeting. Another opportunity.

One of my concerns during the summer was that several differences of opinion existed between the European Union and JUSCANZ countries on key issues. Members of the two groups also shared this concern. From 5 to 8 September, Canada hosted a well-attended WEOG regional meeting in Montebello, Quebec, to provide an opportunity to share views and seek compromises on key issues. This was followed by a meeting in Paris on 6 November, hosted by France (as president of the European Union), immediately before a regular meeting of the OECD Environmental Health and Safety Committee. As the bureau representative for these countries, I attended both meetings to assist in attempts at narrowing the differences between the two groups.

The last major event before INC-5 was a meeting of the GEF Council in Washington from 1 to 3 November. A discussion was scheduled on the possible role of the GEF in facilitating the implementation of the POPs convention, including a proposal to include a new POPs operational program. I was invited to attend to inform the council on the state of the negotiations and the outcome of the Vevey meeting. Given the sensitivity of the unresolved issue of the convention financial mechanism, I described the current situation and the differing points of view, being careful not to give the impression that I was endorsing the GEF merely by attending the meeting. I pointed out that it was entirely up to the INC participants and the GEF council members to take such actions or decisions each thought would be warranted, and I invited the GEF CEO to report to INC-5 on the outcome of the council meeting.

The GEF council agreed to be guided by the decisions of the INC and the future CoP to the convention. It also agreed that, if designated as the financial mechanism for the POPs agreement, the GEF would prepare an initial assessment of the financial resources required to ensure a quick start to providing assistance to eligible countries; make available additional financial resources for this purpose through the third replenishment scheduled to be decided by May 2002; and initiate early action to fund enabling activities with existing resources. The council also approved several steps to streamline the GEF project cycle and improve overall effectiveness and requested the responsible IGOs to ensure the efficient and timely disbursement of funds for enabling activities. These outcomes were influenced by the combination of the state of play in the negotiations (deferment of a decision to set up the POPs operational program until after INC-5) and the report of the Vevey meeting (the streamlining of the GEF procedures). The stage was now set for the showdown at INC-5.

The final negotiation meeting, hosted by the South African government and partially funded by Denmark, took place in Johannesburg from 4 to 10 December 2001. About 550 representatives of 121 governments, 11 IGOs, and 97 NGOs attended the meeting with the fresh memory of the failure of the climate change negotiations in The Hague a few weeks earlier. The pessimists were talking about a sixth INC, feeling that the time might not be right for a successful outcome, while the optimists felt that the negotiators would work even harder to avoid repeating the experience of The Hague.

There were many issues yet to resolve and final text was needed to complete the work. The major issues for INC-5 related to

- control measures for intentionally produced POPs (especially the mechanism to implement the measures for DDT and 'specific exemptions'), unintentionally produced POPs (minimization vs. elimination and targeted national reductions vs. national action plans), and stockpiles and wastes (relationship to other regimes, such as the Basel Convention);
- measures for addition of new POPs (completing the criteria and process for reviewing nominations);
- the inclusion of precaution in the convention; and
- securing agreement on financial and technical assistance, especially the decision on a financial mechanism.

The controls were a high priority for the WEOG countries and they were well prepared with proposals to resolve the remaining issues in this area. The highest priority for the G-77 and China was the decision on financial and technical assistance. The WEOG and CEE countries supported a GEF mechanism while G-77 and China strongly supported a stand-alone mechanism. The issue of 'precaution' promised to be difficult to resolve and the EU and JUSCANZ countries were expected to be most active in debating this issue.

I had informed delegates ahead of time that provision had been made for three sessions per day beginning on Tuesday, but that I was prepared to relax this requirement if sufficient progress was made during the week. This was not to be the case and, with the exception of Monday, most delegates were in the convention centre from eight in the morning until nearly midnight, with regional group meetings and contact groups in operation in addition to the plenary that met throughout the week. The week started off well, with delegates agreeing to accept the chair's draft text as the basis for the week's negotiations. As a result of the heroic efforts of co-chairs of the contact groups, some getting only a few hours of sleep each night, success was at hand by mid-week on the control measures, although agreement on the final aspects was not obtained until the early morning hours of 10 December. Issues began to be resolved on such matters as pesticides and industrial chemicals, trade, screening of new and existing chemicals and pesticides, and other provisions. However, several issues remained near the end of the week. These included the policy option for unintentionally produced POPs (minimization vs. elimination), the relationship of the POPs convention with the Basel Convention, how precaution would be reflected in the convention, and the interrelated issues of technical assistance, technology transfer, and the financial mechanism.

On the latter issues, I proposed to the bureau that a contact group be set up to develop draft proposals for the plenary and was surprised when informed that this would not be acceptable, that the G-77 and China insisted that I preside over the discussion. With far too many delegates in the plenary to have the necessary type of exchange, this would prove a difficult challenge. I was also concerned that there needed to be transparency in this discussion to ensure that, whatever proposal resulted, there would be no holdouts over the nature of the process used. The other problem was interpretation: with an issue of this importance to the developing countries, we needed to ensure that all could follow the discussion. The solution for this was to hold a meeting of a smaller contact group in the centre of the plenary room with interpretation to allow the group's discussion to be witnessed by anyone who was interested. The room was reconfigured and, in the space of one three-hour session, we had an appropriate exchange of views and were able to mandate a smaller group to work on a proposal for the INC to consider. This group laboured until late Saturday evening, when it struck a compromise proposal.

One of the last main issues to be addressed was the role of precaution in the convention. The main protagonists in this case were the EU and JUSCANZ countries, which spent many hours in discussion looking for a solution. At times the debate got quite intense and, on Saturday evening, during a break between plenary sessions, I was invited into the meeting to listen to both sides, give my reaction to their positions, and make suggestions on moving towards a solution. After giving my reaction, I reminded the group that the fate of the convention was in their hands and that as everyone agreed that the convention should include references to precaution, it would be a pity indeed to lose the deal over how to word such references in the convention. The two sides came to an agreement-in-principle shortly after I left to rejoin the plenary and, early Sunday morning, agreed on exact language.

Beginning Saturday morning, we held four three-hour plenary sessions and many smaller groups met to resolve remaining issues; however, we had yet to agree on articles on technical assistance and a few other matters. At about 5:30 Sunday morning, we lost our interpreters, who had done a wonderful job in staying with us as long as they did. I needed a break at this point but I wanted to know whether there was a will to finish the convention and I gauged this by asking if there would be any objection to finishing the negotiations in English only. At this moment, any single delegation could have refused and halted the process; however, no delegation objected and for the first time we all knew for sure that we were going to complete the convention. We resumed at about 6:30 A.M. and at 7:28 A.M., according to the staff of the *Earth Negotiations Bulletin*, we concluded the last article of the convention. An hour later, after the closing formalities, the South African hosts treated the survivors of the long night to a champagne breakfast.

JOURNEY'S END: CELEBRATION IN STOCKHOLM

About a week after my return home from Johannesburg, I received the first compilation of the convention text for review. I met with the secretariat and head of the LDG in Geneva in mid-January to scrutinize the text and sort out some problems. The convention text was then made available in February for all to see. There proved to be only two minor problems, which were easily rectified, and the convention was then distributed to delegates who would attend the Stockholm Conference of Plenipotentiaries.

As a result of the late finish of the negotiations at INC-5, several resolutions were not completed and participants agreed to hold a one-day meeting before the Conference of Plenipotentiaries to complete them. This occurred on 21 May 2001 and brought agreement on draft resolutions to continue the INC process to prepare for a fast start for Conference of the Parties 1 (CoP1) and to provide oversight on the international co-operative action on POPs, with particular attention to addressing the needs of developing countries. On 22–23 May, the Stockholm Convention and the resolutions were adopted and the convention signed by ninety-one countries and the European Union. Immediately following this signing ceremony, Canada became the first country to ratify the convention – a particularly satisfying moment for me.

CHECKING THE REAR-VIEW MIRROR

The road to Stockholm was a long and winding one, indeed. In reflecting on the period from 1995 to 2001, I recall the following factors that influenced the process and led to the development of a widely supported convention that is acknowledged by all stakeholders as a significant advance in international environmental law. These are grouped into factors that influenced either the process or the content of the convention and followed by a brief summary of some lessons learned that may be of interest in future activities of a similar nature.

Factors that influenced the process:

- Sustained involvement of stakeholders throughout the process
  Stakeholder involvement began with the UNEP/IFCS ad hoc Working Group on POPs, continued with the UNEP/IFCS sub-regional workshops, and was maintained during the INC process, despite limitations that were necessary due to the nature of an intergovernmental negotiation process. Convinced of the importance of maintaining this approach, I repeatedly challenged participants in the process to keep it open, inclusive, and transparent, and I added another element to this: accountability. I committed to keeping myself accountable to the INC delegates for my actions related to the negotiation process, and in turn challenged them to be accountable to their colleagues in the INC process for their actions.

- Sustained publicity for the POPs process
  The peaceful and effective demonstrations at each INC by the environmental NGOs helped to attract publicity for the negotiation process. The Inuit carving of a mother and child presented to Dr Klaus Töpfer by the ICC was an-

other publicity coup. The carving was displayed from INC-2 throughout the remainder of the negotiations as a constant reminder to the delegates of the significance and importance of the task and became an icon for the negotiations. The secretariat also undertook to publicize aspects of the negotiation process using print and electronic media. Public attention and political support are valuable in demonstrating the importance and relevance of the issue and in maintaining priority for the activity at the senior and political levels of government.

- Open and transparent communication throughout the process
  The need for and benefits of open communication became apparent during the IFCS working group activities. Following INC-2, I noted the need to assist delegates in preparing for INC meetings by giving a sense of what to expect to happen and when and setting goals for each meeting so that we would conclude by INC-5. I found that distributing scenario notes helped in this regard as it served to "manage the expectations" of delegates. I received positive feedback on this and would recommend the practice to others.

  I must also acknowledge the team behaviour exhibited by the bureau members throughout the process, including the extended bureau members from the Legal Drafting Group, Implementation Aspects Group, and contact groups, who met before each INC and early each morning before they attended regional group meetings. They were always willing to provide their input in planning and organizing meetings, conveying messages to their groups, informing me of developments within their groups that I should be aware of, and identifying problems and possible solutions. This two-way communication was an essential ingredient in the planning and conduct of the negotiations and became increasingly important during the latter stages, when advance information on unexpected developments allowed time for making necessary adjustments.

  Finally, we were able to successfully conclude negotiations in plenary, without having to resort to a small group (e.g., "Friends of the Chair") to resolve the issues remaining in the final twenty-four to forty-eight hours of INC-5. Previous experience had shown me the risks of smaller groups trying to make a deal on behalf of a large body of negotiators, and I was particularly pleased that small groups were used only when mandated by the plenary. Several delegates have since remarked to me that they were pleased that we completed our work in plenary and that the process maintained full transparency right through until the end.

- Strong and reliable secretariat support throughout the process
  A key factor in the overall success of this process was the strong support by the IFCS Secretariat during the working group activities and the UNEP Secretariat before and during the negotiations. Dr Klaus Töpfer attended all the

INC meetings and constantly supported the negotiations, including advancing funds when donations were slow to come, thus allowing meetings to be scheduled at six- to eight-month intervals to maintain the momentum of the negotiations. I must acknowledge the steadfast and professional performance delivered by the UNEP Secretariat, headed by Jim Willis. The experience gained during the negotiation of the Rotterdam Convention prepared the secretariat to support the bureau, the INC, and other related meetings to ensure timely planning and organization and the distribution of documents sufficiently in advance of meetings to enable delegates to conduct consultations before the meetings. Throughout the INC process, secretariat staff were at my side at all times for advice and other assistance and this allowed me to focus on the issues and discussions, session after session. I believe that the solid performance of the secretariat throughout the negotiations also earned the trust and appreciation of the delegates.

Factors that influenced the process or content of the convention:

- Early agreement on the scientific basis for international action
  The first turning point in the overall process was the March 1996 IFCS meeting in Canberra, where it was concluded that there was sufficient scientific evidence available to warrant international action on POPs. Getting this multi-stakeholder agreement early on allowed subsequent efforts to be directed towards elaborating *when* and *how* to deal with POPs and we did not revisit the scientific justification for the issue. This would be a key factor in any similar activity.

- Adherence to agreements in the IFCS report
  The agreements that were reflected in the IFCS report were tenaciously held to during the development of the mandate at UNEP Governing Council in 1997 and throughout the preparation for and conduct of the negotiations. It is noteworthy that the agreements reflected in the IFCS report were developed during a relatively short period in 1995–96 (as required to complete the mandate in UNEP GC Decision 18/32) and accounted for a large proportion of the provisions in the final text of the convention.

- Scientific basis for adding new POPs
  Establishing an Experts Group that maintained stakeholder involvement was instrumental in developing a proposal that included scientific and socio-economic factors that was quickly adopted by the INC and served as an excellent basis for the final convention text. The Experts Group was also instrumental in including a provision in the convention to prevent the introduction into commerce of new industrial chemicals or pesticides that have POPs properties.

- Separate discussions on policies and technical and financial assistance
  There was early agreement to separate the discussions of policy content for the convention from the discussions on technical and financial aspects. This enabled progress to be made in each area, while recognizing that resolution was required in both areas to complete the convention. I do not know how the negotiators could have addressed all the issues in the time that was available for the negotiations without this approach.

- Steady pacing in developing the convention text
  INC-1 began with a list of possible articles. INC-2 considered an expanded outline for the convention based on text from other existing multilateral environmental agreements but left blank articles for which there was no prior experience. This outline was accepted at INC-2 as the basis for negotiations and the text was developed and modified at INC-3 and INC-4. Based on a request from delegates at INC-4, a chair's text was offered at INC-5 as an alternative to the text as it stood at the end of INC-4. The chair's draft text was adopted as the basis for negotiation at INC-5. This pacing allowed a good amount of general policy consensus to be reached before attempting the actual drafting of the legal text. It also allowed negotiators to buy into the process and the product in a stepwise manner along the way.

The following lessons learned may be of interest in future similar activities:

- An early agreement on the scientific basis for action is key to shifting from a discussion of *whether* action is needed to a discussion that addresses *when* and *how* to take action.
- A process that features open and transparent communication and includes participation by all stakeholders should yield a balanced product that will enjoy widespread support.
- The chair and bureau members should serve as models for effective two-way communication throughout the process.
- Publicity should be encouraged concerning the negotiation process. The public and politicians should be kept aware of the issues and the need for actions that will involve all sectors of society.
- Strong and professional secretariat support for both the chair and the delegates is a key ingredient in the negotiation process.
- Early consideration should be given to separating the discussions on policies from those on financial and technical aspects. This would allow progress to be made in both areas at the same time, recognizing that both areas must be satisfactorily addressed to successfully conclude negotiations.
- Pacing is important in developing the convention text. It is worthwhile to begin with a general framework and then allow delegates to fill it in as knowledge and understanding grow. It is not advisable to offer a full draft text too

early in the process. The chair can be invited to submit a draft text for consideration at the last negotiating session as an alternative to the ongoing draft.

THE ROAD AHEAD

The long journey to Stockholm required a lot of hard work by thousands of people. They produced a major new convention that includes policies that will be implemented in a long-term process of change to protect human health and the environment. Much work remains ahead to implement the convention.

To ensure that the momentum built during the negotiations is not lost, the participants at the Stockholm conference agreed to continue the INC process to oversee preparations for the first meeting of the CoP and to coordinate the ongoing actions of stakeholders to address POPs issues. Annual meetings will be held, beginning with INC-6 in Geneva (June 2002), where delegates are to address a wide range of issues relating to implementation of the convention, including national implementation plans, technical assistance and financing issues, organizational and operational aspects of the future CoP, and the development of guidance for those involved in taking action on POPs. This interim process will continue until the convention enters into force, and then the CoP will become the body responsible for overseeing implementation of the convention. However, by the time that the signature period closed on 22 May 2002, 150 countries and the European Community had signed the Stockholm Convention and 8 had ratified it. More than 108 countries were already taking actions on POPs before the completion of the convention and this number is expected to grow before the entry into force of the convention as stakeholders around the world address POPs problems. I suspect that the long and winding road to Stockholm was but a prelude to a longer journey ahead. Hopefully, it will not all be uphill.

ACKNOWLEDGMENT

The success of the negotiations was a result of the efforts of the thousands of individuals who participated in the development of this convention. I met hundreds of people during the negotiations, only a few of whom I have mentioned by name in this chapter, and their individual and collective efforts and contributions are gratefully acknowledged. I must acknowledge the support and encouragement that I received from my colleagues at Environment Canada and others within the Government of Canada, beginning with my initial role in 1995, when I was authorized to become involved in the working group, right through to the Stockholm conference. Without this unwavering support, I would not have been able to continue in my role as chair. Particular thanks are due to my colleagues on the Canadian delegations to the various meetings, who supported and encouraged me, especially at those times when I began to doubt my ability to carry on to the conclusion.

NOTES

1 The Group of 77 (G-77) was established in 1964 by seventy-seven developing countries at the end of the first session of the United Nations Conference on Trade and Development (UNCTAD) in Geneva. From the group's first ministerial meeting in Algiers in 1967, a permanent institutional structure gradually developed. The membership of the G-77 has increased to 133 countries; however, the original name was retained because of its historic significance. The G-77 seeks to harmonize the negotiating positions of its members. See also http: //www.g77.org.

2 The CEE group comprises central and eastern European countries, most with economies in transition.

3 The JUSCANZ group of countries refers to a bloc of developed nations that, when named, included Japan, United States, Canada, Australia, and New Zealand and now includes Iceland, Liechtenstein, Norway, the Republic of Korea, and Switzerland. (Mexico occasionally participates because of its membership in NAFTA; however, it did not participate as a part of JUSCANZ in the POPs negotiations.)

12 The Inuit Journey towards a POPs-Free World

SHEILA WATT-CLOUTIER

My public role in the POPs process began with an invitation to speak at a forum organized by Physicians for Social Responsibility in Montreal, the day before the first United Nations negotiating session towards a legally binding agreement to eliminate persistent organic pollutants (POPs) at their source.

However, my personal involvement started when I was a child growing up in a traditional Inuit world: hunting, fishing, and gathering; travelling on the land, ice, and water by dog team in the winter and by canoe in the summer. I was the youngest of a small family born to a single mother and grandmother (in those days the white fathers did not stay). My mother worked very hard to provide for our family, working for the Hudson Bay Company and later at the nursing stations. Because she travelled a fair amount with her work, we were raised by our grandmother during most of our early childhood. We lived quite traditionally; my two brothers were the only males in the house and they learned to become hunters at a young age, feeding the family with our precious "country food," mainly caribou, ptarmigan, fish, goose, seal, and whale. To this day, I am an avid country-food eater and have a very close connection to my Inuit heritage.

My perspectives on this POPs process will be from the heart: it is my spiritual journey. I will start by saying that although my role in these negotiations has been that of a person elected by my people to work on global issues – including the protection of our environment – there can be no doubt that it led my personal and professional journeys to mirror each other. As I work to help the world rid itself of its man-made toxins, a parallel process is happening in my own spirit, as deep-rooted legacies of generational wounds that have left their mark deep in my soul are now, through my own cleansing, leaving my body and spirit.

Anyone who knows me well knows how spiritual I am and that I am more so since the loss of three close family members in the last three years and the near loss of one of my own children. I buried my beloved only sister, who without warning died at age forty-eight of a massive heart attack, a week after the second session in Nairobi. A few days before leaving for the fifth session in Johannesburg, we buried an aunt who had played a key role in my childhood. Just days after my return from Johannesburg, one of my children was airlifted out of the Arctic by air ambulance and went into emergency surgery moments before potential death. Less than three months after the signing of the Stockholm Convention, I also buried my mother. My personal losses are mirrored by those of many families in our Inuit world who have come to know personal loss all too well.

I am part of a generation that has experienced tumultuous change in a short period of time, coming from a very traditional way of life to a modern, high-tech world. In fewer than fifty years I have come from travelling only by dog team and canoe to flying in jumbo jets all over the world. This change has come at great cost to our society: we Inuit are known to have the highest suicide rates in North America, higher than any of our Indian brethren, and the speed of change has been a contributor. This change from a strong, independent way of life – living and learning from the land under our own education, judicial, social, and economic systems – to a highly dependent way of life, often involving alcohol, drugs, institutions, and processes, has shaken the very foundation of our families and communities.

More recently though, we are stopping, thinking, and, more importantly, feeling: we have begun to recognize what has happened in our Inuit world and we have started to understand the importance of regaining control over the health of our communities that has been lost over the last few decades. We have begun to appreciate the importance of the *wisdom of the land* in regaining the health of our families and communities. The land teaches not only the technical skills of aiming a gun or harpoon or skinning a seal, but also what is required to survive, giving confidence to our people. It builds the character skills of judgement, courage, patience, strength under pressure, and withstanding stress, which together is the wisdom that will help our young people to change, to choose life over self-destruction.

As we gradually revert to a more sustaining way of life, the last thing we need is to think that our cultural way of life – including our precious country food – is adding to our turmoil. Imagine the shock, confusion, and rage that we initially felt when evidence of high levels of POPs was discovered in our cord blood and nursing milk in the mid-1980s. In fact, levels of many of these substances in the breast milk of Inuit women have been found to be higher than in women anywhere else in the world. Published studies confirm the presence of more than 200 chemicals – including DDT, PCBs, dioxin, lead, mercury, toluene, benzene, and xylene. These were found in mothers' milk, not in some hazardous-waste site. We were being poisoned – not of our doing but from afar.

Faced with this stark reality, I brought a sense of responsibility, commitment, urgency, and passion to my work – both on the issue of POPs and on other issues, such as climate change, that could bring disastrous negative impacts to our very way of life. The issue is not just the contaminants on our plate: our entire way of life is at stake. Traditional Inuit wisdom, the teachings of the land, the power of the hunt, and the consumption of our country food all contain answers to the problems we are facing. They will lead us to meet our challenges and help us to survive in this modern world that has come so rapidly to our Arctic doorstep.

This chapter will include my perspectives on the Inuit Circumpolar Conference's (ICC) role in the work of the negotiations and, as importantly, on some of the side events and important players with whom I have had the privilege of becoming a comrade-in-arms in this fight.

The first negotiating session in Montreal must be considered a strong opening for our message and our struggle. I addressed the forum organized by the Physicians for Social Responsibility the day before the UN negotiations began. The audience's standing ovation was a significant demonstration of support for the clear message that those most affected, Inuit, would become involved in the process and would bring the collective weight of a people to the fight against POPs. This was when I first met Barry Commoner, a well-known American scientist and advocate for the environment, who urged me to bring forward Inuit perspectives in the POPs negotiations, and Tom Goldtooth, of the U.S.-based Indigenous Environmental Network (IEN), who was committed to injecting the views of Indigenous peoples into the POPs negotiations. We subsequently worked closely with Tom and his colleagues.

Then, the next day, my short intervention on the first day of the UN negotiating session was equally well received. Christine Stewart, then Canadian minister of Environment, had delivered a welcoming speech in which she mentioned the high levels of POPs Inuit mothers had in their nursing milk. My initial reaction could be described as cautiously optimistic: with this kind of support and recognition at the onset of discussions, we just might have a crack at getting the world to understand our plight. These were hopeful moments for me in my newly acquired role as spokesperson for the coalition that had been established to voice the concerns and interests of Inuit, Métis, Dene, and Yukon First Nations in northern Canada. (First called the Northern Aboriginal Peoples' Coordinating Committee on POPs, the coalition later became known as Canadian Arctic Indigenous Peoples Against POPs (CAIPAP).) As ICC president for Canada and vice-president for ICC International, I was representing the Inuit of Greenland, Alaska, and Russia, as well as Canada on this particular issue.

At the second session, in Nairobi, Kenya, through the International POPs Elimination Network (IPEN), we were able to make connections with Indigenous peoples of Africa who were also struggling with the issue of POPs. It was a parallel IPEN/World Wildlife Fund-sponsored evening reception that proved, in the long term, most important to Inuit. During that reception I presented to Dr Klaus Töpfer, Executive Director of UNEP, an Inuit carving of a mother and child that became a symbol for the duration of the negotiations.

Klaus Töpfer was a man who, from the outset of this process, gave attention to and connected the POPs issue with the Inuit and the Arctic. As our host for the week in Nairobi, he said that the ultimate goal of this treaty must be the elimination of the production and use of POPs, not simply better management. ICC's position was also that the POPs convention, when eventually ratified, should reflect the goal of elimination rather than risk management. Our objective had been to use ICC's UN observer status to press for a comprehensive, rigorous, and verifiable global POPs treaty. We knew that even if the taps were shut off at their source today, it would take fifty years to clean up the Arctic sink. Simply managing these toxins was not going to be enough for those who live in the Arctic and live with this reality. Klaus Töpfer was consistent in his demonstration of respect and his understanding of our plight throughout the entire process of the negotiations, and I will be forever grateful to him for that.

In appreciation for that respect I presented him with a soapstone carving of a mother and child carved by an Inuk woman from my region of Nunavik. He in turn presented the carving to John Buccini, chair of the sessions, and said that it would sit in front of the chair through the rest of the negotiations. It did indeed. John Buccini told me once that when fatigue threatened, he would only have to look to the carving for further energy and strength. (John did a remarkable job of chairing these challenging negotiations. His calm, organized, and inclusive demeanour reminded me of the qualities of our Inuit Elders; it radiated throughout the negotiations and led to the signing of the convention.)

The third session of negotiations, held in Geneva, Switzerland, was perhaps the most challenging to me. My staff and I would often finalize most of the text of my speeches only when we were at the negotiation session. I always felt that feeling the pulse of the place, the players, and the issues was important before finalizing the text. Inuit are few in number and we need to be as strategic and focused as possible – just as the hunter must be wise and focused when going out for a hunt – when we make our interventions. This is another level of hunting, after all, and our people expect their leaders to come home with something of substance. In fact, it was the case that Inuit – only 150,000 strong in the entire world – exerted influence beyond our numbers in the negotiations; our "hunt" was successful.

The importance of DDT to malaria control was the "hot" issue in Geneva and the timing of my presentation – the day after most of the NGOs and others had intervened – gave us the opportunity to reflect and respond in a way that would be appropriate and helpful to the bigger picture. I remember quite clearly a person from The Malaria Project suggesting, in his presentation, that if we continued with the goal of eliminating POPs, including DDT, we would be bringing death to many thousands and perhaps millions of people. He used images of jumbo jets filled with people to show the numbers that would die each year if DDT were not available to help fight malaria. As I was listening to this presentation I started to question how 150,000 Inuit could compete with those stark figures of potential deaths. Our team, which included Shirley Adamson of the Yukon First Nations, had to develop a strategy to circumvent seeing this solely in terms of numbers; if numbers were going to be part of the equation then the Inuit and Aboriginal peoples of the circumpolar world would surely lose.

We told the negotiators that in the circumpolar world we know about disease and death; whole families and communities had been virtually annihilated by smallpox in the last century and some Indigenous peoples in Russia stand now on the brink of extinction. I stressed that the issue of POPs was about the survival of entire peoples, potentially a loss to the whole world. We chose the "high moral ground," empathizing with people who felt they needed DDT to preserve life and health. We told the delegation that we would not be party to any agreement that threatens others as it was not our way, and we continued to push the delegation not to see the DDT issue in either/or terms as we felt it would lead to inappropriate choices.

I stated in my speech that "I cannot believe that a mother in the Arctic should have to worry about contaminants in the life-giving milk she feeds her infant. Nor can I believe that a mother in the South has to use these very chemicals to protect her babies from disease. Surely we must commit ourselves to finding and using alternatives. While adopting elimination, not perpetual management, as an ultimate goal, the POPs convention must simultaneously ensure that cost-effective alternatives, particularly for DDT, are made available in the developing world." I also reminded delegates of what I said in Montreal on the first session: that a "poisoned Inuk child, a poisoned Arctic, and a poisoned planet are all one in the same," and that we were all in this together.

The fourth negotiating session was in Bonn, Germany, and in the months preceding we had planned with the head of the host German delegation to bring Aboriginal artists from Canada to perform at a reception that the German government would host. Performances by Aqsarniit ("The Northern Lights"), an Inuit drum dancing and throat singing group, and the Tagish Nation Dancers (composed of Tagish and Tlingits peoples) as well as the country food we offered to those at the reception helped sensitize the world to our culture. It was a proud time for us and certainly for me, as my daughter, Sylvia, was one of the performers promoting our way of life.

However, alarm bells began to ring that week as we heard that some delegations were starting to believe that the Arctic was getting too much attention.

Surprisingly, even a person who had intimate knowledge and perspectives of the Arctic, having worked for an Inuit organization, asked one of my staff "What are you guys up to on Thursday night at the reception anyway?" and suggested that we shouldn't "be too repetitive in [our] intervention." We listened to these alarm bells, and knowing that perceptions – or misperceptions – have potential to do a lot of damage, accepted that not everyone understood or sympathized fully with us. However, I also knew that because the priorities of others could easily dilute the reality of this issue to the Arctic, it would be up to us to keep our strategy on track; we had to maintain our focus to keep the Arctic face visible on the map.

As an individual, my emotions may be triggered from time to time; however, I must work through those quickly as it is crucial for the larger scheme of things that I not operate from a position of fear. As an elected leader, it is my responsibility to remain open and to lead with wisdom rather than fear; in light of what was happening we knew we had to bridge the gap between North and South. My intervention in Bonn reflected this and we called upon the collective wisdom of all countries to find ways and means to achieve a strong agreement. We recognized that POPs posed a significant human and environmental health threat both at the origin of their use and, through long-range transport, to Arctic Indigenous peoples. We also recognized that people from the Arctic to Mexico and India as well as everyone in between are at risk and would either bear the burden of a weak agreement or reap the benefits of a good global convention.

We also pressed for financial commitments to implement an agreement, for without a financing mechanism the developing world and countries with economies in transition would not be able to move from a dependence on POPs to safer, more effective alternatives and cleaner processes. Although we in the Arctic were net recipients of these POPs, our message was always to work with the world. As partners, not victims, we once again asked the people of the world for their help, courage, and wisdom to build a convention that would provide safety to people in the countries still using these chemicals and to Arctic peoples who were receiving them.

By this time there was much media interest in this issue. The world generally knows more about the Arctic's wildlife than its people. We believed it was crucial that the world appreciate that the Arctic is a real, majestic land with a resilient people who, to survive, are closely linked to the land upon which we hunt, fish, and gather. Between the fourth session, in Bonn, and the fifth, in Johannesburg, I hosted the UNEP communications team and BBC World television crews who came to Iqaluit, where I presently live, to film how our world is being affected by POPs. They went seal hunting and got to experience our culture at its best. They ate country food and, importantly, got to see just what it is we are up against. The film that UNEP produced was shown at the fifth and final session in Johannesburg, and the BBC piece aired on BBC World television several times before Johannesburg. While all this was very helpful, we could never assume that everyone would "get it." UNEP started showing its video in the corridors once the sessions started and a UNEP employee commented disparagingly on one of the scenes showing a group of us eating our country food raw as we Inuit so often do. I incorporated that into my intervention the next day, reminding the delegation that we eat what we hunt; that we eat animals raw is the reality we have presented to you for the last two years. I added that I hoped our message – ensuring our health and cultural survival – was getting through.

The final negotiating session was held in South Africa, the great country that gave us the leading figure of our time, Nelson Mandela. Feeling certain his participation would help reinforce our cause, I had invited Paul Okalik, the first premier of the new Canadian territory of Nunavut, to join us as part of the coalition delegation in the final negotiations. Not only was Premier Okalik a great delegate, he and I had the honour and privilege of meeting Nelson Mandela on the seventh of December during the week of my forty-seventh birthday. What a birthday gift! Through his life, Premier Okalik had revered and learned from the experiences of Nelson Mandela; our meeting with him was memorable, to say the least, and he left us hopeful of meeting again.

By the fifth session we had become impressed with the commitment of nations around the globe to address POPs and, reflecting these sentiments, I stated that we were firmly convinced that the job could be completed by the end of the week. We also continued to press for the guarantee of significant funding and technical support to the developing world and economies in transition. We were encouraged by comments from the head of the Global Environment Facility (GEF), Mr Mohamed El-Ashry, about his commitment to making the GEF an accessible funding agency to deal with financing for implementation. I first met Mr El-Ashry at the Arctic Parliamentarians meeting in Finland, where, in my speech, I pressed him publicly to listen to what the developing world was perceiving about the GEF's inaccessibility and also invited him to attend the final POPs negotiation session in Johannesburg to show his commitment to the process. My forthright invitation to him was taken somewhat as an offence to the American chair of the Arctic Council, who felt it was the right of the chair to extend such an invitation. We needed not only a strong convention but one that could be implemented and although I did not feel apologetic to the chair of the Arctic Council, I did have an amicable "chat" with him on the matter and all was smoothed over. Mohamed El-Ashry came to South Africa and we became friends, finding common ground in being "grandparents against POPs." As he announced the birth of his second grandchild in his presentation, he set the context of cleaning our world for the future generations.

Prior to the Johannesburg meetings, Finland had assumed the chair of the Arctic Council and made great efforts to bring all Arctic nations and the council's permanent participants together by first hosting a luncheon and then presenting the political declaration signed that autumn by ministers representing all eight Arctic nations, in Barrow, Alaska. The declaration highlighted the need for a comprehensive and verifiable global POPs convention, committing the Arctic states to "coordinate closely in international fora on environmental and sustainable development matters of importance to the Arctic." The declaration was, in our opinion, a real step towards an effective convention. I remember clearly another hopeful moment during that ministerial meeting in Barrow, when, after my presentation on the POPs issue, the American spokesperson, Frank Loy, Under Secretary of State for Global Affairs in the State Department, commended Klaus Töpfer for his work on POPs and wished Klaus well as he was truly doing "God's work." Hearing that statement from Frank Loy and listening to a great speech from Fran Ulmer, Lieutenant Governor of Alaska, I remember thinking, "by jove we just might have these tough Americans on our side." In my experience with the Americans at the political level, it is always difficult to get anything through with the American delegation unless they were the initiators. Knowing that the Americans can exert great influence on others, I became more hopeful now that an American in that position felt it was "God's work" to try to rid the world of toxins.

Throughout the week in Johannesburg there were stressful times, with the possibility that we were not going to get a convention. Tensions were high at all levels and for the first time I felt the strain on our chair, John Buccini, as some unexpected divisive partnership issues surfaced among what were thought to be certain allies. We knew from the recently concluded, less-than-successful negotiations on climate change in The Hague, that any global negotiations can be fraught with difficulties and that mistrust and misunderstandings are common. Attitudes and objectives within and between North and South are often very different. The pressure was on not to repeat the history of the climate change negotiations and, after an all-nighter on the last day, the world had a convention, a convention that we at ICC and the coalition could live with.

Reaching the day of the signing in Stockholm, Sweden, had truly been a team effort. How proud I was as an Inuk to have participated with the world in this monumental effort to rid the world of POPs. My last speech in Stockholm reflected my appreciation to all the players involved throughout this process, including all members of CAIPAP: the Council of Yukon First Nations, the Dene Nation, the Inuit Circumpolar Conference, the Inuit Tapirisat of Canada (now known as Inuit Tapiriit Kanatami), and, during the final three sessions, the Russian Association of Indigenous Peoples of the North.

The process also nurtured working links with many environmental organizations and, through the exceptional efforts of the members of IPEN, we made many friends and developed relationships even with industry representatives. We learned to understand each other's perspectives through much dialogue and I feel we together accomplished a great feat. Not only was I proud as an Inuk, I felt proud as a Canadian, watching as our minister of Environment, David Anderson, signed and ratified the convention on the spot. I remember feeling at the time that the world was still compassionate and loving and humanity was still good. I felt that the human element all of the players brought to the negotiations had led to the success of the agreement. Each country's team had approached the negotiations with optimism, working within their countries and among countries in a true spirit of compromise. I thanked them all from my heart and noted special thanks to Chair John Buccini; Klaus Töpfer, Executive Director of UNEP; and Jim Willis of UNEP with his committed staff.

I ended my speech by saying:

> As Indigenous peoples, we have always sought balance in our interactions with the environment. The Stockholm Convention must set the standard that we rise together to restore the balance. Thank you all for your openness and wisdom. Your combined efforts have brought us an important step closer to fulfilling the basic human right of every person, to live in a world free of toxic contamination. For Inuit and Indigenous peoples, this means not only a healthy and secure environment but also the survival of a people. For that I am grateful. *Nakurmiik*. Thank you.

As an ending to this chapter from the "heart," thanks are due to other players without whom I could not have accomplished my work as an elected leader of our people and as a spokesperson for the coalition. As an elected Inuit leader I am acutely aware that the Inuit of the world are a minority within their own countries and that relations between Indigenous peoples and their governments can often become strained. Respect is a key criterion Inuit use when evaluating others, and both John Buccini and Klaus Töpfer certainly embodied this quality throughout the entire journey.

Throughout this process we worked with and fostered a relationship with our Canadian delegation comprising various players from Foreign Affairs, Health Canada, Industry Canada, INAC, CIDA, and others. While it took some effort initially, and was very challenging at times, the delegation and northern Indigenous peoples developed a partnership that soon became a model for other countries. At the request of the coalition, the Government of Canada included on its delegation Carol Mills, a Dene woman well-versed in contaminant issues. History was being re-written, as government and Indigenous peoples singing from the same song sheet is a rarity. By speaking reasonably, wisely, and professionally and always from the "high moral ground," we were able to strengthen and support the position of the Canadian delegation and help bridge the sometimes wide gulf between developed and developing worlds. Although all of the Canadian delegation figured importantly in the negotiations, from where I stood, the support of David Stone of the Northern Contaminants Program of INAC and Ken McCartney of Foreign Affairs certainly was instrumental to me both on and off the delegation.

On the domestic front we welcomed and appreciated the support of federal ministers responsible for Environment, Foreign Affairs and International Trade, and Finance. We also benefited greatly from the support and commitment of Clifford Lincoln, Karen Kraft Sloan, Charles Caccia, Rick Laliberté, John Herron, and additional members of the House of Commons Standing Committee on Environment and Sustainable Development. It is not often that MPs attend international negotiations to encourage their own national delegation or for standing committees to scrutinize and oversee a delegation during the process, but this is exactly what happened during the global POPs negotiations.

Absolutely without a doubt I would not have been able to accomplish my work in the manner that I did without the incredibly loyal technical and political advice from Stephanie Meakin and Terry Fenge. They added an exceptional calibre of commitment and value to this work and, on behalf of the Arctic peoples of Canada, I thank them immensely for their part in the work. Terry Fenge, with no holds barred, did what he is good at and what I hired him for: keeping the Canadian government and other governments committed to reaching a convention that would work for the Arctic. Stephanie Meakin, either with child in womb or nursing her newborn, adding to the symbolism of our message, was with us through all the negotiations and worked with our Canadian delegation at home between sessions. A biologist, Stephanie worked continuously to give us the assurance that the unfolding convention policy reflected the Arctic focus we needed. Terry and Stephanie never let down their guard for a second and their loyalty to me has had more far-reaching benefits than one would, in simple terms, understand. Together we were re-writing history. The significant level of trust between me, an Inuit politician, and Stephanie and Terry, two non-Inuit advisors, that developed during the POPs negotiations was important, relatively unusual, and most welcome.

Finally, I should speak to the coalition of which I, perhaps because of the ICC observer status to the UN, became spokesperson. I attended and spoke at every session in all five countries – Canada, Kenya, Switzerland, Germany, and South Africa – and it was an honour and a pleasure to serve as a spokesperson for the coalition. The coalition worked quite well as a team throughout the negotiations, yet I felt relationships strain as the media started to really pick up on the issue and focus on me and my Inuit world. With ICC's history of involvement in these issues it seemed quite natural that Inuit/Arctic would become the focus of the attention. In fact I felt it to be good strategy to encourage this perspective since the science-based reality is that we Inuit, being avid consumers of marine mammals, are indeed the greatest net recipients of these POPs. Harnessing the attention, in my opinion, was playing good politics. Shirley Adamson and John Burdeck, of Yukon First Nations and Will Mayo, a member of the Tanana Tribe of the interior of Alaska, were elected leaders who understood and respected the politics of this without feeling threatened or undermined by the Inuit exposure. Shirley and Will helped draft the texts of our speeches and their words are reflected in the session in Geneva and at the signing session in Stockholm. Shirley and I also spoke together at an IPEN forum in Geneva. John and I, along with Minister David Anderson, did a press conference together in Stockholm at the signing. I was grateful for the maturity, wisdom, and constructive input offered by Shirley, John, and Will and for their help during the process. I would be remiss in not acknowledging the participation of Cindy Dickson and Chief Bob Charlie of Yukon First Nations in many of the negotiating sessions and of John Crump and Clive Tesar of the Canadian Arctic Resources Committee for their assistance, especially in Johannesburg. As I have said on many occasions, the hunter must be strategic and focused. There is too much at stake to expend energies on anything other than the ultimate goal. In our world, opportunities to gain attention for the Arctic are few and it is important to keep the focus. In my role as spokesperson for all the Aboriginal peoples of Arctic Canada – and the circumpolar world – I strove at all times to be respectful and inclusive in my deliberations.

The gala night in Stockholm at the signing of the convention was a night filled with excitement, accomplishment, and affirmation as we were hosted by Sweden in a magnificent museum with the largest wooden boat that I have ever seen – right in the middle of the reception hall! That night I presented John Buccini and Klaus Töpfer with painted ostrich eggs. Stephanie Meakin and I first got the idea while in South Africa where we saw these beautifully painted ostrich eggs in many stores. I picked up three unpainted eggs and a young Inuk artist and mother painted Arctic scenes on them. The gesture symbolized a bridging of North and South and the eggs were well received. During the evening, UNEP presented me with an award for ICC, "in recognition of contributions to the 'POPs Club' leading to the successful completion of the Stockholm convention on POPs."

I have a habit of picking up pennies as I walk. I call them my pennies from heaven or angels that keep me connected to source and reassure me that I am on the right path. As Stephanie Meakin and I were transferring planes at Heathrow Airport, on our way to Stockholm, we were racing down the hall to catch our flight and I was, as usual, picking up pennies. Stephanie, who was hurrying me along as I picked up these pennies, was somewhat puzzled until, once on the plane, I explained what they represented for me. So you can well imagine the amazed reaction from Stephanie at the gala in Stockholm when Klaus Töpfer presented me with what he called a "lucky penny," which he had given to John Buccini at the onset of these negotiations to ensure luck in his new role as chair. John returned the penny to Klaus that evening and in turn Klaus presented my "lucky global penny" to me – directly from my angels, I say.

As challenging as this work has been, it has been uplifting and rewarding because we achieved a convention that will eventually make our Arctic environment and eating our country food safe once again and because the teachings and lessons of the journey were invaluable for me and for our people. Because we engaged in the politics of influence rather than the politics of protest, we were able to exert influence out of all proportion to our numbers.

Though small in numbers, we became equal partners in a world of millions and helped to re-awaken a conscience. Though small in numbers and up against vested interests seeking the status quo, we were neither intimidated nor afraid in our attempt to save our cultural way of life, for the power itself was in the attempt. I have been changed forever just by the energy of that attempt.

APPENDIX ONE

Contributors

SECTION I PERSISTENT ORGANIC POLLUTANTS: GLOBAL POISONS THREATEN THE NORTH

POPs, THE ENVIRONMENT, AND PUBLIC HEALTH

ERIC DEWAILLY, a specialist in public health (France and Canada) with an M.D. and a Ph.D. in human toxicology, is a professor in the faculty of Medicine of Université Laval. Internationally recognized for his work among fishing communities, especially in the Arctic, dealing with risk assessment and management issues as well as nutritional outcomes, Dr Dewailly is currently chairman of the Environmental Health Panel of the International Union for Circumpolar Health and scientific director of the WHO Collaborating Centre for Environmental Health and Impact Assessment. He is also head of the Ocean and Human Health Program of the centre and acts as scientific advisor for the World Health Organization (WHO), the Food and Agriculture Organization (FAO) of the United Nations, the United Nations Environment Programme (UNEP), the U.S. Food and Drug Administration (FDA), the U.S. National Institute of Environmental Health Sciences (NIEHS), and various Canadian government ministries. He is director of the Public Health Research Unit of the Université Laval Medical Centre and director of the medical division of the International Center for Ocean and Human Health at the Bermuda Biological Station for Research.

CHRISTOPHER FURGAL is a researcher in the Public Health Research Unit, CHUL Research Centre, Université Laval, Quebec, conducting work in biological and social sciences on environmental contaminants, climate change, and other environmental health-related issues, their management, and their impacts on people in the circumpolar North. He completed his Ph.D. at the

University of Waterloo in 1999 and holds a Canadian Institutes of Health Research Post-doctoral Fellowship for his work. Much of Dr Furgal's current research is conducted in co-operation with Aboriginal organizations in the Canadian and circumpolar North. He is a member of the Nunavik Nutrition and Health Committee, co-lead author of the International Arctic Science Committee (IASC)-directed *Arctic Climate Impact Assessment* chapter on health impacts in circumpolar Arctic regions, and co-lead author of the *Canadian Arctic Contaminants Assessment Report II*, which deals with the communications, education, and process aspects of the Northern Contaminants Program.

CANADIAN ARCTIC INDIGENOUS PEOPLES, TRADITIONAL FOOD SYSTEMS, AND POPs

HARRIET V. KUHNLEIN is founding director of the Centre for Indigenous Peoples' Nutrition and Environment (CINE) and professor of human nutrition at McGill University. She received her higher education from Pennsylvania State University, Oregon State University, and the University of California, Berkeley, and joined McGill University as director of the School of Dietetics and Human Nutrition in 1985. Dr Kuhnlein is a registered dietitian in Canada, holds membership in several nutrition societies and the Canadian Society for Circumpolar Health, and was the 2001–2002 recipient of McGill University's Earl Crampton Award for Distinguished Service in Nutrition. She participated in the Sub-committee on Interpretation and Uses of the Dietary Reference Intakes of the U.S. National Academy of Sciences and is chair of the Task Force on Indigenous Peoples' Food and Nutrition of the International Union of Nutritional Sciences.

LAURIE H.M. CHAN is a toxicologist and an associate professor based at the Centre for Indigenous Peoples' Nutrition and Environment (CINE) at McGill University. He holds a Northern Research Chair established by the Natural Sciences and Engineering Research Council. He has studied exposure to contaminants from traditional food systems and their role in the health of Indigenous populations in the North for ten years, and his contributions to CINE's research have been widely used to empower northerners to make healthy and informed diet choices. Dr Chan has assisted more than twenty northern communities to address their concerns about local environmental problems.

GRACE EGELAND holds a Canada Research Chair in Environment, Nutrition, and Health and is an associate professor based at the Centre for Indigenous Peoples' Nutrition and Environment (CINE), McGill University. After completing a Ph.D. in chronic disease epidemiology in 1989 at the University of Pittsburgh, School of Public Health, she joined the U.S. Centers for Disease Control, working for the National Institute for Occupational Safety and Health and the National Center for Environmental Health. Subsequently, she worked with the State of Alaska Department of Health and Social Services and the University of Bergen in Bergen, Norway. Dr Egeland joined McGill University in 2002.

OLIVIER RECEVEUR is an associate professor in the faculty of Medicine at the University of Montreal. He received a master's degree in epidemiology and a Ph.D. in nutrition from the University of California, Berkeley, and joined McGill University and the Centre for Indigenous Peoples' Nutrition and Environment (CINE) in 1993. His contributions as CINE field coordinator and epidemiologist have led to greater understanding of Arctic Aboriginal peoples' dietary benefits and risks. Dr Receveur's interests include nutrition in developing countries, and he is the chair-elect of the Programme for Appropriate Technology in Health (PATH) Canada.

CANADIAN RESEARCH AND POPs: THE NORTHERN CONTAMINANTS PROGRAM

RUSSEL SHEARER is manager of the inter-agency Northern Contaminants Program under the Northern Science and Contaminants Research Directorate of Indian and Northern Affairs Canada. He chairs the intergovernmental and inter-agency technical working group, which provides a scientific evaluation of funding proposals. Mr Shearer was educated at the University of British Columbia, University of Guelph, and Dalhousie University in marine biology and oceanography. From 1980 to 1988 he worked for Environment Canada in Halifax and Yellowknife in environmental regulation of hazardous materials and ocean dumping activities and conducted research and monitoring related to contaminants and the offshore Arctic oil and gas sector. He was an editor and author of the *Canadian Arctic Contaminants Assessment Report* (*CACAR*) and the AMAP Assessment Report. He is currently coordinating the preparation of the *CACAR II*.

SIU-LING HAN completed a master's degree in biology at the University of Waterloo and worked in private industry before joining the Northern Contaminants Program in 1992. From 1998 to the present, Ms Han has been the manager of the Wildlife Research Program for the Government of Nunavut's Department of Sustainable Development in Iqaluit, Nunavut.

CIRCUMPOLAR PERSPECTIVES ON POPs: THE ARCTIC MONITORING AND ASSESSMENT PROGRAMME

LARS-OTTO REIERSEN was educated as a marine biologist at the University of Oslo, Norway, where he conducted research on the basic processes in marine life and the effects of oil and other contaminants on such processes. He has worked in environmental regulation of shipping and oil and gas activities and in the monitoring and assessment of pollution in the North Sea and adjacent areas. Mr Reiersen was involved in the establishment of the Arctic Environmental Protection Strategy/Arctic Council and since 1992 has served as executive secretary for AMAP.

SIMON WILSON earned his bachelor of science in environmental sciences and a Ph.D. in trace metal environmental chemistry at Lancaster University, U.K. He worked as an environmental data scientist at the International Council for the Exploration of the Sea in Copenhagen and was involved in the preparation of the North Sea Quality Status Report. In 1993 he joined AMAP, where his responsibilities as deputy executive secretary include coordination of AMAP data handling and preparation of the AMAP assessments.

VITALY KIMSTACH received a Ph.D. in analytical chemistry from Rostov State University and a doctor of science degree in chemistry at the Institute of Rare Metals, Moscow. He is certified as a professor in chemistry by the USSR Supreme Certification Commission and is an academician of the Russian Academy of Natural Sciences. Dr Kimstach has held a variety of professorships at Rostov State University and was deputy director of the Hydrochemical Institute and deputy head of the Russian Federal Service for Hydrometeorology and Environmental Monitoring, responsible for the Russian state environmental monitoring network. In 1993 he joined AMAP as deputy executive secretary.

THE DEPOSITION OF AIRBORNE DIOXIN EMITTED BY NORTH AMERICAN SOURCES ON ECOLOGICALLY VULNERABLE RECEPTORS IN NUNAVUT

BARRY COMMONER served as director of the Center for the Biology of Natural Systems (CBNS), Queens College, City University of New York from its inception in 1966 to 2000. Dr Commoner has directed studies of air transport and deposition of dioxins, furans, and hexachlorobenzene in the Great Lakes; of atrazine in the Midwest; of dioxins and furans on dairy farms in Wisconsin and Vermont and in the polar regions of North America; and of PCBs from the Hudson River to the New York City watersheds. He now serves as senior scientist on the staff of CBNS.

PAUL WOODS BARTLETT is CBNS's principal atmospheric modeller of pollutants, chiefly pesticides, PCBs, and dioxins. A Ph.D. candidate at the New School University in New York, he participates in working groups of AMAP and the Co-operative Programme for Monitoring and Evaluation of the Long-range Transmission of Air Pollutants in Europe (EMEP) and is on the steering committee of the international inter-comparison program for modelling persistent organic pollutants for the Task Force on Measurements and Modelling of EMEP.

KIMBERLY COUCHOT is a research associate at CBNS. Specializing in geographic information systems (GIS), she updates inventories of various pollutants, including pesticides, dioxins, and PCBs, and maps the results of the atmospheric modelling of those pollutants.

HOLGER EISL is a senior research associate at CBNS, conducting research in alternative energy concepts, pollution prevention, and environmental economics. He received his bachelor of science equivalent in physics and MRP equivalent from the Technical University of Berlin. In the United States he completed master's degrees in regional planning and energy management and policy and a Ph.D. in city and regional planning at the University of Pennsylvania. Dr Eisl is chair of the Intermediate Technology Development Group of North America, which works on issues of sustainability and technology choice.

SECTION II REGIONAL AND GLOBAL POPs POLICY

REGIONAL POPs POLICY: THE UNECE CLRTAP POPs PROTOCOL

HENRIK SELIN is a Wallenberg Fellow in Environment and Sustainability at the Massachusetts Institute of Technology. He has authored numerous scholarly publications on both regional and global attempts at handling hazardous chemicals. Dr Selin has been following actions on persistent organic pollutants under the Convention on Long-range Transboundary Air Pollution (CLRTAP) since 1996 and has attended several CLRTAP POPs sessions.

GLOBAL POPs POLICY: THE 2001 STOCKHOLM CONVENTION ON PERSISTENT ORGANIC POLLUTANTS

DAVID LEONARD DOWNIE is director of Educational Partnerships for the Earth Institute at Columbia University, where he has taught courses in international environmental politics since 1994. Educated at Duke University and the University of North Carolina, Dr Downie is author of numerous scholarly publications on the Stockholm Convention, the Montreal Protocol, the United Nations Environment Programme, and other topics in global environmental politics. From 1994 to 1999, Dr Downie served as director, Environmental Policy Studies at the School of International and Public Affairs, Columbia University.

THE STOCKHOLM CONVENTION IN THE CONTEXT OF INTERNATIONAL ENVIRONMENTAL LAW

NIGEL BANKES teaches in the faculty of Law at the University of Calgary. He has a law degree from the University of Cambridge and an LLM from the University of British Columbia. Mr Bankes spent the academic year of 1999-2000 as the professor-in-residence in the Legal Bureau at the Department of Foreign Affairs and International Trade in Ottawa.

POPs AND INUIT: INFLUENCING THE GLOBAL AGENDA

TERRY FENGE is president of Terry Fenge Consulting Incorporated and strategic counsel to the chair of the Inuit Circumpolar Conference. Born and raised in the United Kingdom, he was educated at the universities of Wales, Victoria, and Waterloo, where in 1982 he took a Ph.D. in the faculty of Environmental Studies. Dr Fenge has been both research director and executive director of the Canadian Arctic Resources Committee, an Ottawa-based public-interest organization. From 1986 to 1992, he was research director for the Tunngavik Federation of Nunavut, the Inuit organization that negotiated the Nunavut Land Claims Agreement.

POPs IN ALASKA: ENGAGING THE USA

HENRY P. HUNTINGTON is an independent researcher in Eagle River, Alaska. He received his Ph.D. in polar studies from the University of Cambridge in 1991 and has worked for the Alaska Eskimo Whaling Commission and the Inuit Circumpolar Conference. His research has examined Arctic policy, traditional ecological knowledge, conservation, and other aspects of human interactions with the Arctic environment. Dr Huntington is a member of the U.S. Polar Research Board and currently serves as board president of the Arctic Research Consortium of the United States.

MICHELLE SPARCK works part-time as public relations liaison for the Association of Village Council Presidents (Bethel, Alaska) and part-time as liaison for the Alaska Native Health Board (Anchorage, Alaska). Educated at the American University in Washington, D.C., she has also worked for the Coastal

Villages Region Fund, in Senator Ted Stevens' office, and for the U.S. House of Representatives Committee on Resources, Subcommittee on Fisheries Conservation, Wildlife, and Oceans.

THE LONG AND WINDING ROAD TO STOCKHOLM: THE VIEW FROM THE CHAIR

JOHN ANTHONY BUCCINI served as chair of the various intergovernmental bodies involved in developing the Stockholm Convention on Persistent Organic Pollutants. He graduated from the University of Manitoba with a Ph.D. in chemistry in 1970. Following post-doctoral studies at Carleton University, Ottawa, Dr Buccini joined Health Canada in 1972 and Environment Canada in 1982, where he served as a program manager until his retirement in October 2000. He continues to provide consultant services on environmental issues.

THE INUIT JOURNEY TOWARDS A POPS-FREE WORLD

SHEILA WATT-CLOUTIER was elected chair of the Inuit Circumpolar Conference (ICC) in August 2002. Born in Kuujjuaq, northern Quebec, Ms Watt-Cloutier was raised traditionally. She has taken a counselling programme at McGill University as well as occupational and training sessions dealing with education and human development. In the mid-1970s she worked as an Inuktitut interpreter for the Ungava hospital and from 1991 to 1995 as an advisor in a review of the education system in northern Quebec that resulted in the ground-breaking report *Silatunirmut: The Path to Wisdom*. As corporate secretary of Makivik Corporation from 1991 to 1995, she oversaw the administration of the corporation established pursuant to the 1975 James Bay and Northern Quebec Agreement. She was elected president of ICC Canada in 1995 and re-elected to this office in 1998.

APPENDIX TWO

POPs Science and Policy: A Brief Northern Lights Timeline

DAVID DOWNIE AND MIKE KRAFT
COLUMBIA UNIVERSITY[1]

1774 Chlorine discovered. Chlorine gives properties of persistence and lipophilicity to POPs.

1873–74 DDT synthesized.

1929 PCBs first produced in commercial mixtures. Eventually PCBs would be manufactured in the USA, Austria, France, Germany, Italy, Spain, United Kingdom, Russian Federation, China, and Japan and exported to virtually every country.

1933 Hexachlorobenzene (HCB) introduced commercially as a fungicide for wheat.

1937–9 DDT's insecticide properties discovered and commercial production begins.

1948 Paul Müller received the Nobel Prize for Medicine and Physiology for the discovery of pesticide applications for DDT.

1949 Chlordane is first registered in Canada to control insect pests in crops and forests and for domestic and industrial applications.

1962 Rachel Carson's *Silent Spring* published in the United States. Over time this book contributes tremendously to public awareness concerning the dangers of DDT and other pesticides.[2]

1966 PCBs discovered in Baltic Sea fish. *New Scientist* reported that Sören Jensen expressed concern over PCBs entering the air and resulting in increased environmental concentration levels.[3] Samples taken by Jensen from fish throughout Sweden; an eagle; and from his own, his wife's, and his five-month old daughter's hair all contained PCBs. Testing of feathers from eagles preserved at the Swedish National Museum of Natural History revealed environmental traces of PCBs back to 1944.

1972 USA bans DDT except for uses essential to public health.

1973 OECD Council calls for restrictions on the production and use of some chemicals, including PCBs.

Mid-1970s In response to environmental and safety concerns, most uses of DDT and chlordane in Canada are phased out by the mid-1970s.[4]

1976 European Community places restrictions on the marketing and use of PCBs and certain other chemicals.[5]
USA bans the manufacturing, processing, distribution, and use of PCBs, except in a "totally enclosed manner."

1977 Canada restricts manufacturing, processing, and uses of PCBs.[6]

1978 A "Persistent Toxic Substance" annex is added to the 1972 United States–Canada Great Lakes Water Quality Agreement.
European Community prohibits marketing of plant protection products containing certain substances including the POPs aldrin, chlordane, dieldrin, DDT, endrin, HCH, heptachlor, and hexachlorobenzene (with the exception of some uses).[7]

1985 Canada prohibits all uses of DDT.
European Community prohibits production, marketing, and use of PCBs.
Canadian scientists Kinloch and Kuhnlein initiate separate studies concluding PCBs and other POPs and metals are in high concentration in food species of Indigenous peoples in northern Canada. High PCB blood concentrations are found in some Native people.[8]
PCBs linked to reproductive impairment in seals in the Baltic Sea and in beluga whales in the St. Lawrence River.

1986 Toxaphene banned in the United States.

1989 Governments agree to the London Guidelines for the Exchange of Information on Chemicals in International Trade. These guidelines represent the first steps towards regulating international trade of hazardous chemicals.[9]
A study led by Canadian scientist Eric Dewailly reveals that the blood and breast milk of Inuit women have elevated levels of POPs.

1990 The Canadian Environmental Protection Act (CEPA) declares that dioxins and furans released into the environment as by-products from various manufacturing and industrial processes are toxic. Subsequent regulations passed in 1992 and 1994 lead to significant reductions in releases of dioxins and furans, particularly from the pulp and paper industry.

The Executive Body of the Convention on Long-range Transboundary Air Pollution (CLRTAP) decides to initiate POPs assessments based on Arctic concerns.

1991 Eight Arctic states establish the Arctic Environmental Protection Strategy, including the Arctic Monitoring and Assessment Programme.

June 1992 The United Nations Conference on Environment and Development (UNCED) adopts *Agenda 21*. Chapter 19 focuses on toxic chemicals and calls for the creation of the new international mechanism to address the management of toxic chemicals including the prevention of illegal international trade – providing a foundation for international work on POPs safety to begin in earnest.[10]

1994 Separate scientific studies correlate immune system damages and behavioural impairments in top predator species with certain POPs.

April 1994 Governments create the Intergovernmental Forum on Chemical Safety (IFCS) to facilitate co-operation among governments in the attempt to promote chemical risk assessment and management.[11]

March 1995 The thirty-one-country IFCS Inter-sessional Group discusses priority activities under chapter 19 of *Agenda 21*.

May 1995 UNEP's Governing Council adopts Decision 18/32, which calls for the IOMC, IPCS, and IFCS to conduct an international assessment of twelve POPs (aldrin, chlordane, DDT, dieldrin, dioxins, endrin, furans, hexachlorobenzene, heptachlor, mirex, PCBs, and toxaphene) and develop recommendations on potential international action for consideration by the UNEP Governing Council and World Health Assembly no later than 1997.[12]

November 1995 The Intergovernmental Conference to Adopt a Global Programme of Action for Protection of the Marine Environment from Land-based Activities (GPA) considered POPs as part of its agenda. The "Washington Declaration" issued from the conference calls, in part, for talks on a legally binding treaty to reduce or eliminate the discharge, manufacture, and use of the twelve POPs.[13]

March 1996 The second meeting of the IFCS Inter-sessional Group establishes the ad hoc working group on POPs.

1996 Eight Arctic states establish the Arctic Council, to which the Inuit Circumpolar Conference, Saami Council, and Russian Association of Indigenous Peoples of the North are accorded "permanent participant" status.

June 1996 A special intergovernmental experts group meets to address UNEP Governing Council Decision 18/32, focusing on obtaining the necessary information to provide sufficient evidence for the IFCS to formulate a recommendation on the need for international action.

The IFCS ad hoc working group meets shortly after the experts meeting and concludes that sufficient information exists to demonstrate the need for international action on twelve POPs, including a legally binding instrument. This recommendation by the IFCS provides UNEP with the basis to develop a formal mandate to begin negotiations on a global POPs convention.

1997 Arctic Monitoring and Assessment Programme publishes *Arctic Pollution Issues: A State of the Arctic Environment Report.*

Government of Canada publishes *Canadian Arctic Contaminants Assessment Report.*

January 1997 Formal political negotiations on POPs commence under CLRTAP.

February 1997 UNEP Governing Council adopts Decision 19/13C, which calls on UNEP to convene formal intergovernmental negotiations with a mandate to prepare an international legally binding instrument beginning with controls on the twelve POPs. It further requested that the INC establish an experts group (CEG) to develop criteria and a procedure for identifying additional POPs as candidates for future international action. The decision also included a number of immediate actions to address POPs.[14]

April 1997 The United States and Canada sign the Great Lakes Binational Toxics Strategy (GLBTS), establishing a co-operative process for the two governments and sectors of society to work towards the virtual elimination of persistent and bioaccumulative toxic substances resulting from human activity.[15]

May 1997 The World Health Assembly of the WHO endorses the recommendations of the IFCS and requests that the WHO participate actively in the POPs treaty negotiations.[16]

July 1997 – June 1998 UNEP and the IFCS convene eight regional and sub-regional workshops to raise awareness of POPs issues in developing countries and countries with economies in transition and build support for the global negotiations.[17]

1998 Arctic Monitoring and Assessment Programme publishes *AMAP Assessment Report: Arctic Pollution Issues.*

March 1998 Government representatives complete the final draft of what becomes the Rotterdam Convention on the Prior Informed Consent (PIC) Procedure for Certain Hazardous Chemicals and Pesticides in International Trade.[18]

June 1998 Thirty-three countries and the European Community finalize the regional Århus Protocol on POPs under the framework of the United Nations Economic Commission for Europe (UNECE) Convention on Long-range Transboundary Air Pollution (CLRTAP).[19]

July 1998 Negotiations for a global POPs treaty begin with the first meeting of the Intergovernmental Negotiating Committee (POPs INC-1), in Montreal.[20]

September 1998 Representatives from nearly 100 countries formally adopt the Rotterdam Convention on the Prior Informed Consent (PIC) Procedure for Certain Hazardous Chemicals and Pesticides in International Trade.

January 1999 POPs INC-2 takes place in Nairobi.

September 1999 POPs INC-3 takes place in Geneva.

March 2000 POPs INC-4 takes place in Bonn.

Canada becomes the first country to make a specific funding commitment for implementing a future global POPs convention. Administered by the World Bank, the fund will provide CAN$20 million over five years (2000–2005) to assist developing countries and countries with economies in transition in dealing with their problems and in complying with the convention obligations.[21]

December 2000 Government representatives at INC-5 in Johannesburg agree on a final draft of a global treaty.

May 2001 Representatives from more than 150 countries unanimously approve the Stockholm Convention on Persistent Organic Pollutants.[22]

Canada becomes the first country to ratify the Stockholm Convention.

NOTES

1 This is an indicative rather than an exhaustive chronology and omits both important events and details. More extensive histories, on which this chronology draws, can be found in the chapters in this volume; Rune Lonngren, *International Approaches to Chemicals Control – A Historical Overview* (Sweden: National Chemicals Inspectorate 1992); UNEP Chemicals, POPs INC-1 Information Kit http: //irptc.unep.ch/pops/POPs_Inc/press_releases/infokite.html; David Downie

and Jayne Laipraset, "Declaring War on POPs: Understanding the Stockholm Convention on Persistent Organic Pollutants" (draft journal article under review); "History of POPs Discovery, Use and Ban," http: //193.247.37.2/gpa%5Ftrial/04histo.htm#ddt; Environment Canada website: www.ec.gc.ca; World Wildlife Foundation website: www.wwf.org; and the Environmental Protection Agency website: www.epa.gov.

2 Rachel Carson, *Silent Spring* (Boston: Houghton Mifflin 1962).

3 Anonymous, "Report of a New Chemical Hazard," in *New Scientist* no. 525 (1966): 612. See also Sören Jensen, "The PCB Story," *Ambio* 1, no. 4, (1972): 123–31.

4 North American Commission for Environmental Cooperation, "North American Regional Action Plan on DDT, North American Working Group for the Sound Management of Chemicals, Task Force on DDT and Chlordane," June 1997, http: //www.cec.org/programs_projects/pollutants_health/smoc/ddt.cfm?varlan=english#canada.

5 European Directive 76/769/EEC.

6 For more information on events within Canada, including those listed here, see http: //www.ec.gc.ca/pops/brochure_e.htm and Shalini Gupta and Andrew Gilman, "Canada's Contribution to the International Reduction of Certain Persistent Organic Pollutants," *Environmental Review* 2: 1 (March 2000): 74.

7 European Directive 79/117/EEC, 21 December 1978.

8 Chapter 2 in this volume provides a review of the development of knowledge regarding POPs. See also R.L. Lipnick, J.L.M. Hermens, K.C. Jones, and D.C.G. Muir, "Persistent, Bioaccumulative, and Toxic Chemicals I, Fate and Exposure" (Washington, D.C.: American Chemical Society 2000).

9 For the text of the London Guidelines, see http: //irptc.unep.ch/pic/longuien.htm.

10 For information on UNCED and *Agenda 21* see http: //www.un.org/esa/sustdev/agenda21.htm. For full text, including chapter 19, see http: //www.un.org/esa/sustdev/agenda21text.htm.

11 For more information on the IFCS, see http: //www.who.int/ifcs. For the text of the resolution establishing the IFCS, see http: //www.who.int/ifcs/fs_res1.htm. For meeting reports and other official documents from IFCS meetings, including the IFCS ad hoc working group on POPs, see http: //www.who.int/ifcs/pop_home.htm.

12 For the text of UNEP GC Decision 18/32, see http: //irptc.unep.ch/pops/indxhtms/gc1832en.html. For more information on the IOMC, see http: //www.who.int/iomc. For more information on the IPCS, see http: //www.who.int/pcs.

13 For the text of the Washington Declaration, see http: //www.gpa.unep.org/documents/gpa/wadeclaration/default.htm or http: //www.unep.org/unep/gpa/pol2b12.htm. For the text of the GPA, see http: //www.unep.org/unep/gpa/pol2a.htm. For additional official documents of the GPA and the Washington Conference, see http: //www.gpa.unep.org/documents/default.htm#gpa or http: //irptc.unep.ch/pops/WashConf.html.

14 For the text of UNEP GC Decision 19/13C, see http: //irptc.unep.ch/pops/gcpops_e.html.
15 For the text of the process, see http: //www.epa.gov/glnpo/p2/bns.html.
16 For the text of the WHA resolution, see http: //irptc.unep.ch/pops/POPs_Inc/INC_1/inf6.htm.
17 For official proceedings of regional and sub-regional POPs workshops, see http: //irptc.unep.ch/pops/POPs_Inc/proceedings/coverpgs/procovers.htm. For background on the workshops, see http: //irptc.unep.ch/pops/newlayout/wkshpintro.htm.
18 For the text of the convention and official documents concerning its negotiation and implementation, see http: //www.pic.int/.
19 For the text of the agreement, see http: //www.unece.org/env/lrtap/protocol/98pop.htm.
20 For official reports and other documents from POPs INC-1 to POPs INC-5 and all other sessions of the global POPs negotiations, see http: //www.chem.unep.ch/sc/documents/meetings/. For journalistic summaries and multimedia reports, see http: //www.iisd.ca/linkages/chemical/index.html or http: //irptc.unep.ch/pops/newlayout/press_items.htm.
21 For more information, see http: //www.ec.gc.ca/press/2001/010523_n_e.htm.
22 For the text of the Stockholm Convention, see http: //www.chem.unep.ch/sc/.

APPENDIX THREE

Glossary of Terms and Concepts: POPs and International Negotiations

COMPILED BY DAVID LEONARD DOWNIE AND VICTORIA ELMAN

This section provides brief definitions of selected terms relevant to understanding toxic chemicals – in particular, persistent organic pollutants – and the development of national and international policy to control POPs.[1]

ÅRHUS POPS PROTOCOL. The 1998 Århus Protocol on Persistent Organic Pollutants covers Europe and North America and is part of the 1979 UNECE Geneva Convention on Long-range Transboundary Air Pollution. Adopted on 24 June 1998 in Århus, Denmark, the Århus POPs Protocol focuses on a list of sixteen substances that have been singled out according to agreed-upon risk criteria. The objective is to eliminate any discharges, emissions, and losses of POPs. The protocol bans the production and use of some products outright (aldrin, chlordane, chlordecone, dieldrin, endrin, hexabromobiphenyl, mirex, and toxaphene); schedules others for elimination at a later date (DDT, heptachlor, hexaclorobenzene, and PCBs; and severely restricts the use of DDT, HCH (including lindane), and PCBs. It includes provisions for dealing with the wastes of products that will be banned; obliges parties to reduce their emissions of dioxins, furans, PAHs, and HCB; and establishes specific limit values for emissions from the incineration of municipal, hazardous, and medical waste.[2]

ACUTE EXPOSURE. A single exposure to a hazardous material for a brief length of time.

ALDRIN. An insecticide and POP, one of a group known as "hard pesticides" because of their toxicity (they kill on contact) and persistence in the environment. Chemically related to chlordane and dieldrin, all are chlorinated hydrocarbons and nerve poisons. Formerly used as an agricultural mechanism to control soil pests, aldrin was banned in the United States in 1975.[3]

ARCTIC ATHABASKAN COUNCIL (AAC). An international treaty organization established to represent the interests of U.S. and Canadian Athabaskan member First Nation governments in Arctic Council fora and to foster a greater understanding of the common heritage of all Athabaskan peoples of Arctic North America.[4]

ARCTIC MONITORING AND ASSESSMENT PROGRAMME (AMAP). An international organization (now a program of the Arctic Council), established in 1991, that provides reliable and sufficient information on the status of, and threats to, the Arctic environment and scientific advice on actions to be taken to support Arctic governments in their efforts to take remedial and preventive actions relating to contaminants.[5]

BEST AVAILABLE TECHNIQUES (BAT). Term used in enforcement of certain federal and state environmental laws that require the application of the most effective non-polluting techniques. As part of implementing the action plan, each party to the Stockholm Convention must promote and require the use of BAT for new sources in Part II of Annex C within four years of the entry into force of the convention.[6]

BEST ENVIRONMENTAL PRACTICE (BEP). The application of the most appropriate combination of environmental control measures and strategies to be promoted and used by each party in implementing an action plan of a multilateral environmental agreement.[7]

BIOACCUMULATION. The ability of a chemical to accumulate in living tissue at levels higher than those in the surrounding environment from the build-up or storage of substances (such as contaminants) in the bodies of animals over time. The bioaccumulation of contaminants can result from eating animals or drinking water that contain the contaminants. Contaminants that bioaccumulate are very slow to change or do not change to a chemical form that can be digested and eliminated by the animal.[8] See also bioconcentration factor.

BIOCONCENTRATION FACTOR (BCF). The ratio of a chemical's concentration in an organism to its concentration in the environment or food. It generally indicates the presence of a chemical substance in higher concentrations in an organism than in the direct environment or its food.[9]

BIODEGRADABLE. Refers to any substance that decomposes through the action of microorganisms.[10]

BIOMAGNIFICATION. Occurs when an animal eats a plant or another animal and consumes all the contaminants stored in that food. Contaminants can biomagnify in animals that use other animals for food because the concentration increases with each step from prey to predator.[11]

CANADIAN ARCTIC RESOURCES COMMITTEE (CARC). A citizens' organization, incorporated under federal law, made up of several thousands of people from every province and territory who share an interest in northern Canada and believe that the North should be treated with deliberation and care. CARC is funded mainly by individuals and private foundations.[12]

CARCINOGEN. A chemical, physical, or biological substance that is capable of causing cancer.[13]

CHEMICAL. A substance that does not include living matter. A chemical can exist alone or in a mixture or preparation; it can be manufactured or obtained from nature.[14]

CHLORDANE AND HEPTACHLOR. Pesticides and POPs that have been used mainly for termite control. They disrupt the immune system and are associated with reproductive disorders and blood diseases. They were produced in the United States for export as recently as 1997.[15]

CHRONIC EXPOSURE. Continuous or repeated exposure to a hazardous substance over a long period of time.

CONFERENCE OF THE PARTIES (CoP). The supreme body of the Stockholm Convention (and most global environmental treaties). All parties are members.

CONGENER. A chemical term that refers to one of many variants or configurations of a common chemical structure.[16]

CONSENSUS. An agreement can be adopted by consensus rather than by a vote when there are no stated objections from delegations.

CONTAINMENT. A remediation method that attempts to seal off all possible exposure pathways between a hazardous disposal site and the environment, generally including both capping and institutional controls.

CONTAMINANT. An undesired physical, chemical, biological, or radiological substance.

CONVENTION ON BIOLOGICAL DIVERSITY OF 1992 (CBD). This pact among the vast majority of the world's governments sets out commitments for maintaining the world's ecological underpinnings while engaging in the business of economic development. The convention establishes three main goals: the conservation of biological diversity, the sustainable use of its components, and the fair and equitable sharing of the benefits from the use of genetic resources.[17]

CONVENTION TO COMBAT DESERTIFICATION (CCD). One hundred and seventy-nine governments have joined the United Nations CCD as of March 2002 in an effort to promote effective action to combat desertification through innovative local programs and supportive international partnerships.[18]

CONVENTION ON LONG-RANGE TRANSBOUNDARY AIR POLLUTION (CLRTAP). The first international agreement, signed by thirty-four governments (including the United States and Canada) and the European Community in 1979, to recognize both environmental and health problems caused by the flow of air pollutants across borders and the pressing need for regional solutions.[19]

CO-OPERATIVE PROGRAMME FOR MONITORING AND EVALUATION OF THE LONG-RANGE TRANSMISSION OF AIR POLLUTANTS IN EUROPE (EMEP). A scientifically based and policy-driven instrument under the CLRTAP for international co-operation to solve transboundary air pollution.[20]

CORE INVENTORY OF AIR EMISSIONS (CORINAIR). A project performed since 1995 by the European Topic Centre on Air Emissions under contract to the European Environment Agency. The aim is to collect, maintain, manage, and publish information on emissions into the air by means of a European air emission inventory and database system. CORINAIR concerns emissions from all sources relevant to the environmental problems of climate change, acidification, eutrophication, tropospheric ozone, air quality, and dispersion of hazardous substances. Its geographic scope is the fifteen European Union members, Iceland, Liechtenstein, Norway, and central and eastern European countries.[21]

COUNCIL OF YUKON FIRST NATIONS (CYFN). The central political organization for the First Nations people of the Yukon, existing since 1973. It serves the needs of self-governing First Nations within the Yukon.[22]

COUNTRIES WITH ECONOMIES IN TRANSITION (CEIT). Those central and eastern European countries and former republics of the Soviet Union that are in transition to a market economy.[23]

CRITERIA. Specific quantitative or qualitative values for chosen parameters that are used to identify which chemicals should be classified as POPs.

CUT-OFF VALUES. Specific quantitative values for criteria that are used as thresholds for identifying chemicals as POPs.

DEPOSIT. An accumulation or laying down of a chemical from a specific source.

DESERTIFICATION. The degradation of land in arid, semi-arid, and dry sub-humid areas caused primarily by human activities and climatic variations. It refers not to the expansion of existing deserts, but rather to the degradation of dryland ecosystems, which cover more than one-third of the world's land area and are extremely vulnerable to over-exploitation and inappropriate land use.[24]

DEVELOPING COUNTRY. A country with relatively low levels of industrial capability, technological advancement, and economic productivity.[25]

DEPARTMENT OF INDIAN AFFAIRS AND NORTHERN DEVELOPMENT (DIAND) OF CANADA. See INDIAN AND NORTHERN AFFAIRS CANADA (INAC).

DICHLORODIPHENYL TRICHLOROETHANE (DDT). An organochlorine insecticide, banned in the United States since 1972, used mainly to control mosquito-borne malaria. DDT is passed along the food chain and concentrates in the tissues of predators. Its presence causes eagles to lay eggs whose shells are too fragile to protect embryos. It also is responsible for widespread death of songbirds. Its use on crops has generally been replaced by less persistent insecticides.[26]

DIELDRIN. A chlorinated hydrocarbon pesticide banned in the United States since 1984 and labelled one of the twelve POPs. Dieldrin is a contact poison formerly used for mothproofing as well as for agricultural pest control.[27]

DIFFUSE. Not concentrated; also, the process of dispersing or allowing to mix together by diffusion.

DIOXIN AND FURAN. A shorthand term for a class of chemicals called polychlorinated dibenzo-p-dioxins and polychlorinated dibenzofurans. These POPs have never been intentionally produced on an industrial scale, but are generated as wastes and by-products from combustion sources and certain chemical and industrial processes. They became widespread in the environment following World War II as a consequence of rising production and use of large quantities of chlorine-containing chemicals, which are associated with dioxin generation. Dioxins are the most highly poisonous synthetic substances known.[28]

DIRTY DOZEN. The popular name for twelve toxic substances the UNEP has targeted for early action under a global legally binding convention. The group includes dioxins, furans, polychlorinated biphenyls (PCBs), DDT, chlordane, heptachlor, hexachlorobenzene, toxaphene, aldrin, dieldrin, endrin, and mirex.[29]

ECOSYSTEM. A system that is formed by the interactions of organisms with their non-living environment. The organisms and the environment work together as a unit, called an ecological unit.[30]

EMISSION. The release of a substance into the environment from either a "point" or a "diffuse" source. A point source is one whose dimensions are small compared with the distance to the observation point; e.g., a smokestack seen from afar.

EMISSION LIMIT VALUES (ELV). The concentration or mass of polluting substances in emissions from plants (any establishment or other stationary plant used for industrial or public utility purposes that is likely to cause air pollution) during a specified period which is not to be exceeded.

ENDOCRINE DISRUPTER. A substance that affects the endocrine (hormone-producing) system, thus causing adverse health effects in an organism or its offspring.[31]

ENDRIN. A chlorinated hydrocarbon insecticide, this POP has been used on field crops such as cotton and also for rodent and bird control. Because it is extremely poisonous and long lasting, its use is banned or restricted in many countries.[32]

ENTRY INTO FORCE. Intergovernmental agreements are not legally binding until they have been ratified by a certain number of countries. Recent global environmental treaties have required thirty to fifty-five ratifications to start a ninety-day countdown to entry into force.

ENVIRONMENTALLY SOUND MANNER (ESM). Handling a situation (such as the management of stockpiles, wastes, products, and articles) in a way that does not harm the environment.

EPIDEMIOLOGY. Study of the causes of disease or toxic effects in human populations.[33]

EUROPEAN COMMUNITY (EC). A regional economic integration organization.[34]

EUROPEAN ENVIRONMENT AGENCY (EEA). Established by the European Union in 1993 with the objective to provide the community and the member states with reliable and comparable information at the European level, enabling its member states to take the requisite measures to protect their environment, to assess the result of such measures, and to ensure that the public is properly informed about the state of the environment.[35]

EUROPEAN UNION (EU). A post-World War II initiative, European integration was launched with a proposal by France for a European federation. Fifteen states are now members and thirteen eastern and southern European countries are preparing for accession.[36]

EXPOSURE. The act or fact of exposing or being exposed to a substance. See CHRONIC or ACUTE EXPOSURE.

FOOD AND AGRICULTURE ORGANIZATION OF THE UNITED NATIONS (FAO). Founded in 1945 with a mandate to raise levels of nutrition and standards of living, to improve agricultural productivity, and to better the condition of rural populations.[37]

FOOD CHAIN. The linking together of plants and animals in feeding relationships. At the bottom of food chains are green plants that convert sunlight into food energy for the rest of the chain. Animals that eat the plants are then eaten by another animal, and so on up the chain. The number of animals involved can vary. For example, in the North, the lichen>caribou>human food chain has fewer feeding links and is much shorter than the algae>fish>seal>polar bear>human food chain. In nature, food chains overlap to form food webs.[38]

FRAMEWORK CONVENTION ON CLIMATE CHANGE (FCCC). The international response to climate change took shape with the development of the United Nations FCCC. Agreed to in 1992, it set out a framework for action to control or cut greenhouse gas emissions. Since the FCCC entered into force in 1994, five meetings of the Conference of the Parties have taken place, as well as numerous workshops and meetings of its subsidiary bodies. A protocol to the convention was adopted in 1997 at the Third Conference of the Parties, held in Kyoto. Although it has yet to enter into force, the FCCC's Kyoto Protocol commits industrialized countries to achieve quantified targets for decreasing their emissions of greenhouse gases.[39]

GENOTOXIC. Refers to a chemical's potential to modify the genetic code (DNA).

GLOBAL ENVIRONMENT FACILITY (GEF). Established as a mechanism to promote international co-operation and foster actions in providing new and additional grant and concessional funding to eligible countries to meet the agreed incremental costs of measures designed to achieve global environmental benefits.[40]

GRASSHOPPER EFFECT. Also known as Global Fractionation, Global Distillation. The process by which POPs circulate globally through the atmosphere via repeated release (evaporation) and deposit (e.g., through rainfall).[41]

GROUP OF 77 (G-77). Established in 1964 by seventy-seven developing countries at the end of the first session of the United Nations Conference on Trade and Development (UNCTAD) in Geneva. From its first ministerial meeting in Algiers in 1967, a permanent institutional structure gradually developed. The membership of the G-77 has increased to 133 countries; however, the original name was retained because of its historic significance. The G-77 seeks to harmonize the negotiating positions of its members.[42]

GWICH'IN COUNCIL INTERNATIONAL (GCI). One of six Indigenous peoples' organizations that is a permanent participant in the Arctic Council. The GCI represents the Gwich'in, who make up fifteen villages and towns scattered across northeast Alaska and northwest Canada.[43]

HALF-LIFE. The time required for half of a given quantity of a substance to break down into the environment.[44]

HAZARDOUS SUBSTANCE. A broad term that includes all substances that can be harmful to people or the environment.[45]

HAZARDOUS WASTE. By-products or waste materials of manufacturing and other processes that have some dangerous property; generally categorized as corrosive, ignitable, toxic, reactive, or in some way harmful to people or the environment.[46]

HEAVY METALS. Metals such as lead, chromium, copper, and cobalt that can be toxic at relatively low concentrations.[47]

HELSINKI COMMISSION (HELCOM). A governing body that works to protect the marine environment of the Baltic Sea from all sources of pollution through intergovernmental co-operation between Denmark, Estonia, the European Community, Finland, Germany, Latvia, Lithuania, Poland, Russia, and Sweden.[48]

HEXACHLOROBENZENE (HCB). A POP formed as a by-product while making other chemicals, HCB was widely used as a pesticide to protect the seeds of onions and sorghum, wheat, and other grains against fungus until 1965. It was also used to make fireworks, ammunition, and synthetic rubber. Currently, there are no commercial uses in the United States. A main health effect from eating food contaminated by HCB is liver disease.[49]

HEXACHLOROCYCLOHEXANES (HCHs). A group of manufactured chemicals that do not occur naturally in the environment. The most common form, lindane, was used as an insecticide on fruit and vegetable crops (including greenhouse vegetables and tobacco) and forest crops (including Christmas trees). Lindane is no longer allowed in the United States with the exception of its use as an ointment to treat head and body lice and scabies.[50]

HYPERSENSITIVITY. A state of increased susceptibility, such as an allergic reaction, to an antigen to which the organism has been previously exposed. (An antigen is a substance that stimulates the production of antibodies when introduced into the body.)[51]

IMMUNE SYSTEM. The collection of cells and tissues responsible for recognizing and attacking foreign microbes and substances in the body.[52]

INCINERATOR. A furnace in which wastes are burned.

INDIAN AND NORTHERN AFFAIRS CANADA (INAC). The primary responsibility of INAC is to meet the federal government's constitutional, treaty, political, and legal responsibilities to First Nations, Inuit, and other northerners. To do so, the department works collaboratively with First Nations, Inuit, and other northerners as well as with other federal departments and agencies, provinces, and territories. The department was formerly known as the Department of Indian Affairs and Northern Development (DIAND).[53]

INSECTICIDE. A substance or a mixture of substances for killing or repelling any type of insect.[54]

INTERGOVERNMENTAL FORUM ON CHEMICAL SAFETY (IFCS). Established by the International Conference on Chemical Safety in Stockholm in April 1994 to facilitate co-operation among governments in the attempt to promote chemical risk assessment and management.[55]

INTERGOVERNMENTAL NEGOTIATING COMMITTEE (INC). The UNEP Governing Council provided a mandate for the creation of an INC to conduct the POPs talks. All members of the United Nations or its specialized agencies can participate in the work of an INC as negotiating parties.[56]

INTERGOVERNMENTAL ORGANIZATION (IGO). A public or government organization created by treaty or agreement between states.[57]

INTERGOVERNMENTAL PANEL ON CLIMATE CHANGE (IPCC). Established in 1988 by the World Meteorological Organization (WMO) and the United Nations Environment Programme (UNEP) to assess the available scientific, technical, and socio-economic information in the field of climate change, the

IPCC is organized into three working groups that assess scientific information and the impacts of climate change and formulate response strategies in economic and social dimensions.[58]

INTERNATIONAL ENERGY AGENCY (IEA). An autonomous Paris-based agency linked with the OECD. The IEA Secretariat collects and analyzes energy data, assesses member countries' domestic energy policies and programs, makes projections based on differing scenarios, and prepares studies and recommendations on specialized energy topics.[59]

INTERNATIONAL LABOUR ORGANIZATION (ILO). A UN specialized agency that seeks the promotion of social justice and internationally recognized human and labour rights. Founded in 1919, it is the only surviving major creation of the Treaty of Versailles, which brought the League of Nations into being, and became the first specialized agency of the UN in 1946.[60]

INTERNATIONAL POPS ELIMINATION NETWORK (IPEN). A global network established in 1998 by a number of public-interest NGOs. The mission is to work for the global elimination of POPs, on an expedited yet socially equitable basis. IPEN was formally launched with a public forum at the first session of the UNEP INC-1 in Montreal in June 1998, to start negotiations to develop a global, legal instrument to control or eliminate persistent organic pollutants (POPs). Throughout the course of the five negotiating sessions, the network grew to include more than 350 public health, environmental, consumer, and other non-government organizations in sixty-five countries.[61]

INTERNATIONAL PROGRAMME ON CHEMICAL SAFETY (IPCS). The two main roles of the IPCS are to establish the scientific health and environmental risk assessment basis for safe use of chemicals *(normative functions)* and to strengthen national capabilities for chemical safety (*technical co-operation*).[62]

INTER-ORGANIZATION PROGRAMME FOR THE SOUND MANAGEMENT OF CHEMICALS (IOMC). Called to be established at the UNCED conference in Rio de Janeiro in June 1992 under *Agenda 21*, chapter 19; established in 1995 to serve as a mechanism for coordinating efforts of intergovernmental organizations in the assessment and management of chemicals.[63]

INUIT CIRCUMPOLAR CONFERENCE (ICC). This NGO, which holds consultative status to ECOSOC, has for the past twenty years played a crucial role in asserting the interests of the Inuit (the Arctic people across Russia, Alaska, Canada, and Greenland) and other Indigenous peoples worldwide.[64]

LINDANE. See HCH.[65]

METABOLISM. The sum of all chemical and physical processes within a living organism.

METEOROLOGICAL SYNTHESIZING CENTRE-EAST (MSC-E). An international centre of EMEP established in 1979. For more than fifteen years, the centre dealt with the evaluation of long-range transport of acid compounds; since 1995, MSC-E has focused on researching and modelling the long-range transport of POPs and heavy metals.[66]

METEOROLOGICAL SYNTHESIZING CENTRE-WEST (MSC-W). The MSC-W has been hosted by the Norwegian Meteorological Institute since the beginning of the EMEP in 1979. The centre's main task is to model transboundary fluxes of acidifying air pollution and photochemical oxidants.[67]

MIGRATION. The movement of a contaminant from one place to another.

MIREX. A POP once used as an insecticide for ants and termites and as a flame retardant in plastics, rubber, and paint. Because it is implicated in cancer and reproductive effects, mirex has not been manufactured or used in the United States since 1978.[68]

MULTILATERAL ENVIRONMENTAL AGREEMENT (MEA). An international consensus that addresses transborder or global environmental problems.[69]

NATIONAL IMPLEMENTATION PLAN (NIP). (Often referred to as Implementation Plan.) A requirement, established by the Stockholm Convention, that all parties develop, transmit, and update plans of how they intend to meet their obligations under the treaty.

NATIONAL DELEGATION. One or more officials who are empowered to represent and negotiate on behalf of their government.

NON-GOVERNMENT ORGANIZATION (NGO). Many relevant NGOs attend intergovernmental meetings as observers to interact with delegates and the press and provide information. They can include environmental groups, business groups, research institutions, and urban and local government associations.[70]

OBSERVER. The Conference of the Parties and POPs INCs permit observers to attend their meetings. Observers may include non-party states, the United Nations and its specialized agencies, and other relevant intergovernmental or non-government organizations.

ORGANIC CHEMICALS. Chemicals that are carbon-based.[71]

ORGANISATION FOR ECONOMIC CO-OPERATION AND DEVELOPMENT (OECD). A group of thirty member countries sharing a commitment to democratic government and the market economy, the OECD is best known for its publications and statistics, as its work covers economic and social issues from macroeconomics to trade, education, development, and science and innovation.[72]

PARIS COMMISSION (PARCOM). Established to administer the Paris Convention, which entered into force in 1978, PARCOM regulated and controlled inputs of substances and energy to the sea from land-based sources (via the atmosphere, rivers, or direct discharges) and also from offshore platforms. The commission was involved in a thorough review of the use and manufacture of various substances to establish the best environmental practice or best available techniques to prevent pollution and embarked on a series of measures to protect parts of the convention area affected by high levels of nutrients, which have been linked to the occurrence of abnormal algal blooms. In 1992, ministers of the countries party to the Oslo and Paris conventions met and agreed to merge and modernize the provisions of the two conventions. The new agreement is called the OSPAR Convention.[73]

PARTY. A state (or an REIO such as the European Community) that has ratified a treaty or agrees to be bound by a treaty, and for which the treaty has entered into force.

PENTACHLOROPHENOL (PCP). A manufactured chemical not found naturally in the environment. One of the most heavily used pesticides in the United States, it was used as a biocide and wood preservative. Now only certified applicators can purchase and use PCP.[74]

PERSISTENCE. Refers to a chemical's ability to remain stable and not break down.

PERSISTENT BIOACCUMULATIVE TOXICS (PBTs). A term used by the U.S. EPA to describe what UNEP refers to as persistent organic pollutants (POPs).[75]

PERSISTENT ORGANIC POLLUTANTS (POPs). Chemical substances that persist in the environment, bioaccumulate through the food web, and pose a risk of causing adverse effects to human health and the environment. With the evidence of long-range transport of these substances to regions where they have never been used or produced and the consequent threats they pose to the environment of the whole globe, the international community has called for urgent global actions to reduce and eliminate releases of these chemicals.[76]

PESTICIDE. A substance or a mixture of substances used to kill, destroy, or repel any type of pest, including fungi, insects, and termites. There are two main types of pesticides: insecticides used to kill insects and herbicides used to kill weeds, mold, and fungus.[77]

PHOTOLYSIS. The decomposition of a substance into simpler units due to its absorption of light energy.

PLENARY. An open meeting of the entire INC, where all formal decisions are made.

POLLUTANT RELEASE AND TRANSFER REGISTER (PRTR). An environmental database or inventory of potentially harmful releases to air, water, and soil as well as wastes transported off site for treatment and disposal. The development and implementation of such a system adapted to national needs represents a means for governments to track the generation, release, and fate of various pollutants over time. A PRTR can be an important tool in the total environment policy of a government, providing otherwise difficult to obtain information about the pollution burden, encouraging reporters to reduce pollution, and engendering broad public support for government environmental policies.[78]

POLYCHLORINATED BIPHENYLS (PCBs). Mixtures of up to 209 individual chlorinated compounds (known as congeners). There are no known natural sources of PCBs. They have been used as coolants and lubricants in transformers, capacitors, and other electrical equipment because they do not burn easily and are good insulators. The manufacturing of this POP was stopped in the United States in 1977 because of evidence showing that PCBs build up in the environment and can cause harmful health effects.[79]

POLYCYCLIC AROMATIC HYDROCARBONS (PAHs). A group of more than one hundred different chemicals that are formed during the incomplete burning of coal, oil and gas, garbage, or other organic substances like tobacco or charbroiled meat. They are usually found as a mixture containing two or more of these compounds, such as soot. They are also found in coal tar, crude oil, creosote, and roofing tar, and a few are used in medicines or to make dyes, plastics, and pesticides. Many are carcinogenic and metabolize in the body to form toxic compounds.[80]

POPs REVIEW COMMITTEE (PRC). Established by the Stockholm Convention, this scientific committee evaluates additional chemicals based on their toxicity, persistence, bioaccumulation, and long-range transport for inclusion in the treaty.

PRIOR INFORMED CONSENT (PIC). Jointly administered by the FAO and UNEP, the voluntary PIC procedure is a means for formally obtaining and disseminating the decision of importing countries on whether or not they wish to receive future shipments of a certain chemical and for ensuring that exporting countries comply with these decisions. PIC has covered twenty-two pesticides and five industrial chemicals, of which seven are POPs. Although the voluntary system worked for a while, by the mid-1990s governments saw the need for a legally binding treaty to govern trade.[81]

RATIFICATION. After signing a treaty, a country must ratify it, often with the approval of the parliament or other legislature. The instrument of ratification must be deposited with the depositary, which is often the secretary-general of the United Nations.

REGIONAL GROUPS. The five regional groups – Africa, Asia, Central and Eastern Europe (CEE), Latin America and the Caribbean (GRULAC), and the Western Europe and Others Group (WEOG) – meet privately to discuss issues and nominate bureau members and other officials.

REGIONAL ECONOMIC INTEGRATION ORGANIZATION (REIO). An organization made up of a group of states, such as the European Community, that follow the same economic norms and ratify treaties as one unit.[82]

RELEASE. The transfer of a hazardous substance from a controlled condition to an uncontrolled condition in the air, in water, or on land.

REPRODUCTIVE DISORDERS. Negative effects on the reproductive system and on the production of offspring.[83]

RESIDUAL CONTAMINATION. Amount of a pollutant remaining in the environment after a natural or technological process has taken place.

RISK ASSESSMENT. The qualitative and quantitative study of the risk posed to human health and the environment by the presence or use of specific pollutants.

RULES OF PROCEDURE. The rules that govern the organization and proceedings of the COP, including the procedures for decision-making, voting, and participation.

RUSSIAN ASSOCIATION OF INDIGENOUS PEOPLES OF THE NORTH (RAIPON). An NGO comprising thirty northern regional ethnic associations of Indigenous peoples, formed by the first Congress of Indigenous Peoples of the

North, Siberia, and Far East, which was held in Moscow in March 1990. Its main purpose is to protect the interests and lawful rights of the peoples it represents, including their right to land, natural resources, and self-government in accordance with international standards and Russian legislation, and their right to resolve their own social and economic problems. RAIPON also provides assistance in cultural development and education, promotes international exchange and co-operation, and organizes humanitarian aid.[84]

SECRETARIAT. Generally, a group that is responsible for coordinating and administering the activities of a decision-making body. In the case of the Stockholm Convention, the Geneva-based UNEP Chemicals division coordinated the organization of the POPs INC and will continue that function until the CoP decides to establish another arrangement.

SENSITIZATION. The development of a hypersensitive or allergic reaction upon reexposure to a substance. The reaction may be immediate or delayed, acute or chronic.

SHORT-CHAIN CHLORINATED PARAFFINS (SCCP). A family of complex mixtures that have been widely used through industry for the past forty or more years. The short-chained category consists of chain lengths of ten to thirteen carbons. Used as an extreme pressure additive in lubricating oils, where the requirements for chemical stability are high; for secondary plasticizers in paints and plastics; and for flame retardants in various plastics and textiles. Studies have shown that chronic exposure to high concentrations of SCCPs caused several types of tumours in laboratory rodents and toxicity in certain aquatic species. These data have prompted regulatory scrutiny, first in the United States and more recently in Canada and western Europe.[85]

SIGNATURE. The head of state or government, foreign minister, or another designated official indicates a country's agreement with the adopted text of a treaty and the country's intention to become a party by signing it.

SWEDISH ENVIRONMENTAL PROTECTION AGENCY (SEPA). A central environmental authority under the Swedish government whose tasks involve the development of environmental work, implementation of environmental policy, and the follow-up and assessment of such efforts.[86]

TETRACHLORODIBENZO-P-DIOXIN (TCDD). A dioxin variant, or congener, with four chlorine atoms on each molecule. It is the most potent form of dioxin and is the standard against which toxicity of other dioxin congeners is measured.[87]

TOLERABLE DAILY INTAKE (TDI). The amount that may be consumed every day over a lifetime without causing harm. Following a review in 2001, requested by the (UK) Food Standards Agency, the Committee on Toxicity (COT) concluded that the TDI for dioxins and dioxin-like PCBs should be reduced from 10 picograms (pg) TEQ/kilogram (kg) bodyweight (bw) per day to 2 pg TEQ/kg bw/day.[88]

TOXAPHENE. This POP has been used as an insecticide to control crop worms and weevils. It is now labelled a POP and banned in many regions, but similar substances probably will be produced and used under different names in various regions. The main known effects of toxaphene are on the nervous system, and it is classified by the United States Department of Health and Human Services as a probable human carcinogen.[89]

TOXIC EQUIVALENCY FACTOR (TEF). A numerical index that is used to compare dioxin-like toxicity of different congeners and substances.[90]

TOXICITY. The ability of a chemical to cause injury to humans or the environment. An acute toxic reaction occurs soon after exposure, whereas a chronic reaction is experienced long after the exposure, potentially even from low levels of exposure.[91]

TOXICOLOGY. The scientific study of the effects, chemistry, and treatment of poisonous (toxic) substances.[92]

TOXIC RELEASE INVENTORY (TRI). Under the U.S. EPA, the TRI provides the first comprehensive inventory of toxic chemical pollution from manufacturing facilities in the United States.[93]

UNITED NATIONS (UN). Established in October 1945 by 51 countries (now 189), the United Nations establishes an international treaty (charter) that sets out basic principles of international relations. Its four purposes are to maintain international peace and security, to develop friendly relations among nations, to co-operate in solving international problems and promoting respect for human rights, and to be a centre for harmonizing the actions of nations.[94]

UNITED NATIONS CONFERENCE ON ENVIRONMENT AND DEVELOPMENT (UNCED). Popularly known as the "Earth Summit," the 1992 UNCED held in Rio de Janeiro, Brazil, hosted 178 countries, more than 100 heads of state, more than 1,000 NGOs, and tens of thousands of journalists. The aim of such a conference was to devise strategies to halt and reverse the effects of environmental degradation by strengthening the promotion of environmentally sound development in all countries. The participating leaders signed five major instruments.[95]

UNITED NATIONS DEVELOPMENT PROGRAMME (UNDP). The United Nations' principal provider of development advice, advocacy, and grant support, the UNDP has long enjoyed the trust and confidence of governments and NGOs in many parts of the developing as well as the developed world.[96]

UNITED NATIONS ECONOMIC COMMISSION FOR EUROPE (UNECE). Established by the UN Economic and Social Council (ECOSOC) in 1947, UNECE is one of five regional commissions of the United Nations. Its goal is to encourage greater economic co-operation among its member states, focusing on economic analysis, environment and human settlements, statistics, sustainable energy, trade, industry and enterprise development, timber, and transport. It includes fifty-five member states and the participation of more than seventy international organizations and NGOs.[97]

UNITED NATIONS ENVIRONMENT PROGRAMME (UNEP). Based in Nairobi, UNEP facilitates and conducts assessments, develops guidelines, and otherwise supports intergovernmental processes. Within UNEP, the Geneva-based UNEP Chemicals division coordinated the organization of the POPs INC.[98]

UNITED NATIONS INDUSTRIAL DEVELOPMENT ORGANIZATION (UNIDO). A UN agency specializing in industrial affairs for more than thirty years, UNIDO works with governments, business associations, and individual companies to solve industrial problems and equip them to help themselves by providing a global forum.[99]

UNITED NATIONS INSTITUTE FOR TRAINING AND RESEARCH (UNITAR). An autonomous body within the United Nations, UNITAR seeks to enhance the effectiveness of the UN through training and research and the formation of partnerships with other UN agencies, governments, and NGOs for the development and implementation of training and capacity-building programs that meet countries' needs.[100]

UNITED STATES ENVIRONMENTAL PROTECTION AGENCY (U.S. EPA). The EPA's mission is "to protect human health and to safeguard the natural environment – air, water, and land – upon which life depends." The agency provides leadership in the nation's environmental science, research, education, and assessment efforts and works closely with other federal agencies, state and local governments, and Indian tribes to develop and enforce regulations under existing environmental laws. EPA is responsible for researching and setting national standards for a variety of environmental programs and delegates responsibility for issuing permits and monitoring and enforcing compliance to states and tribes.[101]

WORLD HEALTH ORGANIZATION (WHO). Established on 7 April 1948, the United Nations' specialized agency for health has as an objective the attainment of the highest possible level of health for all peoples. The 191 member states of the World Health Assembly, the supreme decision-making body for WHO, meet in Geneva in May each year to determine the policies of the organization. The Health Assembly elects the director-general, supervises the financial policies of the organization, and reviews and approves the proposed budget. The executive board is composed of thirty-two members technically qualified in the field of health. Members are elected for three-year terms. The main functions of the board are to give effect to the decisions and policies of the Health Assembly, to advise it, and generally to facilitate its work. The secretariat is staffed by some 3,500 health and other experts and support staff on fixed-term appointments, working at headquarters, in the six regional offices, and in countries.[102]

VOLATILITY. The ability of a chemical to evaporate into air.

WORLD WILDLIFE FUND (WWF). The mission of the WWF, founded in 1961, is to stop the degradation of the natural environment and to build a future in which humans live in harmony with nature through conserving biological diversity, ensuring the sustainability of renewable natural resources, and reducing pollution and wasteful consumption.[103]

NOTES

1 For more information, refer to the websites in the endnotes. In compiling this glossary, we consulted the following sources:
Agency for Toxic Substances and Disease Registry, "Glossary," 28 September 2000, http: //www.atsdr.cdc.gov/HEC/PRHS/glossary.html.
Henry W. Art, ed., *The Dictionary of Ecology and Environmental Science (*New York: Henry Holt and Company 1993).
Andy Crump, *Dictionary of Environment and Development: People, Places, Ideas, and Organizations* (Cambridge, Massachusetts: The MIT Press 1991).
EMEP/CORINAIR, *Atmospheric Emission Inventory Guidebook* (Copenhagen: European Environment Agency 2001), http: //reports.eea.eu.int/technical_report_2001_3/en/page007.html
EXTOXNET, EXTOXNET Pesticide Information Notebook, "Glossary," 13 July 2002,

http: //ace.orst.edu/cgi-bin/mfs/01/pips/glossary.htm.
Greenpeace U.S.A., "POPs Glossary of Terms," 1998,
http: //www.greenpeaceusa.org/toxics/popsglossarytext.htm
Mark Reisch and David Michael Bearden, "Superfund Fact Book," 3 March 1997;
"Glossary of Superfund Terms," 20 June 2002,
http: //cnie.org.NLE/CRSreports/Waste/waste-21.cfm
Natural Resources Defense Council, "Glossary of Terms in Environmental Health," 22 May 2001. http: //www.nrdc.org/breastmilk/glossary.asp
United Nations Environment Programme, "The POPs Negotiations: A Glossary," July 1998, UNEP Chemicals and UNEP Information Unit for Conventions,
http: //irptc.unep.ch/pops/POPs_Inc/press_releases/infokite.html#glossary
U.S. Environmental Protection Agency, "EPA New England Superfund Glossary," 28 June 2002, http: //www.epa.gov/Region1/superfund/basics/gloss.htm

2 http: //www.unece.org/env/lrtap/pops_h1.htm
3 http: //www.atsdr.cdc.gov/tfacts1.html
4 http: //www.arcticathabaskancouncil.com/
5 http: //www.amap.no/
6 http: //www.varam.gov.lv/ivnvb/ipnk/Ebat_not.htm and http: //www.chem.unep.ch/sc/documents/convtext/convtext_en.pdf
7 http: //www.chem.unep.ch/sc/experts/foexp.htm#_ftn2 and http: //www.chem.unep.ch/sc/documents/convtext/convtext_en.pdf
8 http: //ace.orst.edu/info/extoxnet/tibs/bioaccum.htm
9 http: //www.acdlabs.com/products/phys_chem_lab/logd/bcf.html
10 http: //www.worldwise.com/biodegradable.html
11 http: //www.cine.mcgill.ca/TF/tfCObm.htm
12 http: //www.carc.org/
13 http: //www.ilpi.com/msds/ref/carcinogen.html
14 http: //sis.nlm.nih.gov/Chem/ChemMain.html
15 http: //www.atsdr.cdc.gov/tfacts31.html
16 http: //www.epa.gov/toxteam/pcbid/table.htm
17 http: //www.biodiv.org/doc/publications/guide.asp
18 http: //www.unccd.int/main.php
19 http: //www.unece.org/env/lrtap/lrtap_h1.htm
20 http: //www.emep.int/index_facts.html or http: //www.unece.org/env/emep/welcome.html
21 http: //www.aeat.com/netcen/corinair/corinair.html
22 http: //www.cyfn.ca/
23 http: //www.uneptie.org/ozonaction/contacts/ceit.html
24 http: //www.unccd.int/convention/text/leaflet.php
25 http: //www.wto.org/english/tratop_e/devel_e/d1who_e.htm
26 http: //pmep.cce.cornell.edu/profiles/extoxnet/carbaryl-dicrotophos/ddt-ext.html

27 http: //www.atsdr.cdc.gov/tfacts1.html
28 http: //pops.gpa.unep.org/1_10dio.htm and
http: //www.atsdr.cdc.gov/tfacts104.html
29 http: //pops.gpa.unep.org/01what.htm
30 http: //www.ecostudies.org/
31 http: //www.cepis.ops-oms.org/enwww/salunino/pestic.html
32 http: //www.atsdr.cdc.gov/tfacts89.html
33 http: //www.epibiostat.ucsf.edu/epidem/epidem.html
34 http: //userpage.chemie.fu-berlin.de/adressen/eu.html
35 http: //www.eea.eu.int/
36 http: //www.epibiostat.ucsf.edu/epidem/epidem.html and http: //europa.eu.int
37 http: //www.fao.org/
38 http: //www.ultranet.com/;sljkimball/BiologyPages/F/FoodChains.html
39 http: //unfccc.int/ and http: //www.iisd.ca/climate/
40 http: //www.gefweb.org/
41 http: //www.abc.net.au/news/features/stories/s299580.htm
42 http: //www.g77.org/
43 http: //www.arcticpeoples.org/ips/background.htm and http: //www.gwichin.nt.ca/
44 http: //einstein.byu.edu/;slmasong/HTMstuff/C24A1.html
45 http: //www.atsdr.cdc.gov/hazdat.html
46 http: //www.epa.gov/epaoswer/osw/
47 http: //www.osha.gov/SLTC/metalsheavy/
48 http: //www.helcom.fi/
49 http: //www.atsdr.cdc.gov/tfacts90.html
50 http: //www.atsdr.cdc.gov/tfacts43.html
51 http: //edcenter.med.cornell.edu/CUMC_PathNotes/Immunopathology/
Immuno_02.html
52 http: //www.bbc.co.uk/health/immune/
53 http: //www.ainc-inac.gc.ca/yt/intro_e.html
54 http: //plantprotection.org/IRAC/
55 http: //www.who.int/ifcs/ and http: //www.nihs.go.jp/DCBI/forum/haines2.html
56 http: //www.chem.unep.ch/sc/documents/meetings/
57 http: //www.lib.msu.edu/publ_ser/docs/igos/igoorg.htm and
http: //www.ll.georgetown.edu/intl/guides/orgs/
58 http: //www.ipcc.ch/
59 http: //www.iea.org/
60 http: //www.ilo.org/
61 http: //ipen.ecn.cz/
62 http: //www.who.int/pcs/
63 http: //www.un.org/esa/sustdev/agenda21.htm or http: //www.who.int/iomc
64 http: //www.inuit.org/
65 http: //www.headlice.org/lindane/index.htm

66 http: //www.msceast.org/EMEP.html
67 http: //www.emep.int/index_mscw.html
68 http: //www.atsdr.cdc.gov/tfacts66.html
69 http: //www.wto.org/english/tratop_e/envir_e/cte01_e.htm
70 http: //www.ngonet.org/ngolocat.htm
71 http: //www.epa.gov/iaq/voc.html
72 http: //www.oecd.org/
73 http: //www.ospar.org/eng/html/welcome.html
74 http: //www.atsdr.cdc.gov/tfacts51.html
75 http: //www.epa.gov/oppt/chemrtk/persbioa.htm
76 http: //www.chem.unep.ch/pops/
77 http: //www.epa.gov/pesticides/
78 http: //www.chem.unep.ch/prtr/default.htm and
http: //www.oecd.org/EN/home/0,,EN-home-540–14-no-no-no-0,00.html
79 http: //www.atsdr.cdc.gov/tfacts17.html
80 http: //www.atsdr.cdc.gov/tfacts69.html
81 http: //irptc.unep.ch/pic/
82 http: //www.jeanmonnetprogram.org/nafta/Units/webguide.html
83 http: //www.cehn.org/cehn/resourceguide/reproductivedisorders.html
84 http: //www.raipon.org/
85 http: //www.nlgi.com/0024.htm
86 http: //www.internat.environ.se/index.php3
87 http: //cfpub.epa.gov/ncea/cfm/dioxin.cfm?ActType=default and
http: //pops.gpa.unep.org/1_10dio.htm
88 http: //www.foodstandards.gov.uk/news/pressreleases/pcbsanddioxins
89 http: //www.atsdr.cdc.gov/tfacts94.html
90 http: //glossary.eea.eu.int/EEAGlossary/T/toxic_equivalency_factor
91 http: //www.doh.gov.uk/cot/index.htm
92 http: //www.toxicology.org/
93 http: //www.epa.gov/tri/
94 http: //www.un.org/
95 http: //www.unep.org/unep/partners/un/unced/home.htm
96 http: //www.undp.org/
97 http: //www.unece.org/Welcome.html
98 http: //www.unep.org/
99 http: //www.unido.org/
100 http: //www.unitar.org/
101 http: //www.epa.gov/
102 http: //www.who.int/home-page/ and http: //www.who.int/ctd/whopes/index.html.
103 http: //www.panda.org/

APPENDIX FOUR

The Stockholm Convention on Persistent Organic Pollutants

The Parties to this Convention,

Recognizing that persistent organic pollutants possess toxic properties, resist degradation, bioaccumulate and are transported, through air, water and migratory species, across international boundaries and deposited far from their place of release, where they accumulate in terrestrial and aquatic ecosystems,

Aware of the health concerns, especially in developing countries, resulting from local exposure to persistent organic pollutants, in particular impacts upon women and, through them, upon future generations,

Acknowledging that the Arctic ecosystems and indigenous communities are particularly at risk because of the biomagnification of persistent organic pollutants and that contamination of their traditional foods is a public health issue,

Conscious of the need for global action on persistent organic pollutants,

Mindful of decision 19/13 C of 7 February 1997 of the Governing Council of the United Nations Environment Programme to initiate international action to protect human health and the environment through measures which will reduce and/or eliminate emissions and discharges of persistent organic pollutants,

Recalling the pertinent provisions of the relevant international environmental conventions, especially the Rotterdam Convention on the Prior Informed Consent Procedure for Certain Hazardous Chemicals and Pesticides in International Trade, and the Basel Convention on the Control of Transboundary Movements of Hazardous Wastes and their Disposal including the regional agreements developed within the framework of its Article 11,

Recalling also the pertinent provisions of the Rio Declaration on Environment and Development and Agenda 21,

Acknowledging that precaution underlies the concerns of all the Parties and is embedded within this Convention,

Recognizing that this Convention and other international agreements in the field of trade and the environment are mutually supportive,

Reaffirming that States have, in accordance with the Charter of the United Nations and the principles of international law, the sovereign right to exploit their own resources pursuant to their own environmental and developmental policies, and the responsibility to ensure that activities within their jurisdiction or control do not cause damage to the environment of other States or of areas beyond the limits of national jurisdiction,

Taking into account the circumstances and particular requirements of developing countries, in particular the least developed among them, and countries with economies in transition, especially the need to strengthen their national capabilities for the management of chemicals, including through the transfer of technology, the provision of financial and technical assistance and the promotion of cooperation among the Parties,

Taking full account of the Programme of Action for the Sustainable Development of Small Island Developing States, adopted in Barbados on 6 May 1994,

Noting the respective capabilities of developed and developing countries, as well as the common but differentiated responsibilities of States as set forth in Principle 7 of the Rio Declaration on Environment and Development,

Recognizing the important contribution that the private sector and non-governmental organizations can make to achieving the reduction and/or elimination of emissions and discharges of persistent organic pollutants,

Underlining the importance of manufacturers of persistent organic pollutants taking responsibility for reducing adverse effects caused by their products and for providing information to users, Governments and the public on the hazardous properties of those chemicals,

Conscious of the need to take measures to prevent adverse effects caused by persistent organic pollutants at all stages of their life cycle,

Reaffirming Principle 16 of the Rio Declaration on Environment and Development which states that national authorities should endeavour to promote the internalization of environmental costs and the use of economic instruments, taking into account the approach that the polluter should, in principle, bear the cost of pollution, with due regard to the public interest and without distorting international trade and investment,

Encouraging Parties not having regulatory and assessment schemes for pesticides and industrial chemicals to develop such schemes,

Recognizing the importance of developing and using environmentally sound alternative processes and chemicals,

Determined to protect human health and the environment from the harmful impacts of persistent organic pollutants,

Have agreed as follows:

ARTICLE 1

Objective

Mindful of the precautionary approach as set forth in Principle 15 of the Rio Declaration on Environment and Development, the objective of this Convention is to protect human health and the environment from persistent organic pollutants.

ARTICLE 2

Definitions

For the purposes of this Convention:

(a) "Party" means a State or regional economic integration organization that has consented to be bound by this Convention and for which the Convention is in force;

(b) "Regional economic integration organization" means an organization constituted by sovereign States of a given region to which its member States have transferred competence in respect of matters governed by this Convention and which has been duly authorized, in accordance with its internal procedures, to sign, ratify, accept, approve or accede to this Convention;

(c) "Parties present and voting" means Parties present and casting an affirmative or negative vote.

ARTICLE 3

Measures to reduce or eliminate releases from intentional production and use

1. Each Party shall:

(a) Prohibit and/or take the legal and administrative measures necessary to eliminate:

(i) Its production and use of the chemicals listed in Annex A subject to the provisions of that Annex; and

(ii) Its import and export of the chemicals listed in Annex A in accordance with the provisions of paragraph 2; and

(b) Restrict its production and use of the chemicals listed in Annex B in accordance with the provisions of that Annex.

2. Each Party shall take measures to ensure:

(a) That a chemical listed in Annex A or Annex B is imported only:

(i) For the purpose of environmentally sound disposal as set forth in paragraph 1 (d) of Article 6; or

(ii) For a use or purpose which is permitted for that Party under Annex A or Annex B;

(b) That a chemical listed in Annex A for which any production or use specific exemption is in effect or a chemical listed in Annex B for which any production or use specific exemption or acceptable purpose is in effect, taking into account any relevant provisions in existing international prior informed consent instruments, is exported only:

(i) For the purpose of environmentally sound disposal as set forth in paragraph 1 (d) of Article 6;

(ii) To a Party which is permitted to use that chemical under Annex A or Annex B; or

(iii) To a State not Party to this Convention which has provided an annual certification to the exporting Party. Such certification shall specify the intended use of the chemical and include a statement that, with respect to that chemical, the importing State is committed to:

a. Protect human health and the environment by taking the necessary measures to minimize or prevent releases;

b. Comply with the provisions of paragraph 1 of Article 6; and

c. Comply, where appropriate, with the provisions of paragraph 2 of Part II of Annex B.

The certification shall also include any appropriate supporting documentation, such as legislation, regulatory instruments, or administrative or policy guidelines. The exporting Party shall transmit the certification to the Secretariat within sixty days of receipt.

(c) That a chemical listed in Annex A, for which production and use specific exemptions are no longer in effect for any Party, is not exported from it except for the purpose of environmentally sound disposal as set forth in paragraph 1 (d) of Article 6;

(d) For the purposes of this paragraph, the term "State not Party to this Convention" shall include, with respect to a particular chemical, a State or regional economic integration organization that has not agreed to be bound by the Convention with respect to that chemical.

3. Each Party that has one or more regulatory and assessment schemes for new pesticides or new industrial chemicals shall take measures to regulate with the aim of preventing the production and use of new pesticides or new industrial chemicals which, taking into consideration the criteria in paragraph 1 of Annex D, exhibit the characteristics of persistent organic pollutants.

4. Each Party that has one or more regulatory and assessment schemes for pesticides or industrial chemicals shall, where appropriate, take into consideration within these schemes the criteria in paragraph 1 of Annex D when conducting assessments of pesticides or industrial chemicals currently in use.

5. Except as otherwise provided in this Convention, paragraphs 1 and 2 shall not apply to quantities of a chemical to be used for laboratory-scale research or as a reference standard.

6. Any Party that has a specific exemption in accordance with Annex A or a specific exemption or an acceptable purpose in accordance with Annex B shall take appropriate

measures to ensure that any production or use under such exemption or purpose is carried out in a manner that prevents or minimizes human exposure and release into the environment. For exempted uses or acceptable purposes that involve intentional release into the environment under conditions of normal use, such release shall be to the minimum extent necessary, taking into account any applicable standards and guidelines.

ARTICLE 4

Register of specific exemptions

1. A Register is hereby established for the purpose of identifying the Parties that have specific exemptions listed in Annex A or Annex B. It shall not identify Parties that make use of the provisions in Annex A or Annex B that may be exercised by all Parties. The Register shall be maintained by the Secretariat and shall be available to the public.

2. The Register shall include:

(a) A list of the types of specific exemptions reproduced from Annex A and Annex B;

(b) A list of the Parties that have a specific exemption listed under Annex A or Annex B; and

(c) A list of the expiry dates for each registered specific exemption.

3. Any State may, on becoming a Party, by means of a notification in writing to the Secretariat, register for one or more types of specific exemptions listed in Annex A or Annex B.

4. Unless an earlier date is indicated in the Register by a Party, or an extension is granted pursuant to paragraph 7, all registrations of specific exemptions shall expire five years after the date of entry into force of this Convention with respect to a particular chemical.

5. At its first meeting, the Conference of the Parties shall decide upon its review process for the entries in the Register.

6. Prior to a review of an entry in the Register, the Party concerned shall submit a report to the Secretariat justifying its continuing need for registration of that exemption. The report shall be circulated by the Secretariat to all Parties. The review of a registration shall be carried out on the basis of all available information. Thereupon, the Conference of the Parties may make such recommendations to the Party concerned as it deems appropriate.

7. The Conference of the Parties may, upon request from the Party concerned, decide to extend the expiry date of a specific exemption for a period of up to five years. In making its decision, the Conference of the Parties shall take due account of the special circumstances of the developing country Parties and Parties with economies in transition.

8. A Party may, at any time, withdraw an entry from the Register for a specific exemption upon written notification to the Secretariat. The withdrawal shall take effect on the date specified in the notification.

9. When there are no longer any Parties registered for a particular type of specific exemption, no new registrations may be made with respect to it.

ARTICLE 5

Measures to reduce or eliminate releases from unintentional production

Each Party shall at a minimum take the following measures to reduce the total releases derived from anthropogenic sources of each of the chemicals listed in Annex C, with the goal of their continuing minimization and, where feasible, ultimate elimination:

(a) Develop an action plan or, where appropriate, a regional or subregional action plan within two years of the date of entry into force of this Convention for it, and subsequently implement it as part of its implementation plan specified in Article 7, designed to identify, characterize and address the release of the chemicals listed in Annex C and to facilitate implementation of subparagraphs (b) to (e). The action plan shall include the following elements:

(i) An evaluation of current and projected releases, including the development and maintenance of source inventories and release estimates, taking into consideration the source categories identified in Annex C;

(ii) An evaluation of the efficacy of the laws and policies of the Party relating to the management of such releases;

(iii) Strategies to meet the obligations of this paragraph, taking into account the evaluations in (i) and (ii);

(iv) Steps to promote education and training with regard to, and awareness of, those strategies;

(v) A review every five years of those strategies and of their success in meeting the obligations of this paragraph; such reviews shall be included in reports submitted pursuant to Article 15;

(vi) A schedule for implementation of the action plan, including for the strategies and measures identified therein;

(b) Promote the application of available, feasible and practical measures that can expeditiously achieve a realistic and meaningful level of release reduction or source elimination;

(c) Promote the development and, where it deems appropriate, require the use of substitute or modified materials, products and processes to prevent the formation and release of the chemicals listed in Annex C, taking into consideration the general guidance on prevention and release reduction measures in Annex C and guidelines to be adopted by decision of the Conference of the Parties;

(d) Promote and, in accordance with the implementation schedule of its action plan, require the use of best available techniques for new sources within source categories which a Party has identified as warranting such action in its action plan, with a particular initial focus on source categories identified in Part II of Annex C. In any case, the requirement to use best available techniques for new sources in the categories listed in Part II of that Annex shall be phased in as soon as practicable but no later than four years after the entry into force of the Convention for that Party. For the identified categories, Parties shall promote the use of best environmental practices. When applying best available techniques and best environmental practices, Parties should take into consideration the general guidance on prevention and release reduction measures in that Annex and guidelines on best available techniques and best environmental practices to be adopted by decision of the Conference of the Parties;

(e) Promote, in accordance with its action plan, the use of best available techniques and best environmental practices:

(i) For existing sources, within the source categories listed in Part II of Annex C and within source categories such as those in Part III of that Annex; and

(ii) For new sources, within source categories such as those listed in Part III of Annex C which a Party has not addressed under subparagraph (d).

When applying best available techniques and best environmental practices, Parties should take into consideration the general guidance on prevention and release reduction measures in Annex C and guidelines on best available techniques and best environmental practices to be adopted by decision of the Conference of the Parties;

(f) For the purposes of this paragraph and Annex C:

(i) "Best available techniques" means the most effective and advanced stage in the development of activities and their methods of operation which indicate the practical suitability of particular techniques for providing in principle the basis for release limitations designed to prevent and, where that is not practicable, generally to reduce releases of chemicals listed in Part I of Annex C and their impact on the environment as a whole. In this regard:

(ii) "Techniques" includes both the technology used and the way in which the installation is designed, built, maintained, operated and decommissioned;

(iii) "Available" techniques means those techniques that are accessible to the operator and that are developed on a scale that allows implementation in the relevant industrial sector, under economically and technically viable conditions, taking into consideration the costs and advantages; and

(iv) "Best" means most effective in achieving a high general level of protection of the environment as a whole;

(v) "Best environmental practices" means the application of the most appropriate combination of environmental control measures and strategies;

(vi) "New source" means any source of which the construction or substantial modification is commenced at least one year after the date of:

a. Entry into force of this Convention for the Party concerned; or

b. Entry into force for the Party concerned of an amendment to Annex C where the source becomes subject to the provisions of this Convention only by virtue of that amendment.

(g) Release limit values or performance standards may be used by a Party to fulfill its commitments for best available techniques under this paragraph.

ARTICLE 6

Measures to reduce or eliminate releases from stockpiles and wastes

1. In order to ensure that stockpiles consisting of or containing chemicals listed either in Annex A or Annex B and wastes, including products and articles upon becoming wastes, consisting of, containing or contaminated with a chemical listed in Annex A, B or C, are managed in a manner protective of human health and the environment, each Party shall:

(a) Develop appropriate strategies for identifying:

(i) Stockpiles consisting of or containing chemicals listed either in Annex A or Annex B; and

(ii) Products and articles in use and wastes consisting of, containing or contaminated with a chemical listed in Annex A, B or C;

(b) Identify, to the extent practicable, stockpiles consisting of or containing chemicals listed either in Annex A or Annex B on the basis of the strategies referred to in subparagraph (a);

(c) Manage stockpiles, as appropriate, in a safe, efficient and environmentally sound manner. Stockpiles of chemicals listed either in Annex A or Annex B, after they are no longer allowed to be used according to any specific exemption specified in Annex A or any specific exemption or acceptable purpose specified in Annex B, except stockpiles which are allowed to be exported according to paragraph 2 of Article 3, shall be deemed to be waste and shall be managed in accordance with subparagraph (d);

(d) Take appropriate measures so that such wastes, including products and articles upon becoming wastes, are:

(i) Handled, collected, transported and stored in an environmentally sound manner;

(ii) Disposed of in such a way that the persistent organic pollutant content is destroyed or irreversibly transformed so that they do not exhibit the characteristics of persistent organic pollutants or otherwise disposed of in an environmentally sound manner when destruction or irreversible transformation does not represent the environmentally preferable option or the persistent organic pollutant content is low, taking into account international rules, standards, and guidelines, including those that may be developed pursuant to paragraph 2, and relevant global and regional regimes governing the management of hazardous wastes;

(iii) Not permitted to be subjected to disposal operations that may lead to recovery, recycling, reclamation, direct reuse or alternative uses of persistent organic pollutants; and

(iv) Not transported across international boundaries without taking into account relevant international rules, standards and guidelines;

(e) Endeavour to develop appropriate strategies for identifying sites contaminated by chemicals listed in Annex A, B or C; if remediation of those sites is undertaken it shall be performed in an environmentally sound manner.

2. The Conference of the Parties shall cooperate closely with the appropriate bodies of the Basel Convention on the Control of Transboundary Movements of Hazardous Wastes and their Disposal to, *inter alia*:

(a) Establish levels of destruction and irreversible transformation necessary to ensure that the characteristics of persistent organic pollutants as specified in paragraph 1 of Annex D are not exhibited;

(b) Determine what they consider to be the methods that constitute environmentally sound disposal referred to above; and

(c) Work to establish, as appropriate, the concentration levels of the chemicals listed in Annexes A, B and C in order to define the low persistent organic pollutant content referred to in paragraph 1 (d)(ii).

ARTICLE 7

Implementation plans

1. Each Party shall:

(a) Develop and endeavour to implement a plan for the implementation of its obligations under this Convention;

(b) Transmit its implementation plan to the Conference of the Parties within two years of the date on which this Convention enters into force for it; and

(c) Review and update, as appropriate, its implementation plan on a periodic basis and in a manner to be specified by a decision of the Conference of the Parties.

2. The Parties shall, where appropriate, cooperate directly or through global, regional and subregional organizations, and consult their national stakeholders, including women's groups and groups involved in the health of children, in order to facilitate the development, implementation and updating of their implementation plans.

3. The Parties shall endeavour to utilize and, where necessary, establish the means to integrate national implementation plans for persistent organic pollutants in their sustainable development strategies where appropriate.

ARTICLE 8

Listing of chemicals in Annexes A, B and C

1. A Party may submit a proposal to the Secretariat for listing a chemical in Annexes A, B and/or C. The proposal shall contain the information specified in Annex D. In developing a proposal, a Party may be assisted by other Parties and/or by the Secretariat.

2. The Secretariat shall verify whether the proposal contains the information specified in Annex D. If the Secretariat is satisfied that the proposal contains the information so specified, it shall forward the proposal to the Persistent Organic Pollutants Review Committee.

3. The Committee shall examine the proposal and apply the screening criteria specified in Annex D in a flexible and transparent way, taking all information provided into account in an integrative and balanced manner.

4. If the Committee decides that:

(a) It is satisfied that the screening criteria have been fulfilled, it shall, through the Secretariat, make the proposal and the evaluation of the Committee available to all Parties and observers and invite them to submit the information specified in Annex E; or

(b) It is not satisfied that the screening criteria have been fulfilled, it shall, through the Secretariat, inform all Parties and observers and make the proposal and the evaluation of the Committee available to all Parties and the proposal shall be set aside.

5. Any Party may resubmit a proposal to the Committee that has been set aside by the Committee pursuant to paragraph 4. The resubmission may include any concerns of the Party as well as a justification for additional consideration by the Committee. If, following this procedure, the Committee again sets the proposal aside, the Party may challenge the decision of the Committee and the Conference of the Parties shall consider the matter at its next session. The Conference of the Parties may decide, based on the screening criteria in Annex D and taking into account the evaluation of the Committee and any additional information provided by any Party or observer, that the proposal should proceed.

6. Where the Committee has decided that the screening criteria have been fulfilled, or the Conference of the Parties has decided that the proposal should proceed, the Committee shall further review the proposal, taking into account any relevant additional

information received, and shall prepare a draft risk profile in accordance with Annex E. It shall, through the Secretariat, make that draft available to all Parties and observers, collect technical comments from them and, taking those comments into account, complete the risk profile.

7. If, on the basis of the risk profile conducted in accordance with Annex E, the Committee decides:

(a) That the chemical is likely as a result of its long-range environmental transport to lead to significant adverse human health and/or environmental effects such that global action is warranted, the proposal shall proceed. Lack of full scientific certainty shall not prevent the proposal from proceeding. The Committee shall, through the Secretariat, invite information from all Parties and observers relating to the considerations specified in Annex F. It shall then prepare a risk management evaluation that includes an analysis of possible control measures for the chemical in accordance with that Annex; or

(b) That the proposal should not proceed, it shall, through the Secretariat, make the risk profile available to all Parties and observers and set the proposal aside.

8. For any proposal set aside pursuant to paragraph 7 (b), a Party may request the Conference of the Parties to consider instructing the Committee to invite additional information from the proposing Party and other Parties during a period not to exceed one year. After that period and on the basis of any information received, the Committee shall reconsider the proposal pursuant to paragraph 6 with a priority to be decided by the Conference of the Parties. If, following this procedure, the Committee again sets the proposal aside, the Party may challenge the decision of the Committee and the Conference of the Parties shall consider the matter at its next session. The Conference of the Parties may decide, based on the risk profile prepared in accordance with Annex E and taking into account the evaluation of the Committee and any additional information provided by any Party or observer, that the proposal should proceed. If the Conference of the Parties decides that the proposal shall proceed, the Committee shall then prepare the risk management evaluation.

9. The Committee shall, based on the risk profile referred to in paragraph 6 and the risk management evaluation referred to in paragraph 7 (a) or paragraph 8, recommend whether the chemical should be considered by the Conference of the Parties for listing in Annexes A, B and/or C. The Conference of the Parties, taking due account of the recommendations of the Committee, including any scientific uncertainty, shall decide, in a precautionary manner, whether to list the chemical, and specify its related control measures, in Annexes A, B and/or C.

ARTICLE 9

Information exchange

1. Each Party shall facilitate or undertake the exchange of information relevant to:

(a) The reduction or elimination of the production, use and release of persistent organic pollutants; and

(b) Alternatives to persistent organic pollutants, including information relating to their risks as well as to their economic and social costs.

2. The Parties shall exchange the information referred to in paragraph 1 directly or through the Secretariat.

3. Each Party shall designate a national focal point for the exchange of such information.

4. The Secretariat shall serve as a clearing-house mechanism for information on persistent organic pollutants, including information provided by Parties, intergovernmental organizations and non-governmental organizations.

5. For the purposes of this Convention, information on health and safety of humans and the environment shall not be regarded as confidential. Parties that exchange other information pursuant to this Convention shall protect any confidential information as mutually agreed.

ARTICLE 10

Public information, awareness and education

1. Each Party shall, within its capabilities, promote and facilitate:

(a) Awareness among its policy and decision makers with regard to persistent organic pollutants;

(b) Provision to the public of all available information on persistent organic pollutants, taking into account paragraph 5 of Article 9;

(c) Development and implementation, especially for women, children and the least educated, of educational and public awareness programmes on persistent organic pollutants, as well as on their health and environmental effects and on their alternatives;

(d) Public participation in addressing persistent organic pollutants and their health and environmental effects and in developing adequate responses, including opportunities for providing input at the national level regarding implementation of this Convention;

(e) Training of workers, scientists, educators and technical and managerial personnel;

(f) Development and exchange of educational and public awareness materials at the national and international levels; and

(g) Development and implementation of education and training programmes at the national and international levels.

2. Each Party shall, within its capabilities, ensure that the public has access to the public information referred to in paragraph 1 and that the information is kept up-to-date.

3. Each Party shall, within its capabilities, encourage industry and professional users to promote and facilitate the provision of the information referred to in paragraph 1 at the national level and, as appropriate, subregional, regional and global levels.

4. In providing information on persistent organic pollutants and their alternatives, Parties may use safety data sheets, reports, mass media and other means of communication, and may establish information centres at national and regional levels.

5. Each Party shall give sympathetic consideration to developing mechanisms, such as pollutant release and transfer registers, for the collection and dissemination of information on estimates of the annual quantities of the chemicals listed in Annex A, B or C that are released or disposed of.

ARTICLE 11

Research, development and monitoring

1. The Parties shall, within their capabilities, at the national and international levels, encourage and/or undertake appropriate research, development, monitoring and cooperation pertaining to persistent organic pollutants and, where relevant, to their alternatives and to candidate persistent organic pollutants, including on their:

(a) Sources and releases into the environment;

(b) Presence, levels and trends in humans and the environment;

(c) Environmental transport, fate and transformation;

(d) Effects on human health and the environment;

(e) Socio-economic and cultural impacts;

(f) Release reduction and/or elimination; and

(g) Harmonized methodologies for making inventories of generating sources and analytical techniques for the measurement of releases.

2. In undertaking action under paragraph 1, the Parties shall, within their capabilities:

(a) Support and further develop, as appropriate, international programmes, networks and organizations aimed at defining, conducting, assessing and financing research, data collection and monitoring, taking into account the need to minimize duplication of effort;

(b) Support national and international efforts to strengthen national scientific and technical research capabilities, particularly in developing countries and countries with economies in transition, and to promote access to, and the exchange of, data and analyses;

(c) Take into account the concerns and needs, particularly in the field of financial and technical resources, of developing countries and countries with economies in transition and cooperate in improving their capability to participate in the efforts referred to in subparagraphs (a) and (b);

(d) Undertake research work geared towards alleviating the effects of persistent organic pollutants on reproductive health;

(e) Make the results of their research, development and monitoring activities referred to in this paragraph accessible to the public on a timely and regular basis; and

(f) Encourage and/or undertake cooperation with regard to storage and maintenance of information generated from research, development and monitoring.

ARTICLE 12

Technical assistance

1. The Parties recognize that rendering of timely and appropriate technical assistance in response to requests from developing country Parties and Parties with economies in transition is essential to the successful implementation of this Convention.

2. The Parties shall cooperate to provide timely and appropriate technical assistance to developing country Parties and Parties with economies in transition, to assist them, taking into account their particular needs, to develop and strengthen their capacity to implement their obligations under this Convention.

3. In this regard, technical assistance to be provided by developed country Parties, and other Parties in accordance with their capabilities, shall include, as appropriate and as mutually agreed, technical assistance for capacity-building relating to implementation of the obligations under this Convention. Further guidance in this regard shall be provided by the Conference of the Parties.

4. The Parties shall establish, as appropriate, arrangements for the purpose of providing technical assistance and promoting the transfer of technology to developing country Parties and Parties with economies in transition relating to the implementation of this Convention. These arrangements shall include regional and subregional centres for capacity-building and transfer of technology to assist developing country Parties and Parties with economies in transition to fulfil their obligations under this Convention. Further guidance in this regard shall be provided by the Conference of the Parties.

5. The Parties shall, in the context of this Article, take full account of the specific needs and special situation of least developed countries and small island developing states in their actions with regard to technical assistance.

ARTICLE 13

Financial resources and mechanisms

1. Each Party undertakes to provide, within its capabilities, financial support and incentives in respect of those national activities that are intended to achieve the objective of this Convention in accordance with its national plans, priorities and programmes.

2. The developed country Parties shall provide new and additional financial resources to enable developing country Parties and Parties with economies in transition to meet the agreed full incremental costs of implementing measures which fulfill their

obligations under this Convention as agreed between a recipient Party and an entity participating in the mechanism described in paragraph 6. Other Parties may also on a voluntary basis and in accordance with their capabilities provide such financial resources. Contributions from other sources should also be encouraged. The implementation of these commitments shall take into account the need for adequacy, predictability, the timely flow of funds and the importance of burden sharing among the contributing Parties.

3. Developed country Parties, and other Parties in accordance with their capabilities and in accordance with their national plans, priorities and programmes, may also provide and developing country Parties and Parties with economies in transition avail themselves of financial resources to assist in their implementation of this Convention through other bilateral, regional and multilateral sources or channels.

4. The extent to which the developing country Parties will effectively implement their commitments under this Convention will depend on the effective implementation by developed country Parties of their commitments under this Convention relating to financial resources, technical assistance and technology transfer. The fact that sustainable economic and social development and eradication of poverty are the first and overriding priorities of the developing country Parties will be taken fully into account, giving due consideration to the need for the protection of human health and the environment.

5. The Parties shall take full account of the specific needs and special situation of the least developed countries and the small island developing states in their actions with regard to funding.

6. A mechanism for the provision of adequate and sustainable financial resources to developing country Parties and Parties with economies in transition on a grant or concessional basis to assist in their implementation of the Convention is hereby defined. The mechanism shall function under the authority, as appropriate, and guidance of, and be accountable to the Conference of the Parties for the purposes of this Convention. Its operation shall be entrusted to one or more entities, including existing international entities, as may be decided upon by the Conference of the Parties. The mechanism may also include other entities providing multilateral, regional and bilateral financial and technical assistance. Contributions to the mechanism shall be additional to other financial transfers to developing country Parties and Parties with economies in transition as reflected in, and in accordance with, paragraph 2.

7. Pursuant to the objectives of this Convention and paragraph 6, the Conference of the Parties shall at its first meeting adopt appropriate guidance to be provided to the mechanism and shall agree with the entity or entities participating in the financial mechanism upon arrangements to give effect thereto. The guidance shall address, *inter alia*:

(a) The determination of the policy, strategy and programme priorities, as well as clear and detailed criteria and guidelines regarding eligibility for access to and utilization of financial resources including monitoring and evaluation on a regular basis of such utilization;

(b) The provision by the entity or entities of regular reports to the Conference of the Parties on adequacy and sustainability of funding for activities relevant to the implementation of this Convention;

(c) The promotion of multiple-source funding approaches, mechanisms and arrangements;

(d) The modalities for the determination in a predictable and identifiable manner of the amount of funding necessary and available for the implementation of this Convention, keeping in mind that the phasing out of persistent organic pollutants might require sustained funding, and the conditions under which that amount shall be periodically reviewed; and

(e) The modalities for the provision to interested Parties of assistance with needs assessment, information on available sources of funds and on funding patterns in order to facilitate coordination among them.

8. The Conference of the Parties shall review, not later than its second meeting and thereafter on a regular basis, the effectiveness of the mechanism established under this Article, its ability to address the changing needs of the developing country Parties and Parties with economies in transition, the criteria and guidance referred to in paragraph 7, the level of funding as well as the effectiveness of the performance of the institutional entities entrusted to operate the financial mechanism. It shall, based on such review, take appropriate action, if necessary, to improve the effectiveness of the mechanism, including by means of recommendations and guidance on measures to ensure adequate and sustainable funding to meet the needs of the Parties.

ARTICLE 14

Interim financial arrangements

The institutional structure of the Global Environment Facility, operated in accordance with the Instrument for the Establishment of the Restructured Global Environment Facility, shall, on an interim basis, be the principal entity entrusted with the operations of the financial mechanism referred to in Article 13, for the period between the date of entry into force of this Convention and the first meeting of the Conference of the Parties, or until such time as the Conference of the Parties decides which institutional structure will be designated in accordance with Article 13. The institutional structure of the Global Environment Facility should fulfill this function through operational measures related specifically to persistent organic pollutants taking into account that new arrangements for this area may be needed.

ARTICLE 15

Reporting

1. Each Party shall report to the Conference of the Parties on the measures it has taken to implement the provisions of this Convention and on the effectiveness of such measures in meeting the objectives of the Convention.

2. Each Party shall provide to the Secretariat:

(a) Statistical data on its total quantities of production, import and export of each of the chemicals listed in Annex A and Annex B or a reasonable estimate of such data; and

(b) To the extent practicable, a list of the States from which it has imported each such substance and the States to which it has exported each such substance.

3. Such reporting shall be at periodic intervals and in a format to be decided by the Conference of the Parties at its first meeting.

ARTICLE 16

Effectiveness evaluation

1. Commencing four years after the date of entry into force of this Convention, and periodically thereafter at intervals to be decided by the Conference of the Parties, the Conference shall evaluate the effectiveness of this Convention.

2. In order to facilitate such evaluation, the Conference of the Parties shall, at its first meeting, initiate the establishment of arrangements to provide itself with comparable monitoring data on the presence of the chemicals listed in Annexes A, B and C as well as their regional and global environmental transport. These arrangements:

(a) Should be implemented by the Parties on a regional basis when appropriate, in accordance with their technical and financial capabilities, using existing monitoring programmes and mechanisms to the extent possible and promoting harmonization of approaches;

(b) May be supplemented where necessary, taking into account the differences between regions and their capabilities to implement monitoring activities; and

(c) Shall include reports to the Conference of the Parties on the results of the monitoring activities on a regional and global basis at intervals to be specified by the Conference of the Parties.

3. The evaluation described in paragraph 1 shall be conducted on the basis of available scientific, environmental, technical and economic information, including:

(a) Reports and other monitoring information provided pursuant to paragraph 2;

(b) National reports submitted pursuant to Article 15; and

(c) Non-compliance information provided pursuant to the procedures established under Article 17.

ARTICLE 17

Non-compliance

The Conference of the Parties shall, as soon as practicable, develop and approve procedures and institutional mechanisms for determining non-compliance with the provisions of this Convention and for the treatment of Parties found to be in non-compliance.

ARTICLE 18

Settlement of disputes

1. Parties shall settle any dispute between them concerning the interpretation or application of this Convention through negotiation or other peaceful means of their own choice.

2. When ratifying, accepting, approving or acceding to the Convention, or at any time thereafter, a Party that is not a regional economic integration organization may declare in a written instrument submitted to the depositary that, with respect to any dispute concerning the interpretation or application of the Convention, it recognizes one or both of the following means of dispute settlement as compulsory in relation to any Party accepting the same obligation:

(a) Arbitration in accordance with procedures to be adopted by the Conference of the Parties in an annex as soon as practicable;

(b) Submission of the dispute to the International Court of Justice.

3. A Party that is a regional economic integration organization may make a declaration with like effect in relation to arbitration in accordance with the procedure referred to in paragraph 2 (a).

4. A declaration made pursuant to paragraph 2 or paragraph 3 shall remain in force until it expires in accordance with its terms or until three months after written notice of its revocation has been deposited with the depositary.

5. The expiry of a declaration, a notice of revocation or a new declaration shall not in any way affect proceedings pending before an arbitral tribunal or the International Court of Justice unless the parties to the dispute otherwise agree.

6. If the parties to a dispute have not accepted the same or any procedure pursuant to paragraph 2, and if they have not been able to settle their dispute within twelve months following notification by one party to another that a dispute exists between them, the dispute shall be submitted to a conciliation commission at the request of any party to the dispute. The conciliation commission shall render a report with recommendations. Additional procedures relating to the conciliation commission shall be included in an annex to be adopted by the Conference of the Parties no later than at its second meeting.

ARTICLE 19

Conference of the Parties

1. A Conference of the Parties is hereby established.

2. The first meeting of the Conference of the Parties shall be convened by the Executive Director of the United Nations Environment Programme no later than one year after the entry into force of this Convention. Thereafter, ordinary meetings of the Conference of the Parties shall be held at regular intervals to be decided by the Conference.

3. Extraordinary meetings of the Conference of the Parties shall be held at such other times as may be deemed necessary by the Conference, or at the written request of any Party provided that it is supported by at least one third of the Parties.

4. The Conference of the Parties shall by consensus agree upon and adopt at its first meeting rules of procedure and financial rules for itself and any subsidiary bodies, as well as financial provisions governing the functioning of the Secretariat.

5. The Conference of the Parties shall keep under continuous review and evaluation the implementation of this Convention. It shall perform the functions assigned to it by the Convention and, to this end, shall:

(a) Establish, further to the requirements of paragraph 6, such subsidiary bodies as it considers necessary for the implementation of the Convention;

(b) Cooperate, where appropriate, with competent international organizations and intergovernmental and non-governmental bodies; and

(c) Regularly review all information made available to the Parties pursuant to Article 15, including consideration of the effectiveness of paragraph 2 (b) (iii) of Article 3;

(d) Consider and undertake any additional action that may be required for the achievement of the objectives of the Convention.

6. The Conference of the Parties shall, at its first meeting, establish a subsidiary body to be called the Persistent Organic Pollutants Review Committee for the purposes of performing the functions assigned to that Committee by this Convention. In this regard:

(a) The members of the Persistent Organic Pollutants Review Committee shall be appointed by the Conference of the Parties. Membership of the Committee shall consist of government-designated experts in chemical assessment or management. The members of the Committee shall be appointed on the basis of equitable geographical distribution;

(b) The Conference of the Parties shall decide on the terms of reference, organization and operation of the Committee; and

(c) The Committee shall make every effort to adopt its recommendations by consensus. If all efforts at consensus have been exhausted, and no consensus reached, such recommendation shall as a last resort be adopted by a two-thirds majority vote of the members present and voting.

7. The Conference of the Parties shall, at its third meeting, evaluate the continued need for the procedure contained in paragraph 2 (b) of Article 3, including consideration of its effectiveness.

8. The United Nations, its specialized agencies and the International Atomic Energy Agency, as well as any State not Party to this Convention, may be represented at meetings of the Conference of the Parties as observers. Any body or agency, whether national or international, governmental or non-governmental, qualified in matters covered by the Convention, and which has informed the Secretariat of its wish to be represented at a meeting of the Conference of the Parties as an observer may be ad-

mitted unless at least one third of the Parties present object. The admission and participation of observers shall be subject to the rules of procedure adopted by the Conference of the Parties.

ARTICLE 20

Secretariat

1. A Secretariat is hereby established.

2. The functions of the Secretariat shall be:

(a) To make arrangements for meetings of the Conference of the Parties and its subsidiary bodies and to provide them with services as required;

(b) To facilitate assistance to the Parties, particularly developing country Parties and Parties with economies in transition, on request, in the implementation of this Convention;

(c) To ensure the necessary coordination with the secretariats of other relevant international bodies;

(d) To prepare and make available to the Parties periodic reports based on information received pursuant to Article 15 and other available information;

(e) To enter, under the overall guidance of the Conference of the Parties, into such administrative and contractual arrangements as may be required for the effective discharge of its functions; and

(f) To perform the other secretariat functions specified in this Convention and such other functions as may be determined by the Conference of the Parties.

3. The secretariat functions for this Convention shall be performed by the Executive Director of the United Nations Environment Programme, unless the Conference of the Parties decides, by a three-fourths majority of the Parties present and voting, to entrust the secretariat functions to one or more other international organizations.

ARTICLE 21

Amendments to the Convention

1. Amendments to this Convention may be proposed by any Party.

2. Amendments to this Convention shall be adopted at a meeting of the Conference of the Parties. The text of any proposed amendment shall be communicated to the Parties by the Secretariat at least six months before the meeting at which it is proposed for adoption. The Secretariat shall also communicate proposed amendments to the signatories to this Convention and, for information, to the depositary.

3. The Parties shall make every effort to reach agreement on any proposed amendment to this Convention by consensus. If all efforts at consensus have been exhausted, and no agreement reached, the amendment shall as a last resort be adopted by a three-fourths majority vote of the Parties present and voting.

4. The amendment shall be communicated by the depositary to all Parties for ratification, acceptance or approval.

5. Ratification, acceptance or approval of an amendment shall be notified to the depositary in writing. An amendment adopted in accordance with paragraph 3 shall enter into force for the Parties having accepted it on the ninetieth day after the date of deposit of instruments of ratification, acceptance or approval by at least three-fourths of the Parties. Thereafter, the amendment shall enter into force for any other Party on the ninetieth day after the date on which that Party deposits its instrument of ratification, acceptance or approval of the amendment.

ARTICLE 22

Adoption and amendment of annexes

1. Annexes to this Convention shall form an integral part thereof and, unless expressly provided otherwise, a reference to this Convention constitutes at the same time a reference to any annexes thereto.

2. Any additional annexes shall be restricted to procedural, scientific, technical or administrative matters.

3. The following procedure shall apply to the proposal, adoption and entry into force of additional annexes to this Convention:

(a) Additional annexes shall be proposed and adopted according to the procedure laid down in paragraphs 1, 2 and 3 of Article 21;

(b) Any Party that is unable to accept an additional annex shall so notify the depositary, in writing, within one year from the date of communication by the depositary of the adoption of the additional annex. The depositary shall without delay notify all Parties of any such notification received. A Party may at any time withdraw a previous notification of non-acceptance in respect of any additional annex, and the annex shall thereupon enter into force for that Party subject to subparagraph (c); and

(c) On the expiry of one year from the date of the communication by the depositary of the adoption of an additional annex, the annex shall enter into force for all Parties that have not submitted a notification in accordance with the provisions of subparagraph (b).

4. The proposal, adoption and entry into force of amendments to Annex A, B or C shall be subject to the same procedures as for the proposal, adoption and entry into force of additional annexes to this Convention, except that an amendment to Annex A, B or C shall not enter into force with respect to any Party that has made a declaration with respect to amendment to those Annexes in accordance with paragraph 4 of Article 25, in which case any such amendment shall enter into force for such a Party on the ninetieth day after the date of deposit with the depositary of its instrument of ratification, acceptance, approval or accession with respect to such amendment.

5. The following procedure shall apply to the proposal, adoption and entry into force of an amendment to Annex D, E or F:

(a) Amendments shall be proposed according to the procedure in paragraphs 1 and 2 of Article 21;

(b) The Parties shall take decisions on an amendment to Annex D, E or F by consensus; and

(c) A decision to amend Annex D, E or F shall forthwith be communicated to the Parties by the depositary. The amendment shall enter into force for all Parties on a date to be specified in the decision.

6. If an additional annex or an amendment to an annex is related to an amendment to this Convention, the additional annex or amendment shall not enter into force until such time as the amendment to the Convention enters into force.

ARTICLE 23

Right to vote

1. Each Party to this Convention shall have one vote, except as provided for in paragraph 2.

2. A regional economic integration organization, on matters within its competence, shall exercise its right to vote with a number of votes equal to the number of its member States that are Parties to this Convention. Such an organization shall not exercise its right to vote if any of its member States exercises its right to vote, and vice versa.

ARTICLE 24

Signature

This Convention shall be open for signature at Stockholm by all States and regional economic integration organizations on 23 May 2001, and at the United Nations Headquarters in New York from 24 May 2001 to 22 May 2002.

ARTICLE 25

Ratification, acceptance, approval or accession

1. This Convention shall be subject to ratification, acceptance or approval by States and by regional economic integration organizations. It shall be open for accession by States and by regional economic integration organizations from the day after the date on which the Convention is closed for signature. Instruments of ratification, acceptance, approval or accession shall be deposited with the depositary.

2. Any regional economic integration organization that becomes a Party to this Convention without any of its member States being a Party shall be bound by all the obligations under the Convention. In the case of such organizations, one or more of whose member States is a Party to this Convention, the organization and its member States shall decide on their respective responsibilities for the performance of their obligations under the Convention. In such cases, the organization and the member States shall not be entitled to exercise rights under the Convention concurrently.

3. In its instrument of ratification, acceptance, approval or accession, a regional economic integration organization shall declare the extent of its competence in respect of the matters governed by this Convention. Any such organization shall also inform the depositary, who shall in turn inform the Parties, of any relevant modification in the extent of its competence.

4. In its instrument of ratification, acceptance, approval or accession, any Party may declare that, with respect to it, any amendment to Annex A, B or C shall enter into force only upon the deposit of its instrument of ratification, acceptance, approval or accession with respect thereto.

ARTICLE 26

Entry into force

1. This Convention shall enter into force on the ninetieth day after the date of deposit of the fiftieth instrument of ratification, acceptance, approval or accession.

2. For each State or regional economic integration organization that ratifies, accepts or approves this Convention or accedes thereto after the deposit of the fiftieth instrument of ratification, acceptance, approval or accession, the Convention shall enter into force on the ninetieth day after the date of deposit by such State or regional economic integration organization of its instrument of ratification, acceptance, approval or accession.

3. For the purpose of paragraphs 1 and 2, any instrument deposited by a regional economic integration organization shall not be counted as additional to those deposited by member States of that organization.

ARTICLE 27

Reservations

No reservations may be made to this Convention.

ARTICLE 28

Withdrawal

1. At any time after three years from the date on which this Convention has entered into force for a Party, that Party may withdraw from the Convention by giving written notification to the depositary.

2. Any such withdrawal shall take effect upon the expiry of one year from the date of receipt by the depositary of the notification of withdrawal, or on such later date as may be specified in the notification of withdrawal.

ARTICLE 29

Depositary

The Secretary-General of the United Nations shall be the depositary of this Convention.

ARTICLE 30

Authentic texts

The original of this Convention, of which the Arabic, Chinese, English, French, Russian and Spanish texts are equally authentic, shall be deposited with the Secretary-General of the United Nations.

IN WITNESS WHEREOF the undersigned, being duly authorized to that effect, have signed this Convention.

Done at Stockholm on this twenty-second day of May, two thousand and one.

NOTES

(i) Except as otherwise specified in this Convention, quantities of a chemical occurring as unintentional trace contaminants in products and articles shall not be considered to be listed in this Annex;

(ii) This note shall not be considered as a production and use specific exemption for purposes of paragraph 2 of Article 3. Quantities of a chemical occurring as constituents of articles manufactured or already in use before or on the date of entry into force of the relevant obligation with respect to that chemical, shall not be considered as listed in this Annex, provided that a Party has notified the Secretariat that a particular type of article remains in use within that Party. The Secretariat shall make such notifications publicly available;

(iii) This note, which does not apply to a chemical that has an asterisk following its name in the Chemical column in Part I of this Annex, shall not be considered as a production and use specific exemption for purposes of paragraph 2 of Article 3. Given that no significant quantities of the chemical are expected to reach humans and the environment during the production and use of a closed-system site-limited intermediate, a Party, upon notification to the Secretariat, may allow the production and use of quantities of a chemical listed in this Annex as a closed-system site-limited intermediate that is chemically transformed in the manufacture of other chemicals that, taking into consideration the criteria in paragraph 1 of Annex D, do not exhibit the characteristics of persistent organic pollutants. This notification shall include information on total production and use of such chemical or a reasonable estimate of such information and information regarding the nature of the closed-system site-limited process including the amount of any non-transformed and unintentional trace contamination of the persistent organic pollutant-starting material in the final product. This procedure applies except as otherwise specified in this Annex. The Secretariat shall make such notifications available to the Conference of the Parties and to the public. Such production or use shall not be considered a production or use specific exemption. Such production and use shall cease after a ten-year period, unless the Party concerned submits a new notification to the Secretariat, in which case the period will be extended for an additional ten years unless the Conference of the Parties, after a review of the production and use decides otherwise. The notification procedure can be repeated;

(iv) All the specific exemptions in this Annex may be exercised by Parties that have registered exemptions in respect of them in accordance with Article 4 with the exception of the use of polychlorinated biphenyls in articles in use in accordance with the provisions of Part II of this Annex, which may be exercised by all Parties.

PART II

Polychlorinated biphenyls

Each Party shall:

(a) With regard to the elimination of the use of polychlorinated biphenyls in equipment (e.g. transformers, capacitors or other receptacles containing liquid stocks) by 2025, subject to review by the Conference of the Parties, take action in accordance with the following priorities:

(i) Make determined efforts to identify, label and remove from use equipment containing greater than 10 per cent polychlorinated biphenyls and volumes greater than 5 litres;

(ii) Make determined efforts to identify, label and remove from use equipment containing greater than 0.05 per cent polychlorinated biphenyls and volumes greater than 5 litres;

(iii) Endeavour to identify and remove from use equipment containing greater than 0.005 percent polychlorinated biphenyls and volumes greater than 0.05 litres;

(b) Consistent with the priorities in subparagraph (a), promote the following measures to reduce exposures and risk to control the use of polychlorinated biphenyls:

(i) Use only in intact and non-leaking equipment and only in areas where the risk from environmental release can be minimised and quickly remedied;

(ii) Not use in equipment in areas associated with the production or processing of food or feed;

(iii) When used in populated areas, including schools and hospitals, all reasonable measures to protect from electrical failure which could result in a fire, and regular inspection of equipment for leaks;

(c) Notwithstanding paragraph 2 of Article 3, ensure that equipment containing polychlorinated biphenyls, as described in subparagraph (a), shall not be exported or imported except for the purpose of environmentally sound waste management;

(d) Except for maintenance and servicing operations, not allow recovery for the purpose of reuse in other equipment of liquids with polychlorinated biphenyls content above 0.005 per cent;

(e) Make determined efforts designed to lead to environmentally sound waste management of liquids containing polychlorinated biphenyls and equipment contaminated with polychlorinated biphenyls having a polychlorinated biphenyls content above 0.005 per cent, in accordance with paragraph 1 of Article 6, as soon as possible but no later than 2028, subject to review by the Conference of the Parties;

(f) In lieu of note (ii) in Part I of this Annex, endeavour to identify other articles containing more than 0.005 per cent polychlorinated biphenyls (e.g. cable-sheaths, cured caulk and painted objects) and manage them in accordance with paragraph 1 of Article 6;

(g) Provide a report every five years on progress in eliminating polychlorinated biphenyls and submit it to the Conference of the Parties pursuant to Article 15;

(h) The reports described in subparagraph (g) shall, as appropriate, be considered by the Conference of the Parties in its reviews relating to polychlorinated biphenyls. The Conference of the Parties shall review progress towards elimination of polychlorinated biphenyls at five year intervals or other period, as appropriate, taking into account such reports.

ANNEX A

ELIMINATION

Part I

Chemical	*Activity*	*Specific exemption*
Aldrin* CAS No: 309-00-2	Production	None
	Use	Local ectoparasiticide Insecticide
Chlordane* CAS No: 57-74-9	Production	As allowed for the Parties listed in the Register
	Use	Local ectoparasiticide Insecticide Termiticide Termiticide in buildings and dams Termiticide in roads Additive in plywood adhesives
Dieldrin* CAS No: 60-57-1	Production	None
	Use	In agricultural operations
Endrin* CAS No: 72-20-8	Production	None
	Use	None
Heptachlor* CAS No: 76-44-8	Production	None
	Use	Termiticide Termiticide in structures of houses Termiticide (subterranean) Wood treatment In use in underground cable boxes
Hexachlorobenzene CAS No: 118-74-1	Production	As allowed for the Parties listed in the Register
	Use	Intermediate Solvent in pesticide Closed-system site-limited intermediate
Mirex* CAS No: 2385-85-5	Production	As allowed for the Parties listed in the Register
	Use	Termiticide
Toxaphene* CAS No: 8001-35-2	Production	None
	Use	None
Polychlorinated Biphenyls (PCB)*	Production	None
	Use	Use Articles in use in accordance with the provisions of Part II of this Annex

ANNEX B

RESTRICTION

Part I

Chemical	*Activity*	*Acceptable purpose or specific exemption*
DDT (1,1,1-trichloro-2,2-bis (4-chlorophenyl)ethane) CAS No: 50–29–3	Production	*Acceptable purpose*: Disease vector control use in accordance with Part II of this Annex *Specific exemption*: Intermediate in production of dicofol Intermediate
	Use	*Acceptable purpose*: Disease vector control in accordance with Part II of this Annex *Specific exemption*: Production of dicofol Intermediate

NOTES

(i) Except as otherwise specified in this Convention, quantities of a chemical occurring as unintentional trace contaminants in products and articles shall not be considered to be listed in this Annex;

(ii) This note shall not be considered as a production and use acceptable purpose or specific exemption for purposes of paragraph 2 of Article 3. Quantities of a chemical occurring as constituents of articles manufactured or already in use before or on the date of entry into force of the relevant obligation with respect to that chemical, shall not be considered as listed in this Annex, provided that a Party has notified the Secretariat that a particular type of article remains in use within that Party. The Secretariat shall make such notifications publicly available;

(iii) This note shall not be considered as a production and use specific exemption for purposes of paragraph 2 of Article 3. Given that no significant quantities of the chemical are expected to reach humans and the environment during the production and use of a closed-system site-limited intermediate, a Party, upon notification to the Secretariat, may allow the production and use of quantities of a chemical listed in this Annex as a closed-system site-limited intermediate that is chemically transformed in the manufacture of other chemicals that, taking into consideration the criteria in paragraph 1 of Annex D, do not exhibit the characteristics of persistent organic pollutants. This notification shall include information on total production and use of such chemical or a reasonable estimate of such information and information regarding the nature of the closed-system site-limited process including the amount of any non-transformed and unintentional trace contamination of the persistent organic pollutant-starting material in the final product. This procedure applies except as otherwise specified in this Annex. The Secretariat shall make such notifications available to the Conference of the Parties and to the public. Such production or use shall not be considered a production or use specific exemption. Such production and use shall cease after a ten-year period, unless the Party concerned submits a new notification to the Secretariat, in which case the period will be extended for an additional ten years unless the Conference of the Parties, after a review of the production and use decides otherwise. The notification procedure can be repeated;

(iv) All the specific exemptions in this Annex may be exercised by Parties that have registered in respect of them in accordance with Article 4.

PART II

DDT (1,1,1-trichloro-2,2-bis(4-chlorophenyl)ethane)

1. The production and use of DDT shall be eliminated except for Parties that have notified the Secretariat of their intention to produce and/or use it. A DDT Register is hereby established and shall be available to the public. The Secretariat shall maintain the DDT Register.

2. Each Party that produces and/or uses DDT shall restrict such production and/or use for disease vector control in accordance with the World Health Organization recommendations and guidelines on the use of DDT and when locally safe, effective and affordable alternatives are not available to the Party in question.

3. In the event that a Party not listed in the DDT Register determines that it requires DDT for disease vector control, it shall notify the Secretariat as soon as possible in order to have its name added forthwith to the DDT Register. It shall at the same time notify the World Health Organization.

4. Every three years, each Party that uses DDT shall provide to the Secretariat and the World Health Organization information on the amount used, the conditions of such use and its relevance to that Party's disease management strategy, in a format to be decided by the Conference of the Parties in consultation with the World Health Organization.

5. With the goal of reducing and ultimately eliminating the use of DDT, the Conference of the Parties shall encourage:

(a) Each Party using DDT to develop and implement an action plan as part of the implementation plan specified in Article 7. That action plan shall include:

(i) Development of regulatory and other mechanisms to ensure that DDT use is restricted to disease vector control;

(ii) Implementation of suitable alternative products, methods and strategies, including resistance management strategies to ensure the continuing effectiveness of these alternatives;

(iii) Measures to strengthen health care and to reduce the incidence of the disease.

(b) The Parties, within their capabilities, to promote research and development of safe alternative chemical and non-chemical products, methods and strategies for Parties using DDT, relevant to the conditions of those countries and with the goal of decreasing the human and economic burden of disease. Factors to be promoted when considering alternatives or combinations of alternatives shall include the human health risks and environmental implications of such alternatives. Viable alternatives to DDT shall pose less risk to human health and the environment, be suitable for disease control based on conditions in the Parties in question and be supported with monitoring data.

6. Commencing at its first meeting, and at least every three years thereafter, the Conference of the Parties shall, in consultation with the World Health Organization, evaluate the continued need for DDT for disease vector control on the basis of available scientific, technical, environmental and economic information, including:

(a) The production and use of DDT and the conditions set out in paragraph 2;

(b) The availability, suitability and implementation of the alternatives to DDT; and

(c) Progress in strengthening the capacity of countries to transfer safely to reliance on such alternatives.

7. A Party may, at any time, withdraw its name from the DDT Registry upon written notification to the Secretariat. The withdrawal shall take effect on the date specified in the notification.

ANNEX C

UNINTENTIONAL PRODUCTION

Part I: Persistent organic pollutants subject to the requirements of Article 5

This Annex applies to the following persistent organic pollutants when formed and released unintentionally from anthropogenic sources:

Chemical
Polychlorinated dibenzo-p-dioxins and dibenzofurans (PCDD/PCDF)
Hexachlorobenzene (HCB) (CAS No: 118–74–1)
Polychlorinated biphenyls (PCB)

Part II: Source categories

Polychlorinated dibenzo-p-dioxins and dibenzofurans, hexachlorobenzene and polychlorinated biphenyls are unintentionally formed and released from thermal processes involving organic matter and chlorine as a result of incomplete combustion or chemical reactions. The following industrial source categories have the potential for comparatively high formation and release of these chemicals to the environment:

(a) Waste incinerators, including co-incinerators of municipal, hazardous or medical waste or of sewage sludge;

(b) Cement kilns firing hazardous waste;

(c) Production of pulp using elemental chlorine or chemicals generating elemental chlorine for bleaching;

(d) The following thermal processes in the metallurgical industry:

(i) Secondary copper production;

(ii) Sinter plants in the iron and steel industry;

(iii) Secondary aluminium production;

(iv) Secondary zinc production.

Part III: Source categories

Polychlorinated dibenzo-p-dioxins and dibenzofurans, hexachlorobenzene and polychlorinated biphenyls may also be unintentionally formed and released from the following source categories, including:

(a) Open burning of waste, including burning of landfill sites;

(b) Thermal processes in the metallurgical industry not mentioned in Part II;

(c) Residential combustion sources;

(d) Fossil fuel-fired utility and industrial boilers;

(e) Firing installations for wood and other biomass fuels;

(f) Specific chemical production processes releasing unintentionally formed persistent organic pollutants, especially production of chlorophenols and chloranil;

(g) Crematoria;

(h) Motor vehicles, particularly those burning leaded gasoline;

(i) Destruction of animal carcasses;

(j) Textile and leather dyeing (with chloranil) and finishing (with alkaline extraction);

(k) Shredder plants for the treatment of end of life vehicles;

(l) Smouldering of copper cables;

(m) Waste oil refineries.

Part IV: Definitions

1. For the purposes of this Annex:

(a) "Polychlorinated biphenyls" means aromatic compounds formed in such a manner that the hydrogen atoms on the biphenyl molecule (two benzene rings bonded together by a single carbon-carbon bond) may be replaced by up to ten chlorine atoms; and

(b) "Polychlorinated dibenzo-p-dioxins" and "polychlorinated dibenzofurans" are tricyclic, aromatic compounds formed by two benzene rings connected by two oxygen atoms in polychlorinated dibenzo-p-dioxins and by one oxygen atom and one carbon-carbon bond in polychlorinated dibenzofurans and the hydrogen atoms of which may be replaced by up to eight chlorine atoms.

2. In this Annex, the toxicity of polychlorinated dibenzo-p-dioxins and dibenzofurans is expressed using the concept of toxic equivalency which measures the relative dioxin-like toxic activity of different congeners of polychlorinated dibenzo-p-dioxins and dibenzofurans and coplanar polychlorinated biphenyls in comparison to 2,3,7,8-tetrachlorodibenzo-p-dioxin. The toxic equivalent factor values to be used for the purposes

of this Convention shall be consistent with accepted international standards, commencing with the World Health Organization 1998 mammalian toxic equivalent factor values for polychlorinated dibenzo-p-dioxins and dibenzofurans and coplanar polychlorinated biphenyls. Concentrations are expressed in toxic equivalents.

Part V: General guidance on best available techniques and best environmental practices

This Part provides general guidance to Parties on preventing or reducing releases of the chemicals listed in Part I.

A. General prevention measures relating to both best available techniques and best environmental practices

Priority should be given to the consideration of approaches to prevent the formation and release of the chemicals listed in Part I. Useful measures could include:

(a) The use of low-waste technology;

(b) The use of less hazardous substances;

(c) The promotion of the recovery and recycling of waste and of substances generated and used in a process;

(d) Replacement of feed materials which are persistent organic pollutants or where there is a direct link between the materials and releases of persistent organic pollutants from the source;

(e) Good housekeeping and preventive maintenance programmes;

(f) Improvements in waste management with the aim of the cessation of open and other uncontrolled burning of wastes, including the burning of landfill sites. When considering proposals to construct new waste disposal facilities, consideration should be given to alternatives such as activities to minimize the generation of municipal and medical waste, including resource recovery, reuse, recycling, waste separation and promoting products that generate less waste. Under this approach, public health concerns should be carefully considered;

(g) Minimization of these chemicals as contaminants in products;

(h) Avoiding elemental chlorine or chemicals generating elemental chlorine for bleaching.

B. Best available techniques

The concept of best available techniques is not aimed at the prescription of any specific technique or technology, but at taking into account the technical characteristics of the installation concerned, its geographical location and the local environmental conditions. Appropriate control techniques to reduce releases of the chemicals listed in Part I are in general the same. In determining best available techniques, special consideration should be given, generally or in specific cases, to the following factors, bearing in mind the likely costs and benefits of a measure and consideration of precaution and prevention:

(a) General considerations:

(i) The nature, effects and mass of the releases concerned: techniques may vary depending on source size;

(ii) The commissioning dates for new or existing installations;

(iii) The time needed to introduce the best available technique;

(iv) The consumption and nature of raw materials used in the process and its energy efficiency;

(v) The need to prevent or reduce to a minimum the overall impact of the releases to the environment and the risks to it;

(vi) The need to prevent accidents and to minimize their consequences for the environment;

(vii) The need to ensure occupational health and safety at workplaces;

(viii) Comparable processes, facilities or methods of operation which have been tried with success on an industrial scale;

(ix) Technological advances and changes in scientific knowledge and understanding.

(b) General release reduction measures: When considering proposals to construct new facilities or significantly modify existing facilities using processes that release chemicals listed in this Annex, priority consideration should be given to alternative processes, techniques or practices that have similar usefulness but which avoid the formation and release of such chemicals. In cases where such facilities will be constructed or significantly modified, in addition to the prevention measures outlined in section A of Part V the following reduction measures could also be considered in determining best available techniques:

(i) Use of improved methods for flue-gas cleaning such as thermal or catalytic oxidation, dust precipitation, or adsorption;

(ii) Treatment of residuals, wastewater, wastes and sewage sludge by, for example, thermal treatment or rendering them inert or chemical processes that detoxify them;

(iii) Process changes that lead to the reduction or elimination of releases, such as moving to closed systems;

(iv) Modification of process designs to improve combustion and prevent formation of the chemicals listed in this Annex, through the control of parameters such as incineration temperature or residence time.

C. Best environmental practices

The Conference of the Parties may develop guidance with regard to best environmental practices.

ANNEX D

INFORMATION REQUIREMENTS AND SCREENING CRITERIA

1. A Party submitting a proposal to list a chemical in Annexes A, B and/or C shall identify the chemical in the manner described in subparagraph (a) and provide the information on the chemical, and its transformation products where relevant, relating to the screening criteria set out in subparagraphs (b) to (e):

(a) Chemical identity:

(i) Names, including trade name or names, commercial name or names and synonyms, Chemical Abstracts Service (CAS) Registry number, International Union of Pure and Applied Chemistry (IUPAC) name; and

(ii) Structure, including specification of isomers, where applicable, and the structure of the chemical class;

(b) Persistence:

(i) Evidence that the half-life of the chemical in water is greater than two months, or that its half-life in soil is greater than six months, or that its half-life in sediment is greater than six months; or

(ii) Evidence that the chemical is otherwise sufficiently persistent to justify its consideration within the scope of this Convention;

(c) Bio-accumulation:

(i) Evidence that the bio-concentration factor or bio-accumulation factor in aquatic species for the chemical is greater than 5,000 or, in the absence of such data, that the log Kow is greater than 5;

(ii) Evidence that a chemical presents other reasons for concern, such as high bio-accumulation in other species, high toxicity or ecotoxicity; or

(iii) Monitoring data in biota indicating that the bio-accumulation potential of the chemical is sufficient to justify its consideration within the scope of this Convention;

(d) Potential for long-range environmental transport:

(i) Measured levels of the chemical in locations distant from the sources of its release that are of potential concern;

(ii) Monitoring data showing that long-range environmental transport of the chemical, with the potential for transfer to a receiving environment, may have occurred via air, water or migratory species; or

(iii) Environmental fate properties and/or model results that demonstrate that the chemical has a potential for long-range environmental transport through air, water or migratory species, with the potential for transfer to a receiving environment in locations distant from the sources of its release. For a chemical that migrates significantly through the air, its half-life in air should be greater than two days; and

(e) Adverse effects:

(i) Evidence of adverse effects to human health or to the environment that justifies consideration of the chemical within the scope of this Convention; or

(ii) Toxicity or ecotoxicity data that indicate the potential for damage to human health or to the environment.

2. The proposing Party shall provide a statement of the reasons for concern including, where possible, a comparison of toxicity or ecotoxicity data with detected or predicted levels of a chemical resulting or anticipated from its long-range environmental transport, and a short statement indicating the need for global control.

3. The proposing Party shall, to the extent possible and taking into account its capabilities, provide additional information to support the review of the proposal referred to in paragraph 6 of Article 8. In developing such a proposal, a Party may draw on technical expertise from any source.

ANNEX E

INFORMATION REQUIREMENTS FOR THE RISK PROFILE

The purpose of the review is to evaluate whether the chemical is likely, as a result of its long-range environmental transport, to lead to significant adverse human health and/or environmental effects, such that global action is warranted. For this purpose, a risk profile shall be developed that further elaborates on, and evaluates, the information referred to in Annex D and includes, as far as possible, the following types of information:

(a) Sources, including as appropriate:

(i) Production data, including quantity and location;

(ii) Uses; and

(iii) Releases, such as discharges, losses and emissions;

(b) Hazard assessment for the endpoint or endpoints of concern, including a consideration of toxicological interactions involving multiple chemicals;

(c) Environmental fate, including data and information on the chemical and physical properties of a chemical as well as its persistence and how they are linked to its environmental transport, transfer within and between environmental compartments, degradation and transformation to other chemicals. A determination of the bio-concentration factor or bio-accumulation factor, based on measured values, shall be available, except when monitoring data are judged to meet this need;

(d) Monitoring data;

(e) Exposure in local areas and, in particular, as a result of long-range environmental transport, and including information regarding bio-availability;

(f) National and international risk evaluations, assessments or profiles and labelling information and hazard classifications, as available; and

(g) Status of the chemical under international conventions.

ANNEX F

INFORMATION ON SOCIO-ECONOMIC CONSIDERATIONS

An evaluation should be undertaken regarding possible control measures for chemicals under consideration for inclusion in this Convention, encompassing the full range of options, including management and elimination. For this purpose, relevant information should be provided relating to socio-economic considerations associated with possible control measures to enable a decision to be taken by the Conference of the Parties. Such information should reflect due regard for the differing capabilities and conditions among the Parties and should include consideration of the following indicative list of items:

(a) Efficacy and efficiency of possible control measures in meeting risk reduction goals:

(i) Technical feasibility; and

(ii) Costs, including environmental and health costs;

(b) Alternatives (products and processes):

(i) Technical feasibility;

(ii) Costs, including environmental and health costs;

(iii) Efficacy;

(iv) Risk;

(v) Availability; and

(vi) Accessibility;

(c) Positive and/or negative impacts on society of implementing possible control measures:

(i) Health, including public, environmental and occupational health;

(ii) Agriculture, including aquaculture and forestry;

(iii) Biota (biodiversity);

(iv) Economic aspects;

(v) Movement towards sustainable development; and

(vi) Social costs;

(d) Waste and disposal implications (in particular, obsolete stocks of pesticides and clean-up of contaminated sites):

(i) Technical feasibility; and

(ii) Cost;

(e) Access to information and public education;

(f) Status of control and monitoring capacity; and

(g) Any national or regional control actions taken, including information on alternatives, and other relevant risk management information.